Essentials of Noncoding RNA in Neuroscience

Essentials of Noncoding RNA in Neuroscience

Ontogenetics, Plasticity of the Vertebrate Brain

Edited by

Davide De Pietri Tonelli

*Neurobiology of miRNA Lab, Neuroscience
and Brain Technologies Department,
Istituto Italiano di Tecnologia, Genoa, Italy*

ACADEMIC PRESS

An imprint of Elsevier

Academic Press is an imprint of Elsevier
125 London Wall, London EC2Y 5AS, United Kingdom
525 B Street, Suite 1800, San Diego, CA 92101-4495, United States
50 Hampshire Street, 5th Floor, Cambridge, MA 02139, United States
The Boulevard, Langford Lane, Kidlington, Oxford OX5 1GB, United Kingdom

Notices
Knowledge and best practice in this field are constantly changing. As new research and experience broaden our understanding,
changes in research methods, professional practices, or medical treatment may become necessary.

Practitioners and researchers must always rely on their own experience and knowledge in evaluating and using any information,
methods, compounds, or experiments described herein. In using such information or methods they should be mindful of their own
safety and the safety of others, including parties for whom they have a professional responsibility.

To the fullest extent of the law, neither the Publisher nor the authors, contributors, or editors, assume any liability for any injury
and/or damage to persons or property as a matter of products liability, negligence or otherwise, or from any use or operation of any
methods, products, instructions, or ideas contained in the material herein.

British Library Cataloguing-in-Publication Data
A catalogue record for this book is available from the British Library

Library of Congress Cataloging-in-Publication Data
A catalog record for this book is available from the Library of Congress

ISBN: 978-0-12-804402-5

For Information on all Academic Press publications
visit our website at https://www.elsevier.com/books-and-journals

www.elsevier.com • www.bookaid.org

Publisher: Mara Conner
Acquisitions Editor: Natalie Farra
Editorial Project Manager: Kathy Padilla
Production Project Manager: Chris Wortley
Designer: Victoria Pearson Esser

Typeset by MPS Limited, Chennai, India

Contents

CHAPTER 3 Computational Challenges and -omics Approaches for the Identification of microRNAs and Targets 39

Ioannis S. Vlachos, Georgios Georgakilas, Spyros Tastsoglou,
Maria D. Paraskevopoulou, Dimitra Karagkouni
and Artemis G. Hatzigeorgiou

CHAPTER 4 Methodological Challenges in Functional Investigation and Therapeutic Use of microRNAs 61

Federica Marinaro, Meritxell Pons-Espinal and Davide De Pietri Tonelli

CHAPTER 5 The Cell Biology of Neural Stem and Progenitor Cells and Neocortex Expansion in Development and Evolution 81

Stefania Tavano and Wieland B. Huttner

CHAPTER 6 miRNA-Dependent and Independent Functions of the Microprocessor in the Regulation of Neural Stem Cell Biology 101

Andrea Erni, Chiara Rolando and Verdon Taylor

CHAPTER 15 Comparative Functions of miRNAs in Embryonic Neurogenesis and Neuronal Network Formation **265**

Emma Ristori and Stefania Nicoli

List of Contributors

Daniel G. Abernathy
Washington University in St. Louis School of Medicine, St. Louis, MO, United States

Benedikt Berninger
University Medical Center of the Johannes Gutenberg University Mainz, Mainz, Germany

Pascal Bielefeld
University of Amsterdam, Amsterdam, The Netherlands

Irene Bozzoni
Sapienza University of Rome, Rome, Italy

Murray J. Cairns
University of Newcastle, Callaghan, NSW, Australia; Schizophrenia Research Institute, Sydney, NSW, Australia; Hunter Medical Research Institute, Newcastle, NSW, Australia

Nathalie Coré
Aix-Marseille Université, CNRS, Marseille, France

Harold Cremer
Aix-Marseille Université, CNRS, Marseille, France

Antoine de Chevigny
Aix-Marseille Université, CNRS, Marseille, France

Davide De Pietri Tonelli
Istituto Italiano di Tecnologia, Genoa, Italy

Andrea Erni
University of Basel, Basel, Switzerland

Carlos P. Fitzsimons
University of Amsterdam, Amsterdam, The Netherlands

Michael Geaghan
University of Newcastle, Callaghan, NSW, Australia; Hunter Medical Research Institute, Newcastle, NSW, Australia

Georgios Georgakilas
University of Thessaly, Volos, Greece; Hellenic Pasteur Institute, Athens, Greece

Stefanie Grosswendt
Berlin Institute for Medical Systems Biology (BIMSB), Berlin, Germany

Artemis G. Hatzigeorgiou
University of Thessaly, Volos, Greece; Hellenic Pasteur Institute, Athens, Greece

Wieland B. Huttner
Max Planck Institute of Molecular Cell Biology and Genetics, Dresden, Germany

Dimitra Karagkouni
University of Thessaly, Volos, Greece; Hellenic Pasteur Institute, Athens, Greece

Ivano Legnini
Sapienza University of Rome, Rome, Italy

Yangjian Liu
Washington University in St. Louis School of Medicine, St. Louis, MO, United States

Qing Richard Lu
Cincinnati Children's Hospital Medical Center, Cincinnati, OH, United States; University of Cincinnati, Cincinnati, OH, United States

Federica Marinaro
Istituto Italiano di Tecnologia, Genoa, Italy

Debora Napoli
Scuola Normale Superiore, Pisa, Italy

Stefania Nicoli
Yale School of Medicine, New Haven, CT, United States

Tomasz Jan Nowakowski
University of California, San Francisco, San Francisco, CA, United States

Maria D. Paraskevopoulou
University of Thessaly, Volos, Greece; Hellenic Pasteur Institute, Athens, Greece

Amy E. Pasquinelli
University of California at San Diego, La Jolla, CA, United States

Tommaso Pizzorusso
Institute of Neuroscience, Pisa, Italy; University of Florence, Florence, Italy

Meritxell Pons-Espinal
Istituto Italiano di Tecnologia, Genoa, Italy

Thomas Pratt
University of Edinburgh, Edinburgh, United Kingdom

David Jonathan Price
University of Edinburgh, Edinburgh, United Kingdom

Ben Pustjens
University of Amsterdam, Amsterdam, The Netherlands

Nikolaus Rajewsky
Berlin Institute for Medical Systems Biology (BIMSB), Berlin, Germany

Emma Ristori
Yale School of Medicine, New Haven, CT, United States

Chiara Rolando
University of Basel, Basel, Switzerland

Marijn Schouten
University of Amsterdam, Amsterdam, The Netherlands

William P. Schreiner
University of California at San Diego, La Jolla, CA, United States

Spyros Tastsoglou
University of Thessaly, Volos, Greece; Hellenic Pasteur Institute, Athens, Greece

Stefania Tavano
Max Planck Institute of Molecular Cell Biology and Genetics, Dresden, Germany

Verdon Taylor
University of Basel, Basel, Switzerland

Neha Tiwari
University Medical Center of the Johannes Gutenberg University Mainz, Mainz, Germany

Ioannis S. Vlachos
University of Thessaly, Volos, Greece; Hellenic Pasteur Institute, Athens, Greece

Haibo Wang
Cincinnati Children's Hospital Medical Center, Cincinnati, OH, United States

Andrew S. Yoo
Washington University in St. Louis School of Medicine, St. Louis, MO, United States

Xianghui Zhao
Fourth Military Medical University, Xi'an, China; Cincinnati Children's Hospital Medical Center, Cincinnati, OH, United States

Preface

The brain is not just "any organ." Indeed, (1) it comprises billions of electrogenic and neuromodulatory cells, such as neurons and glia, respectively, (2) which are remarkably heterogeneous and hyperspecialized cells that (3) are interconnected by hundreds of trillions of synapses, forming the most fascinating and perhaps the most important cellular network in our body: the central nervous system. (4) As cell turnover in the human brain is almost negligible, brain cells and its networks are also highly adaptable or "plastic" allowing us to learn and constantly cope with diverse environmental cues, from early postnatal along the whole lifetime.

How are brain development and its plasticity controlled at the molecular level? these are two of the relevant questions, we will try to address in this book.

Neurogenesis can serve as example of both brain development and plasticity. Most of the neurons and glial cells that compose the six layers of mammalian neocortex arise during development. However, in restricted "niches" of the adult brain, stem cells are preserved and generate neurons and glia throughout life. Adult neurogenesis may influence learning and memory and also being regulated by the environment and activity. Thus, it is considered a form of brain's "structural" plasticity.

Interestingly, despite both embryonic and adult "neurogenesis" contribute to the generation of new neurons (and glia), these two processes are very different. In fact, during the intrauterine cortical development, neural stem cells undergo a "program" leading to the orderly generation of hundreds of neurons first, then glial cells. In contrast, adult neural stem cells are "flexible." They can give rise to very few neurons or glia simultaneously, though during the entire lifespan of an individual. Thus, embryonic and adult neurogenesis require distinct (i.e., in space and time) coordination of the intrinsic signaling pathways and extrinsic factors, leading to maintenance of stem cell pools and to the generation of their differentiated progenies.

Following these considerations, other questions that we will discuss in this book are: *Do different key players control brain development or "plasticity"? Or Vice versa, are the same players differently regulated? Which of these mechanisms were leveraged in evolution?*

Several years of studies provided important evidence of the transcriptional cascades and signaling pathways that govern neurogenesis, gliogenesis, and "plasticity" in brain. However, after the "completion" of the human-genome sequencing (though due to the presence of several hundreds of repetitive elements, the exact sequence of our genome is still not entirely known), it became apparent that less than 2% of the human DNA encodes for proteins. In contrast, in evolution, the complexity of the noncoding transcriptome has increased much more than coding one. These and many other postgenomic "surprises" instilled the idea that this 98% of the human genome *must be* "not so junk DNA." Actually, these discoveries open the possibility that the regulation of protein expression might be crucial to understand the molecular control of brain development and function. Moreover, some of these mechanisms might have been leveraged in evolution.

The idea of this book originated in the symposia *"MicroRNAs in brain development and function"* at the 9th FENS meeting in Milan, in July 2014. The seminar room was full, and delegates following the lectures from outside screens largely outnumbered those inside. This confirmed that the topic of noncoding RNAs raises a great interest in the broad field of neuroscience. Having in mind the above questions, in this book, we will focus on noncoding RNAs and epigenetics.

microRNAs (miRNAs) are the "founders" of the large family of small noncoding RNAs. The first miRNA gene was isolated in *Caenorhabditis elegans* in Sidney's Brenner laboratory in the early 1980s and characterized as the first "heterochronic mutant" by Marty Chalfie, John Sulston, and Robert Horvitz, who named this gene "*lin-4*" (lin, cell lineage abnormal). It took more than a decade before Victor Ambros, Gary Ruvkun, and colleagues discovered that the two-heterochronic genes, *lin-4* and *let-7*, did not encode for protein products but rather encode small RNAs. As of today, the field has grown dramatically: miRNAs are known to be present in most plants, insects, animals and are crucially involved in the control of "virtually all" biological pathways and human diseases.

For historical reasons, several chapters of this book, authored by leading scientists in noncoding RNA and neuroscience, focus on miRNAs. We have attempted to highlight key concepts, mechanisms, and challenges for the functional investigation of miRNAs in the framework of the neural system development, plasticity, dysfunction, and evolution. However, we also summarize novel classes of noncoding RNAs, such as piRNAs and circular RNAs, and discuss epigenetics, as well as cross talks between different noncoding RNAs and pathways with relevance to neuroscience.

The first section of this book (*Introduction and Methods*) comprises four chapters on basic biology of animal miRNAs and methods to infer their functions. Schreiner and Pasquinelli introduce miRNA biogenesis and their turnover. Then, Grosswendt and Rajewsky analyze in details miRNA-mediated mechanisms of target repression, relevance of miRNA-dependent control for silencing of "gene expression noise" and methods to directly investigate miRNA: target interactions. Hatzigeorgiou and colleagues highlight the state of the art and challenges in computational approaches to investigate miRNA functions, targets identification, and novel tools for integrative biology. Finally, Marinaro, Pons Espinal, and myself summarize current methodological approaches and unresolved challenges for the functional investigation and therapeutic use of miRNAs and siRNAs, in vivo.

The second section (*Neurogenesis*) comprises five chapters and is focused on cell biology and molecular control of embryonic and adult neurogenesis. Tavano and Huttner write on corticogenesis in rodents and in human brain evolution. Erni, Rolando, and Taylor present miRNA-dependent and -independent regulations in neural stem cells. Finally, the functions and use of epigenetics and noncoding RNAs are discussed by Liu, Abernathy, and Yoo in the context of cell reprograming; by de Chevigny, Cremer, and Coré in adult neurogenesis of the olfactory system, and by Bielefeld, Pustjens, Schouten, and Fitzsimons for the hippocampus.

The third section (*Gliogenesis*) comprises two chapters focusing on molecular and epigenetic control of gliogenesis and relevance of these pathways for diseases. Tiwari and Berninger describe transcriptional, epigenetic, and miRNA-dependent control of astrogliogenesis. Zhao, Wang, and Lu write about miRNA-dependent control of oligodendrocyte development and implications of these regulatory mechanisms in demyelinating diseases and myelin repair.

The fourth section (*Plasticity and Brain dysfunctions*) comprises two chapters discussing functions of small noncoding RNAs in the molecular control of brain physiopathology. Napoli and Pizzorusso present the functions of miRNAs and piRNAs in neuronal networks maturation and plasticity, whereas Gahagan and Cairns focus on small RNA dysregulation in neurocognitive and neuropsychiatric disorders.

The fifth and last section of this book (*Emerging complexity in small RNA classes and functions*) reviews novel types of noncoding RNAs, such as circular RNAs, ceRNA, human accelerated evolutionary genomic regions (HARs), transposable elements, and ALUs in light of comparative development and function of the neural system. Legnini and Bozzoni write on circular RNAs in the context of neural system function and regulation. Ristori and Nicoli discuss on the comparative functions of miRNAs in development of the nervous system in invertebrates versus vertebrates. Finally, in the last chapter of the book Pratt, Nowakowski, and Price write on the functions of non-coding RNAs, in the context of human neocortical evolution.

Through this book, several guiding principles will become apparent.

- Regulatory capacity and diversity of noncoding RNA-dependent control is particularly valuable in the brain. Indeed, in brain, poor cellular turnover and the constant flow of information require specialization and adaptability, also at the level of molecular control.
- Only very few miRNAs (e.g., miR-9, miR-124, miR-134, miR-132, etc.) that here we define as "hubs" seem <u>individually</u> capable of exerting relevant biological functions in brain. In contrast, most miRNAs-dependent regulations are not strong: The degree of protein down regulation by a miRNA is generally less than twofolds.
- To compensate for such low activity, <u>multiple</u> miRNAs act in <u>combinatorial</u> (also known as convergent, synergic, or cooperative) <u>manner</u> to buffer genetic noise. Combinatorial miRNA actions confer robustness to the expression of specific proteins, or to biological functions. This explains why single miRNA knockouts often don't show any phenotype. This also warrants careful consideration of results obtained upon manipulation of a single miRNA, especially in gain of function studies.
- It therefore follows that miRNAs seem to evolve more rapidly than protein coding RNAs. Indeed, de novo miRNAs are well tolerated by evolution and mutations in precursor and mature miRNAs arise more frequently in "young" miRNAs compared to "ancient" miRNAs. Although this exciting evidence indicates that our miRNA repertoire is evolving right now, it also warrants careful consideration for their use as biomarkers of disease.

On the other hand, several open questions and challenges emerging from this book remain unanswered. Even just considering miRNAs, perhaps the best characterized class of noncoding RNAs, we still do not know much. For example, transcriptional control of miRNA in brain is poorly investigated. Most of the miRNA targets have not been identified or experimentally validated. Moreover, with respect to the use of miRNAs for brain therapy, we still miss efficient delivery methods, as well as crucial information about their toxicity and side effects. Finally, we know very little about the integration of miRNAs with epigenetics and other noncoding RNA classes and mechanisms, such as circular RNAs and piRNAs.

Provocative hypotheses are also discussed in this volume. For example, in contrast to lower vertebrates, it seems that sequestering miRNAs is not a general phenomenon for mammalian circular RNAs. This finding therefore opens the intriguing possibility that mammalian circular RNAs might be translated. If just a subset of them will be found to be translated, a complete new set of mammalian proteome would be discovered.

In light of this evidence, it is clear that we still have much work to do. A simple search of "miRNA" or "microRNA" in PubMed (www.ncbi.nlm.nih.gov/pubmed) retrieves more than 12k publications in 2015, and certainly new classes of noncoding RNAs are being discovered, while we are writing this book.

With these 16 chapters, we aim to provide the "essential" knowledge of "noncoding RNA" biology, function(s), and their exploitation in the context of "neuroscience". Indeed, we made an effort to use a language that will be readily intelligible to any student or scientist from different fields and disciplines, thus to reach the widest possible audience. We hope this selection will not reflect too much our personal opinion and biases and apologize to those researchers whose important work we were not able to cite because of space limitations. Our ultimate goal is to equip readers with the instruments to evaluate the functions of noncoding RNAs in healthy and diseased brain and their potential exploitations for therapy. We hope that after reading this book many of you will *take a walk on the wild side* of the transcriptome.

Davide De Pietri Tonelli
Istituto Italiano di Tecnologia, Genoa, Italy

Acknowledgments

I wish to thank all the authors of the chapters for scholarly presentations of current knowledge and visionary interpretations of emerging trends in noncoding RNAs, in their respective fields. I am grateful to *Elsevier's* staff for their advice and diligence in particular to Natalie Farra for triggering the idea of this book and to Kathy Padilla for excellent help in its implementation. Finally, I thank Drs. Luca Berdondini, Paolo Decuzzi (IIT-Genoa), and all members of my laboratory for valuable discussion and Cinzia Nasso for great assistance in the preediting process of this volume.

MAKING AND MAINTAINING MicroRNAs IN ANIMALS

1

William P. Schreiner and Amy E. Pasquinelli
University of California at San Diego, La Jolla, CA, United States

INTRODUCTION

It has now been over 20 years since the first microRNA (miRNA) genes were discovered as critical regulators of developmental transitions in *Caenorhabditis elegans* worms (Lee et al., 1993; Wightman et al., 1993). Since then, thousands of miRNA genes have been documented in plants and animals (Griffiths-Jones, 2004; Kozomara and Griffiths-Jones, 2014). As a class, miRNAs are essential regulators of gene expression in multicellular organisms, and mis-regulation of specific miRNAs can result in abnormal phenotypes in model organisms and disease in humans (Hammond, 2015; Lin and Gregory, 2015; Tuna et al., 2016). Thus understanding how miRNAs are produced and maintained in the endogenous context is paramount for realizing their full biological functions.

MiRNAs are approximately 22 nucleotides (nt) in length and function by guiding Argonaute (AGO) and associated proteins, called the miRNA-induced silencing complex (miRISC), to their messenger RNA (mRNA) targets. Once there, miRISC generally represses gene expression through translational inhibition and mRNA degradation (Jonas and Izaurralde, 2015). In animals, miRNAs use partial base-pairing to recognize their targets (Ha and Kim, 2014). This property allows one miRNA to bind a variety of target site sequences. In fact, it has been estimated that over 50% of the human genome is regulated by miRNAs (Friedman et al., 2009).

The mature, 22 nt form of the miRNA is a product of multiple processing steps (Fig. 1.1). Most animal miRNAs share a common, well-conserved biogenesis pathway, although there are also non-canonical routes for specific miRNAs (Finnegan and Pasquinelli, 2013; Ha and Kim, 2014; Xie and Steitz, 2014). In the most common pathway, RNA polymerase II transcribes much longer primary miRNAs (pri-miRNAs) that contain 5′m7G caps and 3′polyadenosine (polyA) tails. The microprocessor, consisting of Drosha and Digeorge syndrome critical region gene 8 (DGCR8; also known as Pasha), then cleaves the pri-miRNA into a 70 nt precursor miRNA (pre-miRNA). This pre-miRNA is exported from the nucleus to the cytoplasm by Exportin 5. In the cytoplasm, Dicer cleaves the terminal loop generating a 22 nt double stranded miRNA. Only one strand of the duplex is used for targeting (guide) while the other strand (passenger) is discarded.

While each tissue of a multicellular animal expresses its own set of miRNAs, neuronal cells seem particularly reliant on the miRNA pathway. Of the hundreds of miRNAs discovered, nearly half are expressed to some degree in the mammalian brain (Shao et al., 2010). Early miRNA studies in model systems pointed to the functional importance of this pathway in neuronal development

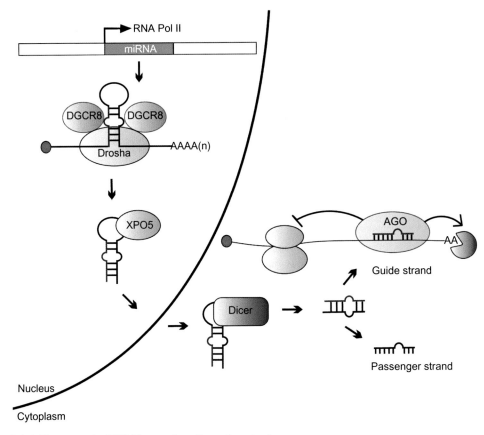

FIGURE 1.1 The general miRNA biogenesis pathway in animals

MiRNAs are transcribed in the nucleus by RNA polymerase II. The resulting primary transcript (pri-miRNA) is processed by the microprocessor complex comprised of Drosha and DGCR8. The excised precursor miRNA (pre-miRNA) is then exported from the nucleus by Exportin 5. Once in the cytoplasm the pre-miRNA is cleaved by Dicer. The guide strand of the transient 22 nt double stranded Dicer product is selected for incorporation into AGO, forming the miRISC. Typically miRISC binds the 3′UTR of a target mRNA and triggers translational inhibition and mRNA destabilization.

and function. For example, brain morphogenesis fails in Zebrafish defective for general miRNA biogenesis (Giraldez et al., 2005). In *C. elegans*, specific miRNAs were found to be essential for controlling the fate of two asymmetric chemosensory neurons (Chang et al., 2004; Johnston and Hobert, 2003). A wide body of research has now shown that miRNAs are important for a variety of neuronal processes, such as neurogenesis, axon guidance, and mature neuron functioning (Schratt, 2009; Shi et al., 2010).

MiRNAs biogenesis is an essential process for effective miRNA-mediated gene silencing. Small perturbations in miRNA biogenesis can alter the mature miRNA sequence or levels and, thus, affect gene expression. Here we provide an overview of the miRNA biogenesis pathway in animals, with an emphasis on examples relevant to neurogenesis.

GENOMIC ORGANIZATION

MiRNAs genes reside in a variety of genomic arrangements that differ in frequency across organisms. In mammals the majority of miRNAs are located within the intronic regions of genes (Chang et al., 2015; Kim and Kim, 2007; Monteys et al., 2010; Rodriguez et al., 2004; Saini et al., 2008). There are also rare examples of exonic sequences, usually 3′ untranslated regions (UTRs), within mRNAs that encode miRNAs (Chang et al., 2015; Kim and Kim, 2007; Rodriguez et al., 2004). MiRNAs that are found in host genes, whether they are exonic or intronic, are often cotranscribed with their protein-coding counterpart (Baskerville and Bartel, 2005; Rodriguez et al., 2004). However, this is not always the case, and some miRNAs that are located within host genes have their own transcriptional regulatory elements (Corcoran et al., 2009; Monteys et al., 2010; Ozsolak et al., 2008). In *C. elegans*, most miRNA genes are intragenic but even the ones located within protein-coding genes typically have their own promoters (Martinez et al., 2008).

MiRNAs are often arranged in clusters, where closely spaced miRNAs are cotranscribed as part of a common primary transcript (Bartel, 2009; Lau et al., 2001; Lee et al., 2002). miRNA clusters sometimes encode members of the same family of miRNAs, where each miRNA has identical 5′ end sequences (Bartel, 2009). This arrangement is thought to facilitate the expression of related miRNAs that are capable of regulating shared targets. However, not all clusters include related miRNAs and some contain multiple family members. For example, the miR-17−92 cluster in humans consists of six miRNAs that belong to four different miRNA families: miR-17 and miR-20a are part of the miR-17 family, miR-19a and miR-19b-1 are part of the miR-19 family, and miR-18 and miR-92a-1 are each part of their own distinct families (Ota et al., 2004). This cluster is often perturbed in cancers, and may also be important for regulating the expression of genes involved in neurodegenerative diseases (He et al., 2005; Mogilyansky and Rigoutsos, 2013).

MicroRNA TRANSCRIPTION

The biogenesis of most miRNAs initiates with transcription by RNA polymerase II (Bracht et al., 2004; Cai et al., 2004; Lee et al., 2004). The resulting pri-miRNAs vary widely in size from hundreds to thousands of nts long (Chang et al., 2015; Saini et al., 2008). For details on the identification of miRNA transcription start sites see Chapter 3, Computational and -Omics Approaches for the Identification of miRNAs and Targets, of this book by Hatzigeorgiou and colleagues. In general the transcription of miRNAs is controlled by the same mechanisms that govern the synthesis of protein coding mRNAs. In addition to regulation by chromatin and DNA modifications, specific transcription factors have been shown to control the expression of particular miRNAs during development and in response to various extrinsic conditions (Liu et al., 2013; Marson et al., 2008; Schanen and Li, 2011). For instance, the well-studied p53 transcriptional regulator not only controls the expression of numerous protein coding genes but also stimulates the transcription of specific miRNAs, including members of the miR-34 family, upon DNA damage (Bommer et al., 2007; Chang et al., 2007; Corney et al., 2007; He et al., 2007; Raver-Shapira et al., 2007; Tarasov et al., 2007; Tazawa et al., 2007). Up-regulation of miR-34 family miRNAs is critical for the tumor suppressive function of p53 because these miRNAs repress the expression of cell proliferation genes and promote apoptosis (Bommer et al., 2007; Chang et al., 2007; Corney et al., 2007; He et al., 2007; Raver-Shapira et al., 2007; Tarasov et al., 2007; Tazawa et al., 2007). Interestingly, miR-34 levels also increase during adulthood in mammals, flies, and worms (Cao et al., 2010; de Lencastre et al., 2010; Ibanez-Ventoso et al., 2006; Li et al., 2011a,b; Liu et al., 2012), and in *Drosophila* this has been shown to have a neuroprotective role (Liu et al., 2012). Whether this induction is also mediated by p53 or other transcription factors is yet to be determined.

A recurring motif in the miRNA pathway is transcriptional regulation through negative feedback loops. One striking example of such regulation involves miR-133b and pituitary homeobox 3 (PITX3) in mouse dopaminergic neurons (Kim et al., 2007). PITX3, a homeodomain transcription factor, promotes the expression of miR-133b. However, PITX3 contains miR-133b binding sites in its 3′UTR. Thus rising levels of miR-133b lead to repression of PITX3, halting its own transcription (Kim et al., 2007). An even more intricate loop exists for miR-273 and lsy-6 in a pair of worm chemosensory neurons. These miRNAs, along with the transcription factors DIE-1 and COG-1, form a double negative feedback loop that is important for determining the left/right fate of two *C. elegans* head neurons (Johnston et al., 2005). In this example the left neuron expresses *die-1*, which induces transcription of lsy-6 miRNA, a repressor of *cog-1*; in the right neuron, this pathway is inhibited through the expression of miR-273, which targets *die-1* for down-regulation (Johnston et al., 2005).

PRIMARY MicroRNA PROCESSING

Pri-miRNA processing occurs in the nucleus and is often cotranscriptional (Ballarino et al., 2009; Morlando et al., 2008; Pawlicki and Steitz, 2008). In most cases an enzyme complex termed the microprocessor recognizes and excises the approximately 70 nt stem loop pre-miRNA (Denli et al., 2004; Gregory et al., 2004; Han et al., 2004; Landthaler et al., 2004). Typically, the miRNA sequence is embedded in a base-paired stem that is flanked by a basal region that consists of an ~ 11 base pair (bp) lower stem that is preceded by unpaired sequences and a terminal loop of ~ 20 nt (Han et al., 2006; Zeng et al., 2005). A set of primary and secondary sequence motifs have been shown to regulate pri-miRNA processing in human cells. First the distance from the junction between the basal single-stranded and double stranded region to the cut site appears to be critical for accurate cleavage, with an 11 bp distance being ideal (Auyeung et al., 2013; Han et al., 2006; Zeng et al., 2005). Second, three important motifs reside in efficiently cleaved miRNAs: a UG at the basal junction, a UGUG in the apical loop, and a CNNC motif downstream of the stem loop (Auyeung et al., 2013). Additionally, base-pairing at all nts except nt 8, as measured from the basal junction, is a feature in the stem structure of efficiently processed primary miRNAs (Fang and Bartel, 2015). While these elements have been proposed to distinguish microprocessor substrates from other structured RNAs in the cell, they are not universally shared by all miRNAs or across species. The consensus seems to be that pri-miRNA processing is modular, with different elements contributing to recognition and cleavage by the microprocessor (Auyeung et al., 2013; Nguyen et al., 2015).

Recent studies have demonstrated that structural remodeling of some miRNA primary transcripts is necessary for efficient recognition by the microprocessor. In *C. elegans* the *let-7* gene encodes primary transcripts that undergo a form of splicing where the 5′ region is removed and replaced with splice leader sequence (Bracht et al., 2004; Mondol et al., 2015). This processing event enables the pri-miRNA to adopt a conformation that makes it a more favorable substrate for Drosha processing (Mondol et al., 2015). The mammalian miR-17−92 clustered miRNA gene is another example where restructuring of the primary transcript is important for microprocessor recognition (Du et al., 2015). The nascent pri-miR-17−92 transcript folds into a structure that blocks processing of all but one of the six miRNAs in this cluster. Cleavage of this transcript to form a progenitor-miRNA removes the sequences that create the inhibitory structure, allowing for efficient processing of all miRNAs in the cluster. Thus upstream processing events that regulate pri-miRNA structure can control their entrance into the miRNA biogenesis pathway.

The microprocessor consists of Drosha and DGCR8 proteins and is essential for the production of most miRNAs. Mice lacking either DGCR8 or Drosha are not viable (Chong et al., 2010; Wang et al., 2007). Monoallelic loss of DGCR8 and other genes on chromosome 22 is associated with Digeorge syndrome (de la Chapelle et al., 1981). Patients with Digeorge syndrome display cognitive defects and have a higher than normal chance of developing schizophrenia (Karayiorgou et al., 1995, 2010). In a mouse model of Digeorge syndrome, haploinsufficiency of Dgcr8 alone led to altered biogenesis of specific miRNAs and various cognitive abnormalities in the animals (Fenelon et al., 2011; Stark et al., 2008). These findings indicate that neuronal function may be particularly sensitive to changes in miRNA dose.

As the catalytic subunit of the microprocessor, Drosha is a ribonuclease (RNase) type III enzyme with two RNase III domains: RIIIDa and RIIIDb (Denli et al., 2004; Gregory et al., 2004; Han et al., 2004; Landthaler et al., 2004). Each of these domains cuts on either side of the RNA stem at ∼11 bp above the basal junction. This cleavage event usually results in a 2 nt overhang on the 3′ end of the product, which is important for subsequent Dicer processing. DGCR8 contains two RNA binding domains and assists in pri-miRNA processing by binding the pri-miRNA near the terminal loop structure. Recent studies have shown that one Drosha and two DGCR8 proteins interact with pri-miRNAs in a heterotrimeric complex with Drosha positioned at the cleavage sites and DGCR8 proteins bound to each RNase III domain (Herbert et al., 2016; Kwon et al., 2016; Nguyen et al., 2015). This arrangement enables the microprocessor to evaluate multiple elements in the pri-miRNA that contribute to accurate and efficient cleavage.

While the microprocessor can cleave pri-miRNAs in vitro a variety of cofactors regulate this step in vivo (Finnegan and Pasquinelli, 2013). An example pertinent to neuronal function is the involvement of several amyotrophic lateral sclerosis (ALS) associated genes in miRNA biogenesis (Eitan and Hornstein, 2016). The fused in sarcoma/translocated in liposarcoma (FUS/TLS) and TDP-43 (TAR DNA-binding protein 43) proteins have been shown to promote the expression of several miRNAs important for neuronal function (Buratti et al., 2010; Gregory et al., 2004; Kawahara and Mieda-Sato, 2012; Ling et al., 2013; Morlando et al., 2012). These proteins interact with the microprocessor and seem to stabilize its recruitment to particular miRNA genes. ALS is a neurodegenerative disease where progressive loss of motor neurons leads to muscle wasting and death. A molecular signature of ALS is the accumulation of pathological aggregates that often include FUS/TLS and TDP-53 (Arai et al., 2006; Deng et al., 2010; Neumann et al., 2006). Thus depletion of functional FUS/TLS and TDP-53 may result in inadequate processing of miRNAs required for neuronal maintenance; for more details on noncanonical functions of the microprocessor, see Chapter 6, MiRNA-Dependent and Independent Functions of the Microprocessor in the Regulation of Neural Stem Cell Biology, of this book.

EXPORT FROM THE NUCLEUS

The newly processed pre-miRNA must be exported from the nucleus to the cytoplasm for further maturation steps. This is accomplished by the nuclear transport factor Exportin 5 and its GTP bound Ran cofactor (Bohnsack et al., 2004; Lund et al., 2004; Yi et al., 2003). Exportin 5 preferentially binds pre-miRNAs with canonical structures, including the 2 nt $3'$ overhang and a paired stem (Okada et al., 2009). Upon nuclear exit, GTP (guanosine triphosphate) hydrolysis by Ran triggers release of the pre-miRNA into the cytoplasm. Depletion of Exportin 5 leads to decreased mature and pre-miRNA levels, indicating that pre-miRNAs trapped in the nucleus may be subject to degradation (Lund et al., 2004; Yi et al., 2003). Inactivating mutations in Exportin 5 have been detected in human colon, stomach, and endometrial tumors (Melo et al., 2010). In cancer cells lines expressing mutant Exportin 5 a subset of miRNAs were significantly down-regulated compared to cells supplemented with wildtype Exportin 5. Since many of these Exportin 5 dependent miRNAs have potential tumor suppressive functions, their depletion in tumor cells with defective Exportin 5 activity may contribute to the oncogenic phenotype. Consistent with the observation that some miRNAs may be more sensitive to the loss of Exportin 5, a recent study that examined the effect of Exportin 5 deletion in human cells found only a modest decrease in the abundance of most miRNAs (Kim et al., 2016). Thus alternative pathways must exist for the delivery of precursor miRNAs to the cytoplasm for final maturation (Xie et al., 2013). The nonessential nature of Exportin 5 for miRNA biogenesis is also consistent with the lack an obvious homolog of this factor in organisms such as *C. elegans* (Bussing et al., 2010).

PRE-MicroRNA PROCESSING

After exiting the nucleus the pre-miRNA needs to be processed further to produce the mature miRNA. Dicer, an RNase III enzyme, catalyzes this step by excising the terminal loop, which results in a ~ 22 nt dsRNA that contains the future mature miRNA. Dicer was initially discovered in *Drosophila* as the enzyme responsible for generating small interfering RNA (siRNA) guides from longer dsRNAs used to induce RNA interference (RNAi) (Bernstein et al., 2001). Subsequent work by many labs determined that an essential function for Dicer is the processing of endogenous miRNA precursors (Grishok et al., 2001; Hutvagner et al., 2001; Ketting et al., 2001; Knight and Bass, 2001). Since Dicer is required for the maturation of almost all miRNAs (Kim et al., 2016), it is not surprising that global loss of this factor is associated with severe pleiotropic phenotypes and eventual lethality (Bernstein et al., 2001; Giraldez et al., 2005; Grishok et al., 2001; Ketting et al., 2001; Knight and Bass, 2001). Directed knock out of Dicer in specific types of neurons has demonstrated the importance of this factor in the differentiation, function, and survival of these cells (Cuellar et al., 2008; Damiani et al., 2008; Davis et al., 2008; Haramati et al., 2010; Schaefer et al.,

2007; Shin et al., 2009; Tao et al., 2011). Tissue specific loss of Dicer, and hence miRNA maturation, has also been used to reproduce disease phenotypes in model systems. For example, ablation of Dicer in mouse postmitotic motor neurons results in defects characteristic of spinal muscular atrophy, including neurodegeneration and muscle wasting (Haramati et al., 2010).

Similar to Drosha, Dicer contains two RNase III domains, and recent structural work suggests an evolutionary connection between these two factors (Kwon et al., 2016). The RIIIDb domain and the RIIIDa domain cleave the miRNA near the terminal loop at the 5′ and 3′ ends, respectively (Lau et al., 2012; Zhang et al., 2004). To position the cleavage sites, it has been postulated that Dicer acts as a molecular ruler, binding to the free 5′ and 3′ ends at the base of the stem and measuring 22 nts toward the loop region (Macrae et al., 2006; Park et al., 2011).

Like the microprocessor, Dicer function is also regulated by a variety of cofactors. In mammals the HIV-1 TAR RNA binding protein (TRBP) interacts with Dicer (Chendrimada et al., 2005; Haase et al., 2005). In addition to stabilizing Dicer protein, TRBP also influences the cleavage size and miRNA strand selected for loading into AGO (Fukunaga et al., 2012; Lee et al., 2013; Paroo et al., 2009; Wilson et al., 2015). Protein activator of PKR (PACT) is another Dicer partner that regulates the fidelity of pre-miRNA maturation (Lee et al., 2006). Structural studies have revealed that PACT and TRBP bind to the same amino acid stretch in Dicer (Wilson et al., 2015). Mutation of this surface in Dicer abolishes interaction with either cofactor or results in altered cleavage sizes and defective strand selection for a subset of miRNAs (Wilson et al., 2015). It is presently unclear why certain miRNAs are more dependent on TRBP or PACT for accurate maturation.

Considering its essential role in the biogenesis of almost all miRNAs, it is intriguing that the expression and function of Dicer is repressed at multiple levels. Through a negative feedback loop, Dicer itself is repressed by the miRNA pathway. In humans the Dicer mRNA contains several let-7 complementary sites that mark it for destabilization and translational inhibition by miRISC (Forman et al., 2008; Tokumaru et al., 2008). During zebrafish hindbrain development, the expression of miR-107 targets Dicer for down-regulation, which results in decreased biogenesis of miR-9 (Ristori et al., 2015). Reduced accumulation of miR-9 was shown to be important for preventing over-proliferation of neurons at the expense of differentiation. The activity of human Dicer protein has been shown to be regulated by its N-terminal helicase domain (Ma et al., 2008). Deletion of this domain greatly increases its catalytic efficiency. Autoregulation of Dicer activity in vivo is presumably mediated by structural rearrangement of the helicase domain, possibly by cofactor binding or modifications.

The best characterized inhibitors of Dicer function act on its pre-miRNA substrates. In one example, general repression of pre-miRNA processing occurs during the macrophage inflammatory response through induction of monocyte chemoattractant protein 1-induced protein (MCPIP1) (Suzuki et al., 2011). MCPIP1 cleaves the loop regions of pre-miRNAs, which reduces their recognition by Dicer and makes them vulnerable to rapid degradation. In contrast to global miRNA down-regulation, processing of specific miRNA precursors by Dicer is regulated by the RNA binding protein LIN28 (Heo et al., 2008; Newman et al., 2008; Rybak et al., 2008; Viswanathan et al., 2008). In mammalian cells, LIN28 recognizes sequences in the loop region of let-7 family miRNAs (Heo et al., 2008; Loughlin et al., 2012; Nam et al., 2011; Newman et al., 2008; Piskounova et al., 2008). While binding of LIN28 to let-7 precursor miRNAs is sufficient to block Dicer, this interaction also recruits RNA uridylyltransferase enzymes that add U tails to the pre-miRNA 3′ ends (Heo et al., 2009; Thornton et al., 2012). The uridylated RNA is then a substrate for rapid decay by

Dis3l2 (Chang et al., 2013; Ustianenko et al., 2013). This regulatory pathway may be integral to the effects of brain-derived neurotrophic factor (BDNF) on neuronal survival, structure, and synapse function in the mammalian brain (Huang et al., 2012). In cultured mammalian neurons, BDNF was found to induce the expression of LIN28, which led to the down-regulation of let-7 miRNAs. The resulting de-repression of let-7 targets proved necessary for the ability of BDNF to stimulate the outgrowth of neuronal dendrites in culture.

MATURE MicroRNA COMPLEX FORMATION

The final step of miRNA maturation involves selection of the single stranded miRNA for loading onto AGO (Dueck and Meister, 2014). During this process, one half of the ∼22 nt duplex is chosen as the guide sequence and stably bound by AGO. The other strand, called the passenger or star, is discarded and presumably degraded through poorly defined mechanisms. MiRNA strand selection is not a random process as only one half of the Dicer product accumulates as the mature miRNA in most cases. Although the choice of one strand over the other is not always predictable, the favored strand often contains certain structural and sequence features. The guide strand typically exhibits looser 5′ base pairing to its partner, which may facilitate its incorporation into AGO (Khvorova et al., 2003; Schwarz et al., 2003). Additionally, miRNA guide strands usually have a 5′ uracil and are purine rich (Hu et al., 2009). The Dicer interacting proteins, TRBP and PACT, also influence strand selection (Lee et al., 2013). The current model suggests that the association of Dicer and TRBP/PACT with AGO aids in the selection of the guide strand from the newly processed miRNA duplex. Occasionally, passenger strands are detected in miRISC and, thus, have the potential to regulate distinct sets of genes (Okamura et al., 2008).

The fidelity of strand selection can have major implications for gene expression since target regulation by miRISC is dependent on the miRNA sequence. Upon binding to AGO the mature miRNA is positioned to search for targets with complementary sequences. Structural studies have shown that initially nts 2−5 in the miRNA, called the seed region, are available for pairing (Elkayam et al., 2012; Nakanishi et al., 2012; Schirle and MacRae, 2012). Binding to a target sequence then induces a structural rearrangement that exposes nts 6−8 for additional interactions with the target (Schirle et al., 2014). Additionally, this conformational change-induced by seed pairing positions sequences in the miRNA 3′ end for possible pairing interactions (Schirle et al., 2014). These studies have reinforced the importance of 5′ seed pairing interactions and raised the possibility of additional sequences contributing to the recognition of specific target sites.

While AGO and its miRNA guide are responsible for binding specific targets, additional factors are required to regulate the expression of the bound mRNA. GW182 proteins bind directly to AGO and recruit factors that trigger translational inhibition and decay of the target mRNA (Pfaff and Meister, 2013). The mechanism of translational repression by miRISC is not well understood but, in most cases, it is also coupled with mRNA destabilization (Jonas and Izaurralde, 2015). In animals, GW182 proteins associated with miRISC recruit deadenylation factors to initiate decay of the mRNA (Behm-Ansmant et al., 2006; Jakymiw et al., 2005; Liu et al., 2005; Meister et al., 2005; Rehwinkel et al., 2005). Shortening of the polyA tail is thought to weaken interactions between the mRNA termini, which exposes the 5′ end for decapping. Upon removal of the 5′m^{7}G cap the mRNA is rapidly degraded by 5′ to 3′ exonucleases. For a detailed examination of how miRISC regulates target gene expression, see Chapter 2, Essentials of MicroRNA-Dependent

Control of mRNA Translation and Decay, MicroRNA Targeting Principles and Methods for Target Identification, of this book.

MicroRNA STABILITY

Compared to mRNAs, miRNAs as a class are exceptionally stable molecules. Half-life studies in mouse fibroblast cells have shown that the average miRNA can persist for several days (Gantier et al., 2011). A general stabilizing factor for miRNAs is their association with AGO. The cellular abundance of mature miRNAs can be directly regulated by the availability of AGO (Diederichs and Haber, 2007; Winter and Diederichs, 2011). When bound by AGO the miRNA termini are likely less accessible to exonucleases. Curiously though, target engagement by miRISC has been found to increase or decrease the stability of the bound miRNA depending on the circumstances. In *C. elegans*, target interactions protect at least some miRNAs from decay (Chatterjee et al., 2011; Chatterjee and Grosshans, 2009). This target-mediated miRNA protection is thought to prevent release of the miRNA from AGO, safeguarding it from the 5′ to 3′ exonucleases XRN-1/-2. Conversely, pairing of miRNAs to target sites in mammalian and *Drosophila* cells has been reported to trigger their destabilization (Cazalla et al., 2010; de la Mata et al., 2015). Extensive base pair interactions were found to induce the addition of nts to the 3′ end of the miRNA, which then serve as a signal for 3′−5′ degradation (Ameres et al., 2010). Recent work in rodent primary neurons has revealed a setting where this pathway is particularly robust (de la Mata et al., 2015). Even in these cells, though, extensive pairing, beyond what is possible with most natural miRNA targets sites, was required for target RNA-directed miRNA degradation. Thus it remains to be determined if target engagement has a prominent role in regulating miRNA stability in vivo.

Several nucleases have been implicated in miRNA degradation. As mentioned above the 5′ to 3′ exonucleases XRN-1/-2 can digest miRNAs released from AGO when target sites are scarce (Chatterjee et al., 2011; Chatterjee and Grosshans, 2009). In worms the decapping scavenger protein stimulates miRNA decay by helping to recruit XRN-1 to these substrates (Bosse et al., 2013). The 3′ to 5′ exoribonuclease, ERI1 appears to be a general regulator of miRNA stability in mammalian cells, as loss of this factor resulted in global up-regulation of mature miRNA levels (Thomas et al., 2014). Other decay enzymes seem to be more selective in their miRNA targets. For example, polynucleotide phosphorylase, an interferon inducible 3′ to 5′ exoribonuclease, specifically targets miR-221 for rapid degradation in human melanoma cells (Das et al., 2010).

While the mechanisms responsible for controlling the turnover of global or select miRNAs are still being elucidated the regulation of miRNA stability seems especially active in neuronal cells. Dynamic miRNA stability has been proposed to be important for learning. In the snail *Aplysia*, exposure to serotonin resulted in down-regulation of the mature levels of several brain-enriched miRNAs (Rajasethupathy et al., 2009). Serotonin is a neurotransmitter released during learning that promotes the activation of CREB (cAMP Response Element Binding), a transcription factor important for pathways involved in long-term memory. One output of serotonin signaling is the rapid decay of miR-124, which relieves CREB from repression by this miRNA.

In another example of induced miRNA decay a specific set of miRNAs become destabilized in response to dark adaptation in mouse retinal cells (Krol et al., 2010). This change in miRNA levels was shown to be relevant for derepression of a target important for photoreceptor function.

Interestingly, this study also demonstrated that miRNAs in general have decreased stability in neuronal cells (Krol et al., 2010).

CONCLUDING REMARKS

In the relatively short amount of time since their discovery, miRNAs have risen to prominence as essential regulators of diverse biological pathways. A key to realizing their roles in cell specification and function is an understanding of how miRNAs are produced and maintained. Neuronal cells in particular seem to engage a myriad of factors and mechanisms to regulate the expression of miRNAs required for their various developmental pathways and activities. The basic steps involved in miRNA biogenesis have been defined. The current challenge is to discover how this process is regulated in different cell types, in response to various external stimuli, and in disease situations.

ACKNOWLEDGEMENTS

We thank members of the Pasquinelli Lab for critical reading of the manuscript. Support for this study was provided by the UCSD Cellular and Molecular Genetics Training Program through an institutional grant from the National Institute of General Medicine (T32 GM007240) to W.P.S. and the National Institutes of Health (NIHGM071654) to A.E.P.

REFERENCES

Ameres, S.L., Horwich, M.D., Hung, J.H., Xu, J., Ghildiyal, M., Weng, Z., et al., 2010. Target RNA-directed trimming and tailing of small silencing RNAs. Science 328, 1534−1539.

Arai, T., Hasegawa, M., Akiyama, H., Ikeda, K., Nonaka, T., Mori, H., et al., 2006. TDP-43 is a component of ubiquitin-positive tau-negative inclusions in frontotemporal lobar degeneration and amyotrophic lateral sclerosis. Biochem. Biophys. Res. Commun. 351, 602−611.

Auyeung, V.C., Ulitsky, I., McGeary, S.E., Bartel, D.P., 2013. Beyond secondary structure: primary-sequence determinants license pri-miRNA hairpins for processing. Cell 152, 844−858.

Ballarino, M., Pagano, F., Girardi, E., Morlando, M., Cacchiarelli, D., Marchioni, M., et al., 2009. Coupled RNA processing and transcription of intergenic primary microRNAs. Mol. Cell. Biol. 29, 5632−5638.

Bartel, D.P., 2009. MicroRNAs: target recognition and regulatory functions. Cell 136, 215−233.

Baskerville, S., Bartel, D.P., 2005. Microarray profiling of microRNAs reveals frequent coexpression with neighboring miRNAs and host genes. RNA 11, 241−247.

Behm-Ansmant, I., Rehwinkel, J., Doerks, T., Stark, A., Bork, P., Izaurralde, E., 2006. mRNA degradation by miRNAs and GW182 requires both CCR4:NOT deadenylase and DCP1:DCP2 decapping complexes. Genes Dev. 20, 1885−1898.

Bernstein, E., Caudy, A.A., Hammond, S.M., Hannon, G.J., 2001. Role for a bidentate ribonuclease in the initiation step of RNA interference. Nature 409, 363−366.

Bohnsack, M.T., Czaplinski, K., Gorlich, D., 2004. Exportin 5 is a RanGTP-dependent dsRNA-binding protein that mediates nuclear export of pre-miRNAs. RNA 10, 185−191.

Bommer, G.T., Gerin, I., Feng, Y., Kaczorowski, A.J., Kuick, R., Love, R.E., et al., 2007. p53-mediated activation of miRNA34 candidate tumor-suppressor genes. Curr. Biol. 17, 1298−1307.

Bosse, G.D., Ruegger, S., Ow, M.C., Vasquez-Rifo, A., Rondeau, E.L., Ambros, V.R., et al., 2013. The decapping scavenger enzyme DCS-1 controls microRNA levels in *Caenorhabditis elegans*. Mol. Cell. 50, 281−287.

Bracht, J., Hunter, S., Eachus, R., Weeks, P., Pasquinelli, A.E., 2004. Trans-splicing and polyadenylation of let-7 microRNA primary transcripts. RNA 10, 1586−1594.

Buratti, E., De Conti, L., Stuani, C., Romano, M., Baralle, M., Baralle, F., 2010. Nuclear factor TDP-43 can affect selected microRNA levels. FEBS J. 277, 2268−2281.

Bussing, I., Yang, J.S., Lai, E.C., Grosshans, H., 2010. The nuclear export receptor XPO-1 supports primary miRNA processing in *C. elegans* and Drosophila. EMBO J. 29, 1830−1839.

Cai, X., Hagedorn, C.H., Cullen, B.R., 2004. Human microRNAs are processed from capped, polyadenylated transcripts that can also function as mRNAs. RNA 10, 1957−1966.

Cao, K., Chen-Plotkin, A.S., Plotkin, J.B., Wang, L.S., 2010. Age-correlated gene expression in normal and neurodegenerative human brain tissues. PLoS ONE 5. doi:10.1371/journal.pone.0013098.

Cazalla, D., Yario, T., Steitz, J.A., 2010. Down-regulation of a host microRNA by a Herpesvirus saimiri noncoding RNA. Science 328, 1563−1566.

Chang, H.M., Triboulet, R., Thornton, J.E., Gregory, R.I., 2013. A role for the Perlman syndrome exonuclease Dis3l2 in the Lin28-let-7 pathway. Nature 97, 244−248.

Chang, S., Johnston Jr., R.J., Frokjaer-Jensen, C., Lockery, S., Hobert, O., 2004. MicroRNAs act sequentially and asymmetrically to control chemosensory laterality in the nematode. Nature 430, 785−789.

Chang, T.C., Wentzel, E.A., Kent, O.A., Ramachandran, K., Mullendore, M., Lee, K.H., et al., 2007. Transactivation of miR-34a by p53 broadly influences gene expression and promotes apoptosis. Mol. Cell. 26, 745−752.

Chang, T.C., Pertea, M., Lee, S., Salzberg, S.L., Mendell, J.T., 2015. Genome-wide annotation of microRNA primary transcript structures reveals novel regulatory mechanisms. Genome Res. 25, 1401−1409.

Chatterjee, S., Grosshans, H., 2009. Active turnover modulates mature microRNA activity in Caenorhabditis elegans. Nature 461, 546−549.

Chatterjee, S., Fasler, M., Bussing, I., Grosshans, H., 2011. Target-mediated protection of endogenous microRNAs in *C. elegans*. Dev. Cell. 20, 388−396.

Chendrimada, T.P., Gregory, R.I., Kumaraswamy, E., Norman, J., Cooch, N., Nishikura, K., et al., 2005. TRBP recruits the Dicer complex to Ago2 for microRNA processing and gene silencing. Nature 436, 740−744.

Chong, M.M., Zhang, G., Cheloufi, S., Neubert, T.A., Hannon, G.J., Littman, D.R., 2010. Canonical and alternate functions of the microRNA biogenesis machinery. Genes Dev. 24, 1951−1960.

Corcoran, D.L., Pandit, K.V., Gordon, B., Bhattacharjee, A., Kaminski, N., Benos, P.V., 2009. Features of mammalian microRNA promoters emerge from polymerase II chromatin immunoprecipitation data. PLoS ONE 4, e5279.

Corney, D.C., Flesken-Nikitin, A., Godwin, A.K., Wang, W., Nikitin, A.Y., 2007. MicroRNA-34b and MicroRNA-34c are targets of p53 and cooperate in control of cell proliferation and adhesion-independent growth. Cancer Res. 67, 8433−8438.

Cuellar, T.L., Davis, T.H., Nelson, P.T., Loeb, G.B., Harfe, B.D., Ullian, E., et al., 2008. Dicer loss in striatal neurons produces behavioral and neuroanatomical phenotypes in the absence of neurodegeneration. Proc. Natl. Acad. Sci. USA 105, 5614−5619.

Damiani, D., Alexander, J.J., O'Rourke, J.R., McManus, M., Jadhav, A.P., Cepko, C.L., et al., 2008. Dicer inactivation leads to progressive functional and structural degeneration of the mouse retina. J. Neurosci. 28, 4878−4887.

Das, S.K., Sokhi, U.K., Bhutia, S.K., Azab, B., Su, Z.Z., Sarkar, D., et al., 2010. Human polynucleotide phosphorylase selectively and preferentially degrades microRNA-221 in human melanoma cells. Proc. Natl. Acad. Sci. USA 107, 11948–11953.

Davis, T.H., Cuellar, T.L., Koch, S.M., Barker, A.J., Harfe, B.D., McManus, M.T., et al., 2008. Conditional loss of Dicer disrupts cellular and tissue morphogenesis in the cortex and hippocampus. J. Neurosci. 28, 4322–4330.

de la Chapelle, A., Herva, R., Koivisto, M., Aula, P., 1981. A deletion in chromosome 22 can cause DiGeorge syndrome. Hum. Genet. 57, 253–256.

de la Mata, M., Gaidatzis, D., Vitanescu, M., Stadler, M.B., Wentzel, C., Scheiffele, P., et al., 2015. Potent degradation of neuronal miRNAs induced by highly complementary targets. EMBO Rep. 16, 500–511.

de Lencastre, A., Pincus, Z., Zhou, K., Kato, M., Lee, S.S., Slack, F.J., 2010. MicroRNAs both promote and antagonize longevity in C. elegans. Curr. Biol. 20, 2159–2168.

Deng, H.X., Zhai, H., Bigio, E.H., Yan, J., Fecto, F., Ajroud, K., et al., 2010. FUS-immunoreactive inclusions are a common feature in sporadic and non-SOD1 familial amyotrophic lateral sclerosis. Ann. Neurol. 67, 739–748.

Denli, A.M., Tops, B.B., Plasterk, R.H., Ketting, R.F., Hannon, G.J., 2004. Processing of primary microRNAs by the microprocessor complex. Nature 432, 231–235.

Diederichs, S., Haber, D.A., 2007. Dual role for argonautes in microRNA processing and posttranscriptional regulation of microRNA expression. Cell 131, 1097–1108.

Du, P., Wang, L., Sliz, P., Gregory, R.I., 2015. A biogenesis step upstream of microprocessor controls miR-17 approximately 92 expression. Cell 162, 885–899.

Dueck, A., Meister, G., 2014. Assembly and function of small RNA–argonaute protein complexes. Biol. Chem. 395, 611–629.

Eitan, C., Hornstein, E., 2016. Vulnerability of microRNA biogenesis in FTD-ALS. Brain Res. 1647, 105–111.

Elkayam, E., Kuhn, C.D., Tocilj, A., Haase, A.D., Greene, E.M., Hannon, G.J., et al., 2012. The structure of human argonaute-2 in complex with miR-20a. Cell 150, 100–110.

Fang, W., Bartel, D.P., 2015. The menu of features that define primary microRNAs and enable de novo design of microRNA genes. Mol. Cell. 60, 131–145.

Fenelon, K., Mukai, J., Xu, B., Hsu, P.K., Drew, L.J., Karayiorgou, M., et al., 2011. Deficiency of Dgcr8, a gene disrupted by the 22q11.2 microdeletion, results in altered short-term plasticity in the prefrontal cortex. Proc. Natl. Acad. Sci. USA 108, 4447–4452.

Finnegan, E.F., Pasquinelli, A.E., 2013. MicroRNA biogenesis: regulating the regulators. Crit. Rev. Biochem. Mol. Biol. 48, 51–68.

Forman, J.J., Legesse-Miller, A., Coller, H.A., 2008. A search for conserved sequences in coding regions reveals that the let-7 microRNA targets Dicer within its coding sequence. Proc. Natl. Acad. Sci. USA 105, 14879–14884.

Friedman, R.C., Farh, K.K., Burge, C.B., Bartel, D.P., 2009. Most mammalian mRNAs are conserved targets of microRNAs. Genome Res. 19, 92–105.

Fukunaga, R., Han, B.W., Hung, J.H., Xu, J., Weng, Z., Zamore, P.D., 2012. Dicer partner proteins tune the length of mature miRNAs in flies and mammals. Cell 151, 533–546.

Gantier, M.P., McCoy, C.E., Rusinova, I., Saulep, D., Wang, D., Xu, D., et al., 2011. Analysis of microRNA turnover in mammalian cells following Dicer1 ablation. Nucleic Acids Res. 39, 5692–5703.

Giraldez, A.J., Cinalli, R.M., Glasner, M.E., Enright, A.J., Thomson, J.M., Baskerville, S., et al., 2005. MicroRNAs regulate brain morphogenesis in zebrafish. Science 308, 833–838.

Gregory, R.I., Yan, K.P., Amuthan, G., Chendrimada, T., Doratotaj, B., Cooch, N., et al., 2004. The microprocessor complex mediates the genesis of microRNAs. Nature 432, 235–240.

Griffiths-Jones, S., 2004. The microRNA registry. Nucleic Acids Res. 32, D109−111.

Grishok, A., Pasquinelli, A.E., Conte, D., Li, N., Parrish, S., Ha, I., et al., 2001. Genes and mechanisms related to RNA interference regulate expression of the small temporal RNAs that control C. elegans developmental timing. Cell 106, 23−34.

Ha, M., Kim, V.N., 2014. Regulation of microRNA biogenesis. Nat. Rev. Mol. Cell. Biol. 15, 509−524.

Haase, A.D., Jaskiewicz, L., Zhang, H., Laine, S., Sack, R., Gatignol, A., et al., 2005. TRBP, a regulator of cellular PKR and HIV-1 virus expression, interacts with Dicer and functions in RNA silencing. EMBO. Rep. 6, 961−967.

Hammond, S.M., 2015. An overview of microRNAs. Adv. Drug. Deliv. Rev. 87, 3−14.

Han, J., Lee, Y., Yeom, K.H., Kim, Y.K., Jin, H., Kim, V.N., 2004. The Drosha-DGCR8 complex in primary microRNA processing. Genes Dev. 18, 3016−3027.

Han, J., Lee, Y., Yeom, K.H., Nam, J.W., Heo, I., Rhee, J.K., et al., 2006. Molecular basis for the recognition of primary microRNAs by the Drosha-DGCR8 complex. Cell 125, 887−901.

Haramati, S., Chapnik, E., Sztainberg, Y., Eilam, R., Zwang, R., Gershoni, N., et al., 2010. miRNA malfunction causes spinal motor neuron disease. Proc. Natl. Acad. Sci. USA 107, 13111−13116.

He, L., Thomson, J.M., Hemann, M.T., Hernando-Monge, E., Mu, D., Goodson, S., et al., 2005. A microRNA polycistron as a potential human oncogene. Nature 435, 828−833.

He, L., He, X., Lim, L.P., de Stanchina, E., Xuan, Z., Liang, Y., et al., 2007. A microRNA component of the p53 tumour suppressor network. Nature 447, 1130−1134.

Heo, I., Joo, C., Cho, J., Ha, M., Han, J., Kim, V.N., 2008. Lin28 mediates the terminal uridylation of let-7 precursor MicroRNA. Mol. Cell 32, 276−284.

Heo, I., Joo, C., Kim, Y.K., Ha, M., Yoon, M.J., Cho, J., et al., 2009. TUT4 in concert with Lin28 suppresses microRNA biogenesis through pre-microRNA uridylation. Cell 138, 696−708.

Herbert, K.M., Sarkar, S.K., Mills, M., Delgado De la Herran, H.C., Neuman, K.C., Steitz, J.A., 2016. A heterotrimer model of the complete microprocessor complex revealed by single-molecule subunit counting. RNA 22, 175−183.

Hu, H.Y., Yan, Z., Xu, Y., Hu, H., Menzel, C., Zhou, Y.H., et al., 2009. Sequence features associated with microRNA strand selection in humans and flies. BMC Genomics 10, 413.

Huang, Y.W., Ruiz, C.R., Eyler, E.C., Lin, K., Meffert, M.K., 2012. Dual regulation of miRNA biogenesis generates target specificity in neurotrophin-induced protein synthesis. Cell 148, 933−946.

Hutvagner, G., McLachlan, J., Pasquinelli, A.E., Balint, E., Tuschl, T., Zamore, P.D., 2001. A cellular function for the RNA-interference enzyme Dicer in the maturation of the let-7 small temporal RNA. Science 293, 834−838.

Ibanez-Ventoso, C., Yang, M., Guo, S., Robins, H., Padgett, R.W., Driscoll, M., 2006. Modulated microRNA expression during adult lifespan in Caenorhabditis elegans. Aging Cell 5, 235−246.

Jakymiw, A., Lian, S., Eystathioy, T., Li, S., Satoh, M., Hamel, J.C., et al., 2005. Disruption of GW bodies impairs mammalian RNA interference. Nat. Cell Biol. 7, 1267−1274.

Johnston, R.J., Hobert, O., 2003. A microRNA controlling left/right neuronal asymmetry in Caenorhabditis elegans. Nature 426, 845−849

Johnston Jr., R.J., Chang, S., Etchberger, J.F., Ortiz, C.O., Hobert, O., 2005. MicroRNAs acting in a double-negative feedback loop to control a neuronal cell fate decision. Proc. Natl. Acad. Sci. USA 102, 12449−12454.

Jonas, S., Izaurralde, E., 2015. Towards a molecular understanding of microRNA-mediated gene silencing. Nat. Rev. Genet. 16, 421−433.

Karayiorgou, M., Morris, M.A., Morrow, B., Shprintzen, R.J., Goldberg, R., Borrow, J., et al., 1995. Schizophrenia susceptibility associated with interstitial deletions of chromosome 22q11. Proc. Natl. Acad. Sci. USA 92, 7612−7616.

Karayiorgou, M., Simon, T.J., Gogos, J.A., 2010. 22q11.2 microdeletions: linking DNA structural variation to brain dysfunction and schizophrenia. Nat. Rev. Neurosci. 11, 402−416.

Kawahara, Y., Mieda-Sato, A., 2012. TDP-43 promotes microRNA biogenesis as a component of the Drosha and Dicer complexes. Proc. Natl. Acad. Sci. USA 109, 3347−3352.

Ketting, R.F., Fischer, S.E., Bernstein, E., Sijen, T., Hannon, G.J., Plasterk, R.H., 2001. Dicer functions in RNA interference and in synthesis of small RNA involved in developmental timing in *C. elegans*. Genes Dev. 15, 2654−2659.

Khvorova, A., Reynolds, A., Jayasena, S.D., 2003. Functional siRNAs and miRNAs exhibit strand bias. Cell 115, 209−216.

Kim, J., Inoue, K., Ishii, J., Vanti, W.B., Voronov, S.V., Murchison, E., et al., 2007. A MicroRNA feedback circuit in midbrain dopamine neurons. Science 317, 1220−1224.

Kim, Y.K., Kim, V.N., 2007. Processing of intronic microRNAs. EMBO J. 26, 775−783.

Kim, Y.K., Kim, B., Kim, V.N., 2016. Re-evaluation of the roles of DROSHA, Exportin 5, and DICER in microRNA biogenesis. Proc. Natl. Acad. Sci. USA 113 (13), E1881−E1889.

Knight, S.W., Bass, B.L., 2001. A role for the RNase III enzyme DCR-1 in RNA interference and germ line development in *Caenorhabditis elegans*. Science 293, 2269−2271.

Kozomara, A., Griffiths-Jones, S., 2014. miRBase: annotating high confidence microRNAs using deep sequencing data. Nucleic Acids Res. 42, D68−73.

Krol, J., Busskamp, V., Markiewicz, I., Stadler, M.B., Ribi, S., Richter, J., et al., 2010. Characterizing light-regulated retinal microRNAs reveals rapid turnover as a common property of neuronal microRNAs. Cell 141, 618−631.

Kwon, S.C., Nguyen, T.A., Choi, Y.G., Jo, M.H., Hohng, S., Kim, V.N., et al., 2016. Structure of human DROSHA. Cell 164, 81−90.

Landthaler, M., Yalcin, A., Tuschl, T., 2004. The human DiGeorge syndrome critical region gene 8 and its *D. melanogaster* homolog are required for miRNA biogenesis. Curr. Biol. 14, 2162−2167.

Lau, N.C., Lim, L.P., Weinstein, E.G., Bartel, D.P., 2001. An abundant class of tiny RNAs with probable regulatory roles in *Caenorhabditis elegans*. Science 294, 858−862.

Lau, P.W., Guiley, K.Z., De, N., Potter, C.S., Carragher, B., MacRae, I.J., 2012. The molecular architecture of human Dicer. Nat. Struct. Mol. Biol. 19, 436−440.

Lee, H.Y., Zhou, K., Smith, A.M., Noland, C.L., Doudna, J.A., 2013. Differential roles of human Dicer-binding proteins TRBP and PACT in small RNA processing. Nucleic Acids Res. 41, 6568−6576.

Lee, R.C., Feinbaum, R.L., Ambros, V., 1993. The *C. elegans* heterochronic gene lin-4 encodes small RNAs with antisense complementarity to lin-14. Cell 75, 843−854.

Lee, Y., Jeon, K., Lee, J.T., Kim, S., Kim, V.N., 2002. MicroRNA maturation: stepwise processing and subcellular localization. EMBO J. 21, 4663−4670.

Lee, Y., Kim, M., Han, J., Yeom, K.H., Lee, S., Baek, S.H., et al., 2004. MicroRNA genes are transcribed by RNA polymerase II. EMBO J. 23, 4051−4060.

Lee, Y., Hur, I., Park, S.Y., Kim, Y.K., Suh, M.R., Kim, V.N., 2006. The role of PACT in the RNA silencing pathway. EMBO J. 25, 522−532.

Li, N., Muthusamy, S., Liang, R., Sarojini, H., Wang, E., 2011a. Increased expression of miR-34a and miR-93 in rat liver during aging, and their impact on the expression of Mgst1 and Sirt1. Mech. Ageing. Dev. 132, 75−85.

Li, X., Khanna, A., Li, N., Wang, E., 2011b. Circulatory miR34a as an RNA based, noninvasive biomarker for brain aging. Aging (Albany, NY) 3, 985−1002.

Lin, S., Gregory, R.I., 2015. MicroRNA biogenesis pathways in cancer. Nat. Rev. Cancer 15, 321−333.

Ling, S.C., Polymenidou, M., Cleveland, D.W., 2013. Converging mechanisms in ALS and FTD: disrupted RNA and protein homeostasis. Neuron 79, 416−438.

Liu, J., Rivas, F.V., Wohlschlegel, J., Yates III, J.R., Parker, R., Hannon, G.J., 2005. A role for the P-body component GW182 in microRNA function. Nat. Cell. Biol. 7, 1261−1266.

Liu, N., Landreh, M., Cao, K., Abe, M., Hendriks, G.J., Kennerdell, J., et al., 2012. The microRNA miR-34 modulates aging and neurodegeneration in Drosophila. Nature 482, 519−523.

Liu, X., Chen, X., Yu, X., Tao, Y., Bode, A.M., Dong, Z., et al., 2013. Regulation of microRNAs by epigenetics and their interplay involved in cancer. J. Exp. Clin. Cancer. Res. 32, 96.

Loughlin, F.E., Gebert, L.F., Towbin, H., Brunschweiger, A., Hall, J., Allain, F.H., 2012. Structural basis of pre-let-7 miRNA recognition by the zinc knuckles of pluripotency factor Lin28. Nat. Struct. Mol. Biol. 19, 84−89.

Lund, E., Guttinger, S., Calado, A., Dahlberg, J.E., Kutay, U., 2004. Nuclear export of microRNA precursors. Science 303, 95−98.

Ma, E., MacRae, I.J., Kirsch, J.F., Doudna, J.A., 2008. Autoinhibition of human dicer by its internal helicase domain. J. Mol. Biol. 380, 237−243.

Macrae, I.J., Zhou, K., Li, F., Repic, A., Brooks, A.N., Cande, W.Z., et al., 2006. Structural basis for double-stranded RNA processing by Dicer. Science 311, 195−198.

Marson, A., Levine, S.S., Cole, M.F., Frampton, G.M., Brambrink, T., Johnstone, S., et al., 2008. Connecting microRNA genes to the core transcriptional regulatory circuitry of embryonic stem cells. Cell 134, 521−533.

Martinez, N.J., Ow, M.C., Reece-Hoyes, J.S., Barrasa, M.I., Ambros, V.R., Walhout, A.J., 2008. Genome-scale spatiotemporal analysis of Caenorhabditis elegans microRNA promoter activity. Genome Res. 18, 2005−2015.

Meister, G., Landthaler, M., Peters, L., Chen, P.Y., Urlaub, H., Luhrmann, R., et al., 2005. Identification of novel argonaute-associated proteins. Curr. Biol. 15, 2149−2155.

Melo, S.A., Moutinho, C., Ropero, S., Calin, G.A., Rossi, S., Spizzo, R., et al., 2010. A genetic defect in exportin-5 traps precursor microRNAs in the nucleus of cancer cells. Cancer Cell. 18, 303−315.

Mogilyansky, E., Rigoutsos, I., 2013. The miR-17/92 cluster: a comprehensive update on its genomics, genetics, functions and increasingly important and numerous roles in health and disease. Cell. Death. Differ. 20, 1603−1614.

Mondol, V., Ahn, B.C., Pasquinelli, A.E., 2015. Splicing remodels the let-7 primary microRNA to facilitate Drosha processing in Caenorhabditis elegans. RNA 21, 1396−1403.

Monteys, A.M., Spengler, R.M., Wan, J., Tecedor, L., Lennox, K.A., Xing, Y., et al., 2010. Structure and activity of putative intronic miRNA promoters. RNA 16, 495−505.

Morlando, M., Ballarino, M., Gromak, N., Pagano, F., Bozzoni, I., Proudfoot, N.J., 2008. Primary microRNA transcripts are processed co-transcriptionally. Nat. Struct. Mol. Biol. 15, 902−909.

Morlando, M., Dini Modigliani, S., Torrelli, G., Rosa, A., Di Carlo, V., Caffarelli, E., et al., 2012. FUS stimulates microRNA biogenesis by facilitating co-transcriptional Drosha recruitment. EMBO J. 31, 4502−4510.

Nakanishi, K., Weinberg, D.E., Bartel, D.P., Patel, D.J., 2012. Structure of yeast Argonaute with guide RNA. Nature 486, 368−374.

Nam, Y., Chen, C., Gregory, R.I., Chou, J.J., Sliz, P., 2011. Molecular basis for interaction of let-7 microRNAs with Lin28. Cell 147, 1080−1091.

Neumann, M., Sampathu, D.M., Kwong, L.K., Truax, A.C., Micsenyi, M.C., Chou, T.T., et al., 2006. Ubiquitinated TDP-43 in frontotemporal lobar degeneration and amyotrophic lateral sclerosis. Science 314, 130−133.

Newman, M.A., Thomson, J.M., Hammond, S.M., 2008. Lin-28 interaction with the Let-7 precursor loop mediates regulated microRNA processing. RNA 14, 1539−1549.

Nguyen, T.A., Jo, M.H., Choi, Y.G., Park, J., Kwon, S.C., Hohng, S., et al., 2015. Functional anatomy of the human microprocessor. Cell 161, 1374−1387.

Okada, C., Yamashita, E., Lee, S.J., Shibata, S., Katahira, J., Nakagawa, A., et al., 2009. A high-resolution structure of the pre-microRNA nuclear export machinery. Science 326, 1275−1279.

Okamura, K., Phillips, M.D., Tyler, D.M., Duan, H., Chou, Y.T., Lai, E.C., 2008. The regulatory activity of microRNA* species has substantial influence on microRNA and 3' UTR evolution. Nat. Struct. Mol. Biol. 15, 354–363.

Ota, A., Tagawa, H., Karnan, S., Tsuzuki, S., Karpas, A., Kira, S., et al., 2004. Identification and characterization of a novel gene, C13orf25, as a target for 13q31-q32 amplification in malignant lymphoma. Cancer Res. 64, 3087–3095.

Ozsolak, F., Poling, L.L., Wang, Z., Liu, H., Liu, X.S., Roeder, R.G., et al., 2008. Chromatin structure analyses identify miRNA promoters. Genes Dev. 22, 3172–3183.

Park, J.E., Heo, I., Tian, Y., Simanshu, D.K., Chang, H., Jee, D., et al., 2011. Dicer recognizes the 5' end of RNA for efficient and accurate processing. Nature 475, 201–205.

Paroo, Z., Ye, X., Chen, S., Liu, Q., 2009. Phosphorylation of the human microRNA-generating complex mediates MAPK/Erk signaling. Cell 139, 112–122.

Pawlicki, J.M., Steitz, J.A., 2008. Primary microRNA transcript retention at sites of transcription leads to enhanced microRNA production. J. Cell. Biol. 182, 61–76.

Pfaff, J., Meister, G., 2013. Argonaute and GW182 proteins: an effective alliance in gene silencing. Biochem. Soc. Trans. 41, 855–860.

Piskounova, E., Viswanathan, S.R., Janas, M., LaPierre, R.J., Daley, G.Q., Sliz, P., et al., 2008. Determinants of microRNA processing inhibition by the developmentally regulated RNA-binding protein Lin28. J. Biol. Chem. 283, 21310–21314.

Rajasethupathy, P., Fiumara, F., Sheridan, R., Betel, D., Puthanveettil, S.V., Russo, J.J., et al., 2009. Characterization of small RNAs in Aplysia reveals a role for miR-124 in constraining synaptic plasticity through CREB. Neuron 63, 803–817.

Raver-Shapira, N., Marciano, E., Meiri, E., Spector, Y., Rosenfeld, N., Moskovits, N., et al., 2007. Transcriptional activation of miR-34a contributes to p53-mediated apoptosis. Mol. Cell 26, 731–743.

Rehwinkel, J., Behm-Ansmant, I., Gatfield, D., Izaurralde, E., 2005. A crucial role for GW182 and the DCP1: DCP2 decapping complex in miRNA-mediated gene silencing. RNA 11, 1640–1647.

Ristori, E., Lopez-Ramirez, M.A., Narayanan, A., Hill-Teran, G., Moro, A., Calvo, C.F., et al., 2015. A Dicer-miR-107 interaction regulates biogenesis of specific miRNAs crucial for neurogenesis. Dev. Cell 32, 546–560.

Rodriguez, A., Griffiths-Jones, S., Ashurst, J.L., Bradley, A., 2004. Identification of mammalian microRNA host genes and transcription units. Genome Res. 14, 1902–1910.

Rybak, A., Fuchs, H., Smirnova, L., Brandt, C., Pohl, E.E., Nitsch, R., et al., 2008. A feedback loop comprising lin-28 and let-7 controls pre-let-7 maturation during neural stem-cell commitment. Nat. Cell. Biol. 10, 987–993.

Saini, H.K., Enright, A.J., Griffiths-Jones, S., 2008. Annotation of mammalian primary microRNAs. BMC Genomics 9, 564.

Schaefer, A., O'Carroll, D., Tan, C.L., Hillman, D., Sugimori, M., Llinas, R., et al., 2007. Cerebellar neurodegeneration in the absence of microRNAs. J. Exp. Med. 204, 1553–1558.

Schanen, B.C., Li, X., 2011. Transcriptional regulation of mammalian miRNA genes. Genomics 97, 1–6.

Schirle, N.T., MacRae, I.J., 2012. The crystal structure of human Argonaute2. Science 336, 1037–1040.

Schirle, N.T., Sheu-Gruttadauria, J., MacRae, I.J., 2014. Structural basis for microRNA targeting. Science 346, 608–613.

Schratt, G., 2009. Fine-tuning neural gene expression with microRNAs. Curr. Opin. Neurobiol. 19, 213–219.

Schwarz, D.S., Hutvagner, G., Du, T., Xu, Z., Aronin, N., Zamore, P.D., 2003. Asymmetry in the assembly of the RNAi enzyme complex. Cell 115, 199–208.

Shao, N.Y., Hu, H.Y., Yan, Z., Xu, Y., Hu, H., Menzel, C., et al., 2010. Comprehensive survey of human brain microRNA by deep sequencing. BMC Genomics 11, 409.

Shi, Y., Zhao, X., Hsieh, J., Wichterle, H., Impey, S., Banerjee, S., et al., 2010. MicroRNA regulation of neural stem cells and neurogenesis. J. Neurosci. 30, 14931−14936.

Shin, D., Shin, J.Y., McManus, M.T., Ptacek, L.J., Fu, Y.H., 2009. Dicer ablation in oligodendrocytes provokes neuronal impairment in mice. Ann. Neurol. 66, 843−857.

Stark, K.L., Xu, B., Bagchi, A., Lai, W.S., Liu, H., Hsu, R., et al., 2008. Altered brain microRNA biogenesis contributes to phenotypic deficits in a 22q11-deletion mouse model. Nat. Genet. 40, 751−760.

Suzuki, H.I., Arase, M., Matsuyama, H., Choi, Y.L., Ueno, T., Mano, H., et al., 2011. MCPIP1 ribonuclease antagonizes dicer and terminates microRNA biogenesis through precursor microRNA degradation. Mol. Cell. 44, 424−436.

Tao, J., Wu, H., Lin, Q., Wei, W., Lu, X.H., Cantle, J.P., et al., 2011. Deletion of astroglial Dicer causes non-cell-autonomous neuronal dysfunction and degeneration. J. Neurosci. 31, 8306−8319.

Tarasov, V., Jung, P., Verdoodt, B., Lodygin, D., Epanchintsev, A., Menssen, A., et al., 2007. Differential regulation of microRNAs by p53 revealed by massively parallel sequencing: miR-34a is a p53 target that induces apoptosis and G1-arrest. Cell Cycle 6, 1586−1593.

Tazawa, H., Tsuchiya, N., Izumiya, M., Nakagama, H., 2007. Tumor-suppressive miR-34a induces senescence-like growth arrest through modulation of the E2F pathway in human colon cancer cells. Proc. Natl. Acad. Sci.USA 104, 15472−15477.

Thomas, M.F., L'Etoile, N.D., Ansel, K.M., 2014. Eri1: a conserved enzyme at the crossroads of multiple RNA-processing pathways. Trends. Genet. 30, 298−307.

Thornton, J.E., Chang, H.M., Piskounova, E., Gregory, R.I., 2012. Lin28-mediated control of let-7 microRNA expression by alternative TUTases Zcchc11 (TUT4) and Zcchc6 (TUT7). RNA 18, 1875−1885.

Tokumaru, S., Suzuki, M., Yamada, H., Nagino, M., Takahashi, T., 2008. let-7 regulates Dicer expression and constitutes a negative feedback loop. Carcinogenesis 29, 2073−2077.

Tuna, M., Machado, A.S., Calin, G.A., 2016. Genetic and epigenetic alterations of microRNAs and implications for human cancers and other diseases. Genes Chromosomes Cancer 55, 193−214.

Ustianenko, D., Hrossova, D., Potesil, D., Chalupnikova, K., Hrazdilova, K., Pachernik, J., et al., 2013. Mammalian DIS3L2 exoribonuclease targets the uridylated precursors of let-7 miRNAs. RNA 19, 1632−1638.

Viswanathan, S.R., Daley, G.Q., Gregory, R.I., 2008. Selective blockade of microRNA processing by Lin28. Science 320, 97−100.

Wang, Y., Medvid, R., Melton, C., Jaenisch, R., Blelloch, R., 2007. DGCR8 is essential for microRNA biogenesis and silencing of embryonic stem cell self-renewal. Nat. Genet. 39, 380−385.

Wightman, B., Ha, I., Ruvkun, G., 1993. Posttranscriptional regulation of the heterochronic gene lin-14 by lin-4 mediates temporal pattern formation in *C. elegans*. Cell 75, 855−862.

Wilson, R.C., Tambe, A., Kidwell, M.A., Noland, C.L., Schneider, C.P., Doudna, J.A., 2015. Dicer-TRBP complex formation ensures accurate mammalian microRNA biogenesis. Mol. Cell. 57, 397−407.

Winter, J., Diederichs, S., 2011. Argonaute proteins regulate microRNA stability: increased microRNA abundance by Argonaute proteins is due to microRNA stabilization. RNA Biol. 8, 1149−1157.

Xie, M., Steitz, J.A., 2014. Versatile microRNA biogenesis in animals and their viruses. RNA Biol. 11, 673−681.

Xie, M., Li, M., Vilborg, A., Lee, N., Shu, M.D., Yartseva, V., et al., 2013. Mammalian 5′-capped microRNA precursors that generate a single microRNA. Cell 155, 1568−1580.

Yi, R., Qin, Y., Macara, I.G., Cullen, B.R., 2003. Exportin-5 mediates the nuclear export of pre-microRNAs and short hairpin RNAs. Genes Dev. 17, 3011−3016.

Zeng, Y., Yi, R., Cullen, B.R., 2005. Recognition and cleavage of primary microRNA precursors by the nuclear processing enzyme Drosha. EMBO J. 24, 138−148.

Zhang, H., Kolb, F.A., Jaskiewicz, L., Westhof, E., Filipowicz, W., 2004. Single processing center models for human Dicer and bacterial RNase III. Cell 118, 57−68.

ESSENTIALS OF miRNA-DEPENDENT CONTROL OF mRNA TRANSLATION AND DECAY, miRNA TARGETING PRINCIPLES, AND METHODS FOR TARGET IDENTIFICATION

Stefanie Grosswendt and Nikolaus Rajewsky
Berlin Institute for Medical Systems Biology (BIMSB), Berlin, Germany

MiRNAs ARE POSTTRANSCRIPTIONAL REGULATORS OF GENE EXPRESSION

Posttranscriptional gene regulation is important for all stages of the life of a (messenger) RNA molecule, including splicing, nuclear/cytoplasmic export/import, transcript stability, as well as subcellular localization and translation (Martin and Ephrussi, 2009; Moore and Proudfoot, 2009; Sonenberg and Hinnebusch, 2009). The factors that mediate these regulatory processes are RNA binding proteins (RBPs) and noncoding RNAs (ncRNAs). In humans, more than 800 RBPs (Baltz et al., 2012; Castello et al., 2012) stably bind to single-stranded RNA sequences or to RNA secondary structures, and either affect the transcript directly, e.g., by stabilizing or cleaving it, or indirectly, by recruiting or preventing the association of other factors. ncRNAs, which are not translated into proteins, include diverse groups of RNAs, like small regulatory RNAs (microRNAs, piRNAs, etc.), long ncRNAs (lncRNAs), transfer RNAs (tRNAs), and ribosomal RNAs (rRNAs) (reviewed in Cech and Steitz, 2014). They often function in conjunction with RBPs in so-called ribonucleoprotein complexes (RNPs). An example hereof is the miRNA-induced silencing complex (miRISC). Its core component Argonaute is bound to a microRNA that guides the complex to its target RNAs to mediate repression of protein production.

Initially discovered in the nematode *Caenorhabditis elegans* (Lee et al., 1993; Wightman et al., 1993; Reinhart et al., 2000), miRNAs were found to form a large group of small RNAs in animals and plants, with some miRNAs being deeply conserved (Pasquinelli et al., 2000; Ibáñez-Ventoso et al., 2008). Currently ~3000 miRNAs have been identified in the human genome, many of them are weakly expressed in most tissues (Friedländer et al., 2014; Kozomara and Griffiths-Jones, 2011). Typically only a few miRNAs are highly expressed in a tissue and are usually conserved in evolution (for more detail on evolutionary conservation of miRNAs see Chapters 15 and 16 of this book). Many of the lowly expressed miRNAs are also only weakly conserved, in contrast. Lowly

Essentials of Noncoding RNA in Neuroscience. DOI: http://dx.doi.org/10.1016/B978-0-12-804402-5.00002-9

expressed miRNAs may be subject to future natural selection (Chen and Rajewsky, 2007). MicroRNAs appear to be involved in nearly all biological processes, as more than 60% of protein coding genes are assumed to be targeted by one or several miRNAs (Friedman et al., 2009). In order to understand the posttranscriptional regulation that is mediated by miRNAs, it is therefore crucial to identify their binding sites.

MiRNA BINDING GUIDES AGO FUNCTION TO TARGET RNAs
THE ROLE OF THE miRNA SEED AND THE KINETICS OF BINDING

Since their discovery, miRNAs present researchers with the challenge of understanding to which transcriptomic sites they bind to exert their function (reviewed in Hausser and Zavolan, 2014). Incorporated into Argonaute proteins, they guide the silencing complex miRISC to transcripts via consecutive Watson−Crick−Pairing and then inhibit translation or destabilize the transcript (Fabian et al., 2010). Consequently, the biological role of a miRNA is mainly specified by its set of targets. To identify miRNA targets is challenging in animals, as target recognition occurs only through partial sequence complementarity (Bartel, 2009). Furthermore, a miRNA typically has hundreds of direct targets and each target might be under the combinatorial regulation of multiple different miRNAs (Brennecke et al., 2005; Krek et al., 2005; Lewis et al., 2005; Xie et al., 2005). Recently it has become clear that miRNA targets may compete for a limited number of miRNAs, requiring to also consider the expression profile of all targets (Jens and Rajewsky, 2015 and references within). Particularly involved in binding is the miRNA seed sequence (or "nucleus" Rajewsky and Socci, 2004), positioned on nucleotides 2−7 from the miRNA 5′end (Bartel, 2009; Lewis et al., 2005; Lai, 2002; Rajewsky, 2006). Complementarity to the seed is most frequently found in experimentally identified Argonaute binding sites and a perfect seed match is often sufficient to elicit a regulatory response (Bartel, 2009).

In addition to such canonical miRNA binding, numerous functional miRNA target sites without a constitutive stretch of complementarity to the miRNA seed have been reported (Bagga et al., 2005; Chi et al., 2012; Didiano and Hobert, 2006; Helwak et al., 2013; Lal et al., 2009; Shin et al., 2010; Vella et al., 2004), amongst them interactions with far-reaching consequences for the organisms' development and homeostasis. Well-known examples are *C. elegans* miRNA let-7 targeting the lin-41 transcript to time transition of larval stages (Reinhart et al., 2000; Vella et al., 2004), miRNA lsy-6 targeting the cog-1 transcript to establish a left/right neuronal asymmetry of the worm (Johnston and Hobert, 2003) or human miRNA-24 noncanonically binding the mRNA of the E2F2 transcription factor to antagonize cell proliferation and allowing cell differentiation (Lal et al., 2009). Noncanonical miRNA interactions are expected to contribute a considerable fraction of Argonaute binding, as experimentally identified AGO sites often lack perfect complementarity to seed sequences of miRNAs that are expressed in the cell (Hafner et al., 2010; Kishore et al., 2011; Corcoran et al., 2011; Zisoulis et al., 2010). However, the range of base pairing patterns and the regulatory impact of noncanonical interactions are currently not well understood because of difficulties in their identification. It is often observed that more extensive complementarities between miRNA and target lead to stronger target regulation but also miRNA binding with complementarities restricted to the seed hexamer or even with imperfect seed complementarity were found to be functional (Brennecke et al., 2005; Vella et al., 2004; Shin et al., 2010; Loeb et al., 2012; Helwak et al., 2013; Grosswendt et al., 2014).

Binding of the miRNA−AGO complex to the target is not only realized by base pairing but also by many contributions of the AGO protein itself. Structural studies of the AGO−miRNA complex have revealed that the bases of the seed nucleotides are well positioned to nucleate the binding of a target RNA (Schirle and Macrae, 2012; Elkayam et al., 2012; Wang et al., 2010). Furthermore, the binding affinity of a small RNA is up to 300-fold increased when it is incorporated into AGO compared to protein-independent annealing. This can be explained by the fact that AGO induces a helical conformation in the miRNA seed, which excludes the possibility that the seed acquires conformations that are less suitable for base pairing. Consequently, the interaction between AGO protein and miRNA already reduces the entropic barrier to target binding (Parker et al., 2009). Interestingly, thermodynamic analysis postulated that base pairing of the seed contributes less then 10% to the energy required for mRNA binding, whereas the largest energy contribution is provided by charges of the protein. The basic nature of the AGO binding channel, a result of numerous arginines and lysines, was estimated to account for about 50% of binding free energy, and another 12% are provided by a strong electronic attraction between the backbone phosphate oxygen of the target nucleotide in position 10 and an Magnesium ion in the PIWI domain of AGO. This underscores the importance of charges contributed by protein domains in target binding efficiency, whereas the seed of the miRNA seems to provide target recognition specificity (Wang et al., 2010). Insights in the time dimension with which miRNA interactions take place came from in vitro studies of the kinetics of Argonaute binding. For mouse Argonaute 2 after $\sim$23 minutes on average half of the miRNA−AGO complexes have dissociated from sites that they had bound via perfect seed complementarity (Wee et al., 2012). The underlying biochemical studies were performed for a relatively small number of miRNA sequences, and it is possible that the binding time is dependent on the type of miRNA, the 3′untranslated regions (3′UTR) context and on the in vivo situation, for instance the binding of other factors nearby that could alter the duration of an interaction.

ALTERNATIVE 3′UTRs USAGE AND THE MOLECULAR MECHANISMS OF miRNA-MEDIATED TRANSLATIONAL REPRESSION

Almost all functionally characterized miRNA target sites are positioned in 3′UTRs and whether a site is actually bound by a miRNA is dependent on many context-specific aspects, for example, coexpression of miRNA and target and the accessibility of a site, which is influenced by the target RNA's secondary structure and by binding of other competing or synergizing trans-regulatory elements. A long 3′UTR is more likely to contain many regulatory sites (such as miRNA sites) that regulate stability and translation of the transcript (Majoros and Ohler, 2007). Consequently, shortening of the 3′UTR can to some extent relieve a gene from miRNA-mediated repression. About half of human genes use alternative 3′UTRs to generate mRNAs that encode the same protein but encounter a different posttranscriptional regulation (Lianoglou et al., 2013). Whether a shorter or a longer 3′UTR is produced can be depended on many cellular conditions, e.g., a widespread reduction in 3′UTR length was found for mRNAs of proliferating cells (Sandberg et al., 2008). Importantly, neural tissues have a very clear tendency to express genes with much longer 3′UTRs than genes expressed in proliferating tissues such as liver. Highly expressed genes in human neural tissues have (on average) 3′UTR of about 1500 nucleotide in length, whereas highly expressed genes in liver have only 500 nucleotide long 3′UTRs (Sood et al., 2006). Interestingly, longer 3′UTR isoforms of genes that show a significantly different 3′UTR usage across tissues tend to contain binding sites for miRNAs that are widely expressed (Lianoglou et al., 2013). This finding demonstrates that not simply miRNA expression but also the choice of the 3′UTR isoform decides to what extend the mRNA is regulated by miRNAs. The close connection between miRNA repression and alternative 3′UTR usage became also obvious for cancer cells, which were found to often express increased numbers of mRNAs with shorter 3′UTR isoforms. These show increased stability and produce more protein, which is in part a result of lost miRNA-mediated repression (Mayr and Bartel, 2009).

The molecular mechanisms of miRNA-mediated translational repression are not fully understood but a wealth of studies concluded that miRNAs inhibit the initiation step of translation and promote deadenlyation of the poly(A)-tail of their targets (reviewed in Fabian et al., 2010). Both were found to occur in conjunction with transcript degradation. Specifically, miRNA binding interferes with recruitment or function of the translation initiation factor elF4F that recognizes the 5′cap of transcripts, and with scanning of the small ribosomal subunit (Ricci et al., 2012) or ribosome assembly at the translation start site. The Argonaute complex not only binds the mRNA but also interacts with GW128 proteins. In turn, GW128 proteins interact with poly (A)-binding proteins (PABPs) thereby pulling the transcript's poly(A)-tail closer, and they recruit the CCR4-NOT deadenylase complex which shortens the poly(A)-tail. This not only decreases the stability of the transcript but also its translatability. It is well established that the poly-A tail and the translation start site are brought together by interactions between the PABPs and translation initiation factors. Is this circularization hampered by shortness of the poly-A tail translating ribosomes are less efficiently recycled, which likely decreases translation (Zekri et al., 2009).

MiRNA-mediated translational inhibition and target degradation are intimately interconnected. A repressive effect of miRNAs on protein synthesis can be detected before any effects on deadenylation and mRNA decay (Djuranovic et al., 2012; Bazzini et al., 2012). Significant repression is already detectable after 2 hours, without signs of deadenylation. However, mRNA decay appears to be the predominant cause of reduced protein synthesis when considering also the late consequences of miRNA targeting (Guo et al., 2010; discussed in Hu and Coller, 2012). Future studies will clarify whether the order of miRNA-mediated silencing effects is dependent on the species or cellular conditions. Independent of the order of events, most targets are repressed by both, translational inhibition and mRNA decay, and how much each mechanism contributes to protein downregulation depends on the individual miRNA−mRNA pair (Selbach et al., 2008).

The degree of protein downregulation mediated by a certain miRNA is mostly less than twofold (Selbach et al., 2008; Baek et al., 2008; Fang and Rajewsky, 2011). However, transcripts are usually bound by multiple miRNAs and this combinatorial targeting actually elicits a more substantial target repression, as seen in the absence of miRNAs in DICER knockout mouse embryonic stem cells (more than twofold derepression for hundreds of genes) (Schmiedel et al., 2015). Also there is a clear linear correlation between number of miRNA target sites and level of repression: the more miRNA sites in the 3′UTR, the stronger the miRNA-mediated repression of this gene (Selbach et al., 2008; Baek et al., 2008).

MiRNAs BUFFER EXPRESSION NOISE TO CONFER ROBUSTNESS TO BIOLOGICAL PATHWAYS AND ACT IN A COMPENSATORY MANNER

In addition to the reduction of protein production, miRNAs are expected to exert an important task in repressing leaky transcription and in buffering aberrant and fluctuating gene expression ("genetic noise") (Schmiedel et al., 2015; see also Siciliano et al., 2013). It was found that miRNAs decrease gene expression noise for lowly expressed genes, whereas they increase the noise for highly expressed genes (Schmiedel et al., 2015). Since miRNAs usually target the more lowly expressed genes, they likely fulfill a role in conveying precision to the protein production of lowly expressed genes. In contrast, highly expressed genes are targeted to a far lower extend and therefore their expression noise can be expected to remain rather unaffected by miRNAs. Also genes targeted by multiple miRNAs have a more-pronounced noise reduction, which might be one of the reasons for the frequently observed combinatorial targeting (Schmiedel et al., 2015).

By buffering expression noise miRNAs confer robustness to genetic pathways (Hornstein and Shomron, 2006; Schmiedel et al., 2015). Hereby, robustness is not limited to the task of keeping a gene expression state stable when facing perturbations but also includes conferring accuracy, uniformity, and irreversibility to developmental transitions and cellular identities. That miRNAs are involved in such a layer of gene-expression regulation is also indicated by the fact that they are often part of positive and negative feedback loops in which they target transcription factors (reviewed in Ebert and Sharp, 2012). Furthermore, miRNAs can establish target gene expression thresholds below which the target is efficiently repressed and above which it could escape miRNA regulation (Mukherji et al., 2011). Such miRNA-induced threshold effects were discussed in context of developmental switches where a miRNA is upregulated, whereas its target transcripts are downregulated in an anticorrelated manner (reviewed in Ebert and Sharp, 2012). However, the concentrations at which such threshold effects are relevant depend on the abundance of targets sites in the cell (Denzler et al., 2014; Jens and Rajewsky, 2015).

Individual depletion of most miRNAs or miRNA families does not result in obvious phenotypes (Alvarez-Saavedra and Horvitz, 2010; Park et al., 2012) and only in rather modest derepression of most target genes, whereas knockout of the organism's Argonaute proteins causes strong developmental defects (Liu, 2004; Su et al., 2009; Vasquez-Rifo et al., 2012). This indicates that the combinatorial effect of different miRNAs seems to be decisive for many targets and that a certain degree of overlap between the targeting profiles of different miRNAs enables them to act in a compensatory manner. Also, it should be mentioned that miRNA knockout studies are not suited to draw general conclusions on the evolutionary importance of individual miRNAs, as the transgenerational fitness and the animals' adaptivity to nonlaboratory conditions were not assessed.

IDENTIFICATION OF miRNA TARGETS VIA PERMUTATION EXPERIMENTS AND BIOINFORMATICS PREDICTION

The first miRNAs were discovered in the analysis of *C. elegans* genetic mutants, which showed an altered larval development. It was found that the genes lin-4 and let-7 negatively regulate protein-coding heterochronic genes, albeit they themselves do not encode for proteins but instead produce small RNAs with sequence complementarity to sites in the 3′UTRs of their target genes (Lee et al., 1993; Reinhart et al., 2000). Here the miRNA:target interactions were uncovered because of drastic complementary loss-of function phenotypes of miRNA and target, respectively. For instance, a mutation in the protein-coding gene lin-14 causes the worm to prematurely undergo developmental programs normally specific for later larval stages, whereas a mutation in the lin-4 gene causes a reiteration of early fates at inappropriately late stages of development. Simultaneously, worms carrying gain-of-function mutations in the lin-14 3′UTR phenocopy lin-4 mutant worms. The additional notions that the lin-14 transcript level stays constant throughout development and that the lin-14 3′UTR harbors lin-4 complementary sequences, suggested that lin-4 small RNA acts as a posttranscriptional negative regulator of lin-14 via direct binding of complementary sites (Lee et al., 1993; Wightman et al., 1993). However, that the identification of miRNA:target interactions via detection of complementary mutant phenotypes would be possible only for a few cases was suggested by knockdown studies in worm and mouse (Miska et al., 2007; Alvarez-Saavedra and Horvitz, 2010; Park et al., 2012). The individual depletion of most miRNAs often did not result in visible phenotypes, and some might only become detectable upon additional perturbations (stress, depletion of proteins which sensitizes the strain, etc.). In conclusion, the power of genetic screens appears to be rather limited for the detection of miRNA interactions.

To uncover targets of an individual miRNA transcriptome-wide, miRNA perturbation experiments with subsequent analysis of changes in transcript or protein levels have been performed (Fig. 2.1).

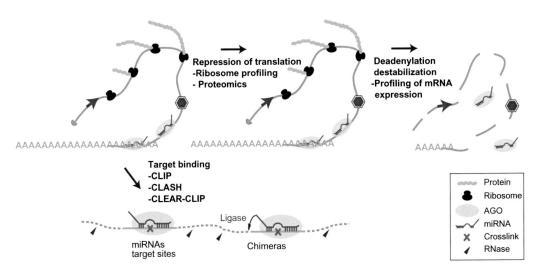

FIGURE 2.1 Methods for miRNA target identification

MiRNA targets can be identified by reading out effects on ribosome-association or protein production (mass spectrometry of pSILAC experiments) as an estimate of translation, and on RNA stability (expression profile determined by microarrays or deep sequencing). Also, binding of the AGO−miRNA complex can be directly investigated at nucleotide resolution via crosslinking and immunoprecipitation (IP) methods that use RNase to specifically recover the AGO-bound target site for deep sequencing. Data of these experiments also contain a very small fraction of chimeric sequences of miRNAs ligated to their targets, generated by an exogenous (CLASH, CLEAR-CLIP) and potentially by an endogenous ligase (CLIP).

The miRNA of interest was either overexpressed or transfected into cell cultures, or to achieve miRNA derepression, knocked out genetically or inhibited by antagonizing small RNAs. To read out the effect of miRNA perturbation on RNA stability, microarrays, and nowadays more prominently deep-sequencing were applied. Also, changes in translation can be investigated by ribosome profiling to learn to which extent the transcript is associated with translating ribosomes (Guo et al., 2010), or by measuring changes in protein synthesis via mass spectrometry of pulsed SILAC experiments (pSILAC, pulsed stable isotope labeling with amino acids in cell culture) (Selbach et al., 2008) (Fig. 2.1). Combined with analyzing potential targets for seed-complementary sites, these approaches led to the discovery of several important interactions. However, they are limited to the investigation of individual miRNAs, influenced by secondary effects and unable to uncover the exact sites of miRNA binding within a transcript.

To overcome experimental limitations, numerous bioinformatic miRNA target prediction algorithms have been developed. To predict miRNA-binding sites, they usually search for canonical seed matches in 3′UTRs and use features like sequence conservation, target site accessibility derived from secondary structure predictions or nucleotide composition, and binding-free energies of potential interactions. They were often successfully applied, but their false positive rates and false negative rates are relatively high (Lewis et al., 2005; Krek et al., 2005; Sethupathy et al., 2006; Selbach et al., 2008; Baek et al., 2008). To counteract the high number of false positive predictions, specificity can be boosted at the expense of sensitivity by, e.g., requiring several seed matches per transcript or a higher degree of conservation. Also, such in silico target identification is unable to account for context-specific aspects that will influence target site binding in vivo, like other RNA-binding proteins that compete with AGO for the same site, expression of 3′UTR isoforms or synergizing cis-regulatory elements that could promote AGO binding. Consequently, computational models to predict miRNA interactions constitute a valuable approach that complements experimental efforts and in recent years both sides work in an increasingly tight interconnection, benefiting from each other's progress (further reading on this topic can be found in Chapter 3: Computational Challenges and -omics Approaches for the Identification of microRNAs and Targets of this book by Hatzigeorgiou and colleagues).

IDENTIFICATION OF miRNA SITES BY RECOVERING AGO-BOUND RNA

CROSSLINKING AND IMMUNOPRECIPITATION TO UNCOVER TARGET SITES OF RBPs

IP studies of AGO proteins enable an experimental and more direct investigation of RNAs that are bound by AGO. Transcripts that are targeted by miRNAs coprecipitate with AGO (RIP, RNA IP) and are subsequently identified by microarrays (RIP-Chip) (Easow et al., 2007; Landthaler et al., 2008) or deep-sequencing (RIP-seq) (Cambronne et al., 2012). However, RIP experiments do not provide a direct way to distinguish target RNAs from background, reassociation of the protein with RNAs that were no original targets can occur during the experiment and RIP also cannot give information on the actual target sites within an AGO-bound transcript.

These issues are largely circumvented in crosslinking and immunoprecipitation experiments (CLIP, Fig. 2.1). Here, protein−RNA interactions are UV-crosslinked and thereby stabilized in living cells, which resolves the problem of reassociation and allows a stringent purification procedure when isolating the protein-bound RNA for sequencing. Additionally the crosslink-induced muta-tions (especially in case of PAR-CLIP) enable a separation of target RNAs from RNAs that were pulled-down nonspecifically. There are different variants of CLIP experiments, which differ amongst other aspects, in the type of UV-crosslinking and the use of the crosslink in bioinformatics analysis. High-throughput sequencing CLIP (HITS-CLIP) and individual-nucleotide-resolution CLIP (iCLIP) make use of the inherent photoreactivity of nucleotides with predominantly aromatic amino acids (phenylalanine, tryptohan, tyrosine, lysine, cysteine) under UV light of 254 nm (Chi et al., 2009; König et al., 2010; Meisenheimer et al., 2000). Frequently, a sequence mutation is found at the site of crosslinking, presumably introduced due to erroneous reverse transcription of crosslinked RNA. In HITS-CLIP, these diagnostic mutations were mostly found to be nucleotide deletions, present in 8%−20% of sequencing reads (Zhang and Darnell, 2011). The difficulty to read over crosslinked nucleotides leads to a frequent drop off of the reverse transcriptase, which is used in the bioinformatics analysis of iCLIP studies.

PAR-CLIP (Fig. 2.2) makes use of photoactivatable thioribonucleotides like 4-thiouridine (4SU) or 6-thioguanosine. Cells grown in the presence of these nucleosides, take them up via nucleoside transporters (Yao et al., 2002), and incorporate them into their nascent RNA. With UV-irradiation of 365 nm, these thioribonucleosides lead to a highly increased crosslinking efficiency compared to 254 nm crosslinking of unlabeled RNA. Furthermore, a very specific nucleotide conversion (T to C mutation in case of 4SU) in sequencing reads was detected at the site of crosslink, which turned out to be very suitable for separating RNAs that were actually bound by the protein from background noise. It is unclear how exactly the characteristic mutations (nucleotide deletions in HITS-CLIP and T to C conversions in PAR-CLIP with 4SU) are introduced at the crosslinked sites. In the case of PAR-CLIP, a guanosine seems to be incorporated into the cDNA when the reverse transcriptase reads over a crosslinked 4-thiouridine, which after PCR amplification renders a C at the sequence position where originally was a T (uridine in RNA). This specific nucleotide change might be the result of an altered base pairing preferences after crosslinking and/or caused by remaining amino acids or small peptides attached at the crosslinked nucleotide (Hafner et al., 2010; Ascano et al., 2012). The latter could also be a plausible explanation for the frequent deletions detected upon 254-nm irradiation, suggesting that the reverse transcriptase skips the crosslinked nucleotide.

HITS-CLIP was performed in a variety of systems, for instance human cell lines, primary mouse T-cells or suspended mouse brain cells, as it only requires UV irradiation and not the prece-dent RNA labeling with thioribonucleosides. The PAR-CLIP protocol, which was so far widely applied in human cell lines, was also adapted for its application in the model system *C. elegans*, enabling the experimental identification of targets of RBPs transcriptome-wide in the living worm (iPAR-CLIP, in vivo PAR-CLIP) (Jungkamp et al., 2011). iPAR-CLIP was later used to investigate the binding of Dicer, an important component of the miRNA processing machinery, and *C. elegans* Argonaute ALG-1 (Rybak-Wolf et al., 2014; Grosswendt et al., 2014). As expected, Dicer was found to specifically and stably bind miRNA precursors but, surprisingly, also a broad target reper-toire beyond miRNAs (Rybak-Wolf et al., 2014). The iPAR-CLIP study of the *C. elegans* Argonaute protein ALG-1 generated a comprehensive map of precisely defined Argonaute binding sites, which show features expected for miRNA binding. To investigate miRNA binding in a living and developing organism has the advantage to capture miRNA interactions in their in vivo context, fulfilling their biological and potentially evolutionary important tasks. At the same time, in vivo studies less likely report miRNA:target pairs which might be theoretically possible based on seed

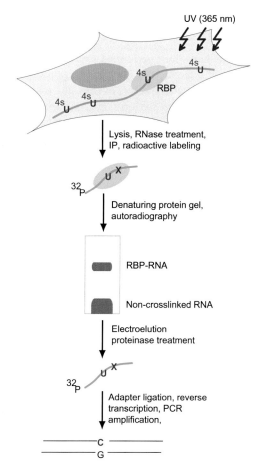

FIGURE 2.2 The concept of PAR-CLIP

Cultured cells are labeled with photoactivatable ribonucleosides (here 4-thiouridine, 4SU). Subsequent UV-irradiation crosslinks labeled RNA to RBP (RNA-binding protein). Upon cell lysis, RNA is partially digested with RNase, radioactively labeled (^{32}P) and covalently connected RBP−RNA complexes can be immunopurified (IP) and size-fractionated using a denaturing gel electrophoresis. Crosslinked (X) RNA is recovered, released from the RBP by proteinase treatment, and converted into a cDNA library for deep-sequencing. Binding sites of the RBP are identified by high read coverage and abundant, diagnostic nucleotide conversions (T to C when using 4SU)

complementarity, but that do not form because of context-specific features. Some sites will simply not be bound by a miRNA because of the secondary structure at the target site, a shorter 3'UTR isoform (Majoros and Ohler, 2007; Mangone et al., 2010) or because of competing trans-acting factors that are temporally and specially coexpressed. The sensitivity for such in vivo features that influence Argonaute binding is absent in purely computational approaches and also affected in experiments using immortalized cell lines, which exhibit a significantly different expression profile compared to primary cells of their kind (Landry et al., 2013; Lin et al., 2014).

ARGONAUTE-CLIP IDENTIFIES SITES OF miRNA BINDING BUT THE TARGETING REMAINS UNCERTAIN

Since 2009 numerous CLIP studies have been performed for Argonaute and identified its binding sites transcriptome-wide at a resolution of a few tens of nucleotides, generating comprehensive maps of sites that are subject to miRNA binding. These studies uncovered and confirmed many features of miRNA regulation, amongst others the importance of the seed, predominant targeting of 3′UTRs over coding sequences (CDS) and 5′UTRs, increased conservation of Argonaute targeted sites and autoregulation of components of the miRNA machinery. Yet, previous analysis of AGO CLIP data did not reveal the identity of the miRNA that guided Argonaute to each individual site. As a consequence the miRNA responsible for binding was predicted on the basis of perfect sequence complementarity to expressed miRNAs. This is highly speculative as a site can contain seed-complementary sites for several miRNAs or could have also been bound by a miRNA in a noncanonical manner. Indeed, AGO CLIP studies usually report a high fraction of confident sites that do not contain perfect complementarity to miRNA seed sequences (Chi et al., 2009; Hafner et al., 2010; Zisoulis et al., 2010; Kishore et al., 2011), indicating a large fraction of AGO-targeting that is mediated by noncanonical miRNA-binding.

More recently AGO CLIP data were directly used to develop bioinformatic tools that aim to predict the miRNA, which most likely mediated AGO-binding to a CLIP-identified site (Majoros et al., 2013; Erhard et al., 2013; Khorshid et al., 2013; Liu et al., 2013). Some of these computational models assume specific modes of interactions and consequently include known features of miRNA binding into their algorithms. MicroMummie for instance, incorporates perfect complementarity to the miRNAs' seed sequence, evolutionary conservation and the spatial position of a potential seed match relative to the crosslink site and within the Ago site into a multivariate Markov Model (Majoros et al., 2013). The miRNA assignment tool PARma not only uses the position of crosslink but in addition includes the cleavage pattern of the RNase used during the experiment, to deduce the k-mer that was most likely bound by a miRNA (PARma, Erhard et al., 2013). In contrast, binding-free energy parameters pivotal for the performance of the biophysical model of miRNA-target interaction "MIRZA," were directly inferred from AGO CLIP data (Khorshid et al., 2013). This enables a less assumption-based analysis of miRNA binding from CLIP data and allows the prediction of noncanonical interactions to some degree.

Albeit these approaches help to more accurately predict miRNA interactions, they all are shaped by the current knowledge on miRNA binding (e.g., making assumptions on the role of the seed or on the importance of binding energies). This impedes the discovery of yet unknown features of miRNA binding. It also renders an immense uncertainty in inferring the identity of a miRNA bound to a target site considering the large number of miRNAs, the brevity of their seeds, and the existence of noncanonical binding modes.

DIRECT IDENTIFICATION OF miRNA:TARGET SITE INTERACTIONS BY LIGATION

LIGASES CONNECT THE miRNAs DIRECTLY TO THEIR BOUND TARGET SITES RESULTING IN CHIMERIC MOLECULES

To experimentally resolve which miRNA guides Argonaute to respective target sites CLIP protocols were adapted to not only recover the sites of AGO-binding but additionally the direct information which miRNA was bound to a site (Helwak et al., 2013; Grosswendt et al., 2014; Moore et al., 2015). This task was accomplished, by adding a ligase to purified AGO−RNA-complexes, which connects the miRNA and its bound target RNA into one chimeric molecule (Fig. 2.3). Helwak et al. adapted a HITS-CLIP protocol in a human cell line for the generation of ligation products of miRNAs and their targets, whereas Grosswendt et al. used the iPAR-CLIP protocol as bases and independently developed a modified version. Albeit the fraction of ligation products in these data is low (<1%), a computational strategy specifically designed to be sensitive and specific enough to detect and map the short sequences of chimeric reads, identified thousands of miRNA chimeras. Then Moore and colleagues accomplished to adapt the HITS-CLIP protocol to significantly increase the fraction of chimeric reads in sequencing data to ∼1.5%−5% (covalent ligation of endogenous Argonaute-bound RNAs (CLEAR)—CLIP).

Surprisingly, Grosswendt et al. also found miRNA chimeras in a control sample, in which no exogenous ligase was added to generate ligation products of miRNA and bound target. It was concluded that an RNA ligase activity, which had been reported to be present in cell lysates (Filipowicz et al., 1983; Perkins et al., 1985; Martinez et al., 2002) likely generated these chimeras in the course of the experiment. Consequently, miRNA chimeras should also be present in data of other AGO CLIP experiments. And indeed, reanalysis of published CLIP experiments performed in other model systems, like human cell lines, mouse brain and virus-infected cells, led to the discovery of thousands of chimeric reads containing a miRNA connected to its target site (Grosswendt et al., 2014). These newly identified miRNA interactions were investigated for known features of miRNA biology, for instance seed complementarity, conservation, and target repression and the results strongly indicated that these miRNA chimeras are derived from endogenous miRNA interactions. A potential candidate responsible for the generation of miRNA chimeras in standard CLIP experiments is the eukaryotic tRNA ligase (Popow et al., 2011), which is termed HSPC117 in humans and has the homologue RTCB in bacteria. It was found to process intron-containing tRNAs and the XBP mRNA as part of its nonconsensus splicing (Jurkin et al., 2014; Lu et al., 2014). The highly conserved ligase requires RNA modifications as produced by the RNases used in most CLIP

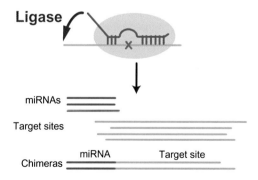

FIGURE 2.3 A ligase connects miRNA and its bound target site into one chimeric molecule

In CLIP experiments, RNA is partially digested. Subsequently the 3′end of the miRNA can be ligated to the 5′end of the bound target site by an experimentally added RNA ligase. Deep sequencing data contain reads of miRNAs, reads of AGO target sites, and to a low percentage also chimeras of miRNAs that are directly connected to their target sites. The crosslink (X) usually occurs between AGO and the target RNA.

experiments, which would also explain why in almost all miRNA chimeras the miRNA is not full-length but truncated from the 3′end and ligated to its target. It remains to be determined, whether this enzyme, which is so far the only known ligase with the required RNA substrate specificities (2′3′cyclic phosphate and 5′hydroxyl), is indeed responsible for chimera-generation in standard CLIP experiments.

Investigation of miRNA binding via analysis of miRNA:target chimeras did not only enable the unambiguous identification of numerous targets for individual miRNAs. It enabled researchers also to confirm previous assumptions and to uncover new features of miRNA binding via analysis that would not have been possible without the comprehensive, experimental data provided by miRNA chimeras.

CHIMERAS RECOVER ENDOGENOUS miRNA INTERACTIONS, THE MAJORITY POSSESSES SEED COMPLEMENTARITY

Studies on miRNA chimeras (Helwak et al., 2013; Grosswendt et al., 2014; Moore et al., 2015) recovered hundreds of previously known and thousands of novel miRNA:target pairs and found various complementarity patterns between the seed of the miRNA and the target site. The frequency of these seed matches and seed match variants strongly in the data exceeds complementarities that could be expected by chance, which was carefully controlled by bioinformatic analysis. Also, the complementarities are not a result of miRNA pairing that occurred after lysis as an experimental artifact. Such false discoveries due to postlysis association of miRNA and random RNA were assessed by species mixing experiments, in which lysate or purified RNA of an unrelated species was added to the lysate during CLIP. Ligation products consisting of miRNA and target from both species were vanishingly few, excluding postlysis association as a grave source of false interaction.

Grosswendt et al. first reported chimeras for mouse brain discovered in data of standard HITS-CLIP experiments. And although the number of miRNA interactions that could be recovered from these data was low, it was possible to identify detailed features of seed pairing. For instance, position 2 and 7 were most tolerant to mismatches compared to the seed positions in between (averaged over all recovered miRNAs), which turned to be an evolutionary conserved feature as it was also found for human and *C. elegans* miRNAs. Later, Moore and colleagues identified 130,000 miRNA interactions in mouse brain neocortex (postnatal day 13) by skillfully adapting the CLIP protocol for the generation and efficient recovery of miRNA chimeras (CLEAR-CLIP). Even though the dataset was dominated by a small number of brain-abundant miRNAs, it was possible to identify targets and preferred base pairing patterns for over 200 miRNAs expressed in the brain.

Similar to the findings by Grosswendt et al., for miRNAs of various species, Moore and colleagues found obvious seed complementarities for 75% of recovered mouse brain miRNA interactions. Specifically, ~30% miRNA interactions had a perfect seed match or an off-set seed match (complementarity to miRNA nt 3−8). An additional fraction of ~35% had an imperfect seed match containing one mismatched nucleotide, and another ~10% had a bulged nucleotide within the seed or the seed match. The remaining ~25% of chimera-identified interactions did not show obvious seed complementarities. It should be emphasized that miRNA sites with an imperfect seed match, i.e., containing a mismatch or a bulge, are extremely difficult to predict by bioinformatic means. The information content of a hexamer containing such a variation is too low to result in predictions with a reasonable false-discovery rate. Since imperfect seed complementarities constitute a large fraction of miRNA binding, direct biochemical binding information in form of chimeras empower their identification and further investigation.

RESEARCH OPPORTUNITIES OPENED UP BY CHIMERAS: CONSERVATION ANALYSIS OF IMPERFECT SEED SITES AND 3′END PAIRING, CLASSIFICATION OF DIFFERENT BINDING MODES, INVESTIGATION OF miRNA FAMILIES

That perfect seed sites are largely conserved was one of the early findings in miRNA research, which not only allowed bioinformatics prediction to estimate the prevalence of miRNA binding in the transcriptome, but it also indicated their biological importance. Now, with the large number of unambiguously identified interactions, analyzing the conservation of imperfect seed complementarity became possible as well. Strikingly, seed matches containing 1 mismatch were significantly conserved, albeit less than perfect matches (Grosswendt et al., 2014). The evolutionary conservation of imperfect seed sites likely indicates that they are functionally relevant. That evolution acted to maintain not only the seed match but the entire miRNA binding site was found when target sequence stretches that pair with the miRNA 3′end were investigated for conservation (Moore et al., 2015). Pairing to the miRNA 3′end appears to be a conserved feature of miRNA binding, albeit these parts of the binding site are less conserved than a perfect 2−8 seed match itself but comparable to the degree of conservation of perfect and imperfect 2−7 seed sites.

The majority of miRNA interactions have some seed complementarity together with pairing in the 3′part of the miRNA. Each miRNA employs various seed complementarities (i.e., 2−7, 2−8, or 3−8 perfect matches or matches with a mismatched or bulged nt) but used them in different relative frequencies. Also the position of the bulged nucleotide (within the seed or the seed match) showed miRNA-dependent preferences. Most of the top expressed miRNAs in the brain do not allow a bulged nucleotide at one or more position, and especially position 5 is intolerant to bulges. Chimeras of miRNA-124 also confirmed the well-known preference of this miRNA to bind with a G bulge between miRNA position 5 and 6 (Chi et al., 2009; Moore et al., 2015). In general, chimeras allow the discovery of several features specific for a miRNA of interest, a strong preference of G-U wobble pairing in position 3 found for miR-30 is another example (Moore et al., 2015).

Many targets had in addition to complementarity to the seed also complementarity to the 3′end of the miRNA, which was not always positioned on miRNA nucleotides 13−16 but included many different patterns. Moore and colleagues identified in total six major modes of binding; five of them with obvious seed complementarities. Most miRNAs employed a combination of seed and 3′end pairing and showed an enrichment or a depletion for pairing patterns in one or more of these major modes, indicating miRNA-specific binding preferences. For brain miRN-124, a strong dependence on seed pairing with different patterns of 3′pairing was found. Some miRNAs, such as miR-9 and miR-181, were also detected to have seed-dependent but also frequent seed-independent pairing. Overall, interactions with weaker seed matches (pairing to miRNA nt 2−7) have slightly more complementarity to the miRNA 3′end, than those with stronger seed matches (pairing to miRNA nt 2−8) and interactions with mismatches or bulges in the seed have the most prevalent binding to the 3′end of miRNAs (Moore et al., 2015).

A longstanding riddle of miRNA biology is the existence and conservation of miRNA families. Members of a miRNA family share the same seed sequence but differ in the sequence of their 3′part, which led to assumptions that they bind many shared but also some distinct sites. To analyze this issue large-scale direct binding data of miRNA family members as provided by chimeras is of extreme value. And indeed, target sites were significantly more often ligated to miRNAs of the same family than expected by chance, reflecting the fact that miRNAs share common sites (Grosswendt et al., 2014). Furthermore, differences in complementarities beyond the seed were noticed. For instance human miR-196a and miR-196b differ in nucleotide position 12. Ligated targets show an enriched complementarity starting either at position 12 or only at position 13, indicating that these family members also have some distinct targeting specificities. Similar complementarity differences were found for targets ligated to human miR-15a and miR-15b (Grosswendt et al., 2014). Also, calculation of duplex energies of miRNA family members and their ligated targets indicated that each family member was usually bound to targets with which they formed the most stable structures, suggesting target preferences between different family members (Moore et al., 2015). However, different complementarities between target sites and miRNA family members do not answer the question whether these sites are indeed differentially bound. Therefore, Moore and colleagues tested differential binding of miRNA family members experimentally, using fluorescent reporters. In several cases, miRNA target sites showed more repression for the identified family member than for the other paralogues. Sites, which were found ligated to several different family members, were similarly regulated by them, indicating functional redundancy. Further investigation of these intrafamily target preferences will hopefully uncover the reasons for the biological necessity of miRNA families, as suggested by their evolutionary conservation.

FUNCTIONALITY OF CHIMERA-IDENTIFIED INTERACTIONS

Studies that employed miRNA:target chimeras to investigate miRNA binding (Helwak et al., 2013; Grosswendt et al., 2014; Moore et al., 2015) also tested the functionality of identified interactions by analysis of miRNA perturbation data and by reporter assays. In miRNA perturbation experiments, the level of a miRNA is downregulated or increased (e.g., by antagomirs or miRNA mimics, respectively). Subsequently, effects on mRNAs and proteins are monitored by microarrays, mRNA sequencing, quantitative PCR, ribosome association analysis, or mass spectrometry. Multiple studies have produced such datasets, which were now reanalyzed focusing specifically on the targets that were found ligated to the miRNA of interest. The targets found in chimeras showed significant responsiveness, confirming that chimeras recover endogenous, biologically relevant miRNA interactions. For instance, in transfection experiments with miR-124, miR-9, and miR-181a mimics in mouse neuroblastoma cells most targets identified via analysis of chimeras were indeed downregulated (Moore et al., 2015). As was seen in several studies, interactions with perfect seed sites showed usually stronger repression than those with imperfect seed matches.

The functionality of interactions with imperfect or no seed matches is difficult to investigate using data of perturbation experiments. A large fraction of mRNAs that contain an interaction with an imperfect seed match for a miRNA also contains additional perfect seed matches for this miRNA. Consequently, the effect mediated by perfect and imperfect sites cannot be disentangled. To analyze mRNAs that contain only an imperfect site and no perfect sites do not seem to be feasible, as especially those mRNA with several sites for a miRNA are likely targets that experience strong and biologically relevant regulation. In contrast, functionality of imperfect seed sites can be adequately tested with reporter assays, in which the site is mutated and reporter expression is compared to the expression of a reporter that contains the nonmutated site. By this comparison, the effect of a site can be investigated in isolation of the remaining 3′UTR context.

It seems likely that miRNAs or groups of miRNAs fulfill a certain biological task in a cell type or tissue-specific manner. The biological processes a miRNA is involved in could be deciphered having numerous in vivo identified targets at hand. Gene ontology analysis for chimera-identified targets of the neuronal miRNAs miR-124 and miR-9 indicated functions in neuron development, synapse formation, and axon guidance (Moore et al., 2015). The functional profile of targets of less-characterized brain miRNAs also indicated neuronal regulatory functions. For instance, targets of miR-26 suggest its involvement in axon development and locomotion, targets of miR-138 indicate its role in neurotransmitter transport/secretion and calcium transport and miR-9* appears to be involved in cell migration and motility. Furthermore, analysis of chimera-identified targets of miR-124, miR-9, and miR-26 in the Kyoto Encyclopedia of Genes and Genomes database reconfirmed their role in glioma (Moore et al., 2015).

PERSPECTIVES

We start with discussing limitations of using chimera-based approaches for identifying miRNA target sites. First of all, chimeras are recovered from ligations naturally occurring in the lysate of CLIP experiments with low frequency (<1% of sequencing reads). Second, the endogenous RNA ligases as well as the experimentally added RNA ligase likely prefer certain structural features of the RNA-duplex (Liu et al., 1997). Thus, chimera-based miRNA target discovery is bound to miss a very large fraction of target sites. Adding a ligase to promote ligation reactions or cofactors to boost the activity of the endogenous ligase could be a strategy to increase the fraction of chimeras generated in the lysate during CLIP experiments (Popow et al., 2014). Moreover, and very obviously, identifying stably bound target sites does not mean that these target sites infer regulation by the respective bound miRNA. Although other criteria such as sequence conservation can help to tease out functionally important target sites, in practice one of the greatest challenges in miRNA target identification is the pruning of these sites for the most relevant interactions to the biological question at hand.

Neural systems pose particular challenges for the understanding of miRNAs. For example, mammalian brain tissues consist of a very large number of different cell types, and even individual neurons have strong differences in gene expression when comparing for example RNAs expressed in synapses to cytoplasmic RNA. miRNAs differ in expression between different cell types and also display large variability within neurons. Moreover, RNA in neurons has by far the largest degree of splicing/isoform diversity compared to RNA expression in other tissues. Not only the average length of 3′UTRs is drastically higher compared to other tissues (Sood et al., 2006) also the number of 3′UTRs isoforms per mRNA is highly diverse and linked to subcellular localization (Taliaferro et al., 2016). Very likely, all these phenomena must be studied on a subcellular level to understand how they are linked to function. Current biochemical approaches for identifying miRNA targets require large input material and therefore usually average over many different cell types or (in the case of FACS sorted cells) at least over all subcellular localization. We believe that further technology development will help to tackle some of these fundamental issues. For example, microfluidics-based droplet capture of single cells allows one to assess the transcriptome of tens of thousands of cells in parallel (Macosko et al., 2015; Klein et al., 2015). Knockdown of a miRNA and applying this single-cell sequencing approach before and after might be able to reveal disregulated miRNA targets within specific cell types which are otherwise invisible because they are masked by all cell types in which these target interactions do not occur or are not regulated.

REFERENCES

Alvarez-Saavedra, E., Horvitz, H.R., 2010. Many families of *C. elegans* microRNAs are not essential for development or viability. Curr. Biol.: CB 20 (4), 367–373.

Ascano, M., et al., 2012. Identification of RNA-protein interaction networks using PAR-CLIP. Wiley Interdiscip. Rev. RNA 3 (2), 159–177.

Baek, D., et al., 2008. The impact of microRNAs on protein output. Nature 455 (7209), 64–71.

Bagga, S., et al., 2005. Regulation by let-7 and lin-4 miRNAs results in target mRNA degradation. Cell 122 (4), 553–563.

Baltz, A.G., et al., 2012. The mRNA-bound proteome and its global occupancy profile on protein-coding transcripts. Mol. Cell. 46 (5), 674–690.

Bartel, D.P., 2009. MicroRNAs: target recognition and regulatory functions. Cell 136 (2), 215–233.

Bazzini, A.A., Lee, M.T., Giraldez, A.J., 2012. Ribosome profiling shows that miR-430 reduces translation before causing mRNA decay in zebrafish. Science (New York, NY) 336 (6078), 233–237.

Brennecke, J., et al., 2005. Principles of microRNA-target recognition. PLoS Biol. 3 (3), e85.

Cambronne, X.A., et al., 2012. Capturing microRNA targets using an RNA-induced silencing complex (RISC)-trap approach. Proc. Natl. Acad. Sci. USA 109 (50), 20473–20478.

Castello, A., et al., 2012. Insights into RNA biology from an atlas of mammalian mRNA-binding proteins. Cell 149 (6), 1393–1406.

Cech, T.R., Steitz, J.A., 2014. The noncoding RNA revolution—trashing old rules to forge new ones. Cell 157 (1), 77–94.

Chen, K., Rajewsky, N., 2007. The evolution of gene regulation by transcription factors and microRNAs. Nat. Rev. Genet. 8 (2), 93–103.

Chi, S.W., et al., 2009. Argonaute HITS-CLIP decodes microRNA-mRNA interaction maps. Nature. 460 (7254), 479–486.

Chi, S.W., Hannon, G.J., Darnell, R.B., 2012. An alternative mode of microRNA target recognition. Nat. Struct. Mol. Biol. 19 (3), 321–327.

Corcoran, D.L., et al., 2011. PARalyzer: definition of RNA binding sites from PAR-CLIP short-read sequence data. Genome. Biol. 12 (8), R79.

Denzler, R., et al., 2014. Assessing the ceRNA hypothesis with quantitative measurements of miRNA and target abundance. Mol. Cell. 54 (5), 766–776.

Didiano, D., Hobert, O., 2006. Perfect seed pairing is not a generally reliable predictor for miRNA-target interactions. Nat. Struct. Mol. Biol. 13 (9), 849–851.

Djuranovic, S., Nahvi, A., Green, R., 2012. miRNA-mediated gene silencing by translational repression followed by mRNA deadenylation and decay. Science (New York, NY) 336 (6078), 237–240.

Easow, G., Teleman, A.A., Cohen, S.M., 2007. Isolation of microRNA targets by miRNP immunopurification. RNA (New York, NY) 13 (8), 1198−1204.

Ebert, M.S., Sharp, P.A., 2012. Roles for MicroRNAs in conferring robustness to biological processes. Cell 149 (3), 515−524.

Elkayam, E., et al., 2012. The structure of human argonaute-2 in complex with miR-20a. Cell 150 (1), 100−110.

Erhard, F., et al., 2013. PARma: identification of microRNA target sites in AGO-PAR-CLIP data. Genome Biol. 14 (7), R79.

Fabian, M.R., Sonenberg, N., Filipowicz, W., 2010. Regulation of mRNA translation and stability by microRNAs. Annu. Rev. Biochem. 79, 351−379.

Fang, Z., Rajewsky, N., 2011. The impact of miRNA target sites in coding sequences and in 3′UTRs. PLoS ONE 6(3) e18067.

Filipowicz, W., et al., 1983. RNA 3′-terminal phosphate cyclase activity and RNA ligation in HeLa cell extract. Nucleic. Acids. Res. 11 (5), 1405−1418.

Friedländer, M.R., et al., 2014. Evidence for the biogenesis of more than 1000 novel human microRNAs. Genome Biol. 15 (4), 1−17.

Friedman, R.C., et al., 2009. Most mammalian mRNAs are conserved targets of microRNAs. Genome Res. 19 (1), 92−105.

Grosswendt, S., et al., 2014. Unambiguous identification of miRNA:target site interactions by different types of ligation reactions. Mol. Cell 1−13.

Guo, H., et al., 2010. Mammalian microRNAs predominantly act to decrease target mRNA levels. Nature. 466 (7308), 835−840.

Hafner, M., et al., 2010. Transcriptome-wide identification of RNA-binding protein and microRNA target sites by PAR-CLIP. Cell 141 (1), 129−141.

Hausser, J., Zavolan, M., 2014. Identification and consequences of miRNA-target interactions—beyond repression of gene expression. Nat. Publ. Group 15 (9), 599−612.

Helwak, A., et al., 2013. Mapping the human miRNA interactome by CLASH reveals frequent noncanonical binding. Cell 153 (3), 654−665.

Hornstein, E., Shomron, N., 2006. Canalization of development by microRNAs. Nat. Genet. 38 (Suppl), S20−S24.

Hu, W., Coller, J., 2012. What comes first: translational repression or mRNA degradation? The deepening mystery of microRNA function. Cell. Res. 22 (9), 1322−1324.

Ibáñez-Ventoso, C., Vora, M., Driscoll, M., 2008. Sequence relationships among C. elegans, D. melanogaster and human microRNAs highlight the extensive conservation of microRNAs in biology. PLoS ONE. 3 (7), e2818.

Jens, M., Rajewsky, N., 2015. Competition between target sites of regulators shapes post-transcriptional gene regulation. Nat. Publ. Group 16 (2), 113−126.

Johnston, R.J., Hobert, O., 2003. A microRNA controlling left/right neuronal asymmetry in Caenorhabditis elegans. Nature. 426 (6968), 845−849.

Jungkamp, A.-C., et al., 2011. In vivo and transcriptome-wide identification of RNA binding protein target sites. Mol. Cell. 44 (5), 828−840.

Jurkin, J., et al., 2014. The mammalian tRNA ligase complex mediates splicing of XBP1 mRNA and controls antibody secretion in plasma cells. EMBO J. Dec 17, 2014;33(24):2922−36. doi: 10.15252/embj.201490332. Epub 2014 Nov. 6.

Khorshid, M., et al., 2013. A biophysical miRNA−mRNA interaction model infers canonical and noncanonical targets. Nat. Methods. 10 (3), 253−255.

Kishore, S., et al., 2011. A quantitative analysis of CLIP methods for identifying binding sites of RNA-binding proteins. Nat. Methods. 8 (7), 559−564.

Klein, A.M., et al., 2015. Droplet barcoding for single-cell transcriptomics applied to embryonic stem cells. Cell 161 (5), 1187−1201.

Kozomara, A., Griffiths-Jones, S., 2011. miRBase: integrating microRNA annotation and deep-sequencing data. Nucleic Acids. Res. 39 (Database issue), D152−D157.

König, J., et al., 2010. iCLIP reveals the function of hnRNP particles in splicing at individual nucleotide resolution. Nat. Struct. Mol. Biol. 17 (7), 909−915.

Krek, A., et al., 2005. Combinatorial microRNA target predictions. Nat. Genet. 37 (5), 495−500.

Lai, E.C., 2002. Micro RNAs are complementary to 3′ UTR sequence motifs that mediate negative post-transcriptional regulation. Nat. Genet. 30 (4), 363−364.

Lal, A., et al., 2009. miR-24 Inhibits cell proliferation by targeting E2F2, MYC, and other cell-cycle genes via binding to "seedless" 3′UTR microRNA recognition elements. Mol. Cell. 35 (5), 610−625.

Landry, J.J.M., et al., 2013. The genomic and transcriptomic landscape of a HeLa cell line. G3 (Bethesda, MD) 3 (8), 1213−1224.

Landthaler, M., et al., 2008. Molecular characterization of human Argonaute-containing ribonucleoprotein complexes and their bound target mRNAs. RNA (New York, NY) 14 (12), 2580−2596.

Lee, R.C., Feinbaum, R.L., Ambros, V., 1993. The *C. elegans* heterochronic gene lin-4 encodes small RNAs with antisense complementarity to lin-14. Cell 75 (5), 843−854.

Lewis, B.P., Burge, C.B., Bartel, D.P., 2005. Conserved seed pairing, often flanked by adenosines, indicates that thousands of human genes are microRNA targets. Cell 120 (1), 15−20.

Lianoglou, S., et al., 2013. Ubiquitously transcribed genes use alternative polyadenylation to achieve tissue-specific expression. Genes Dev. 27 (21), 2380−2396.

Lin, Y.-C., et al., 2014. Genome dynamics of the human embryonic kidney 293 lineage in response to cell biology manipulations. Nat. Commun. 5, 4767.

Liu, C., et al., 2013. CLIP-based prediction of mammalian microRNA binding sites. Nucleic Acids. Res. 41 (14), e138.

Liu, J., 2004. Argonaute2 is the catalytic engine of mammalian RNAi. Science (New York, NY) 305 (5689), 1437−1441.

Liu, J., Jin, Y., Wang, D., 1997. Exon structure requirements for yeast tRNA ligase. Sci. China. Ser. C, Life Sci./Chin. Acad. Sci. 40 (6), 665−669.

Loeb, G.B., et al., 2012. Transcriptome-wide miR-155 binding map reveals widespread noncanonical microRNA targeting. Mol. Cell. 48 (5), 760−770.

Lu, Y., Liang, F.-X., Wang, X., 2014. A synthetic biology approach identifies the mammalian UPR RNA ligase RtcB. Mol. Cell. 55 (5), 758−770.

Macosko, E.Z., et al., 2015. Highly parallel genome-wide expression profiling of individual cells using nanoliter droplets. Cell 161 (5), 1202−1214.

Majoros, W.H., Ohler, U., 2007. Spatial preferences of microRNA targets in 3′ untranslated regions. BMC Genomics 8, 152.

Majoros, W.H., et al., 2013. MicroRNA target site identification by integrating sequence and binding information. Nat. Methods. 10 (7), 630−633.

Mangone, M., et al., 2010. The landscape of *C. elegans* 3′UTRs. Science (New York, NY) 329 (5990), 432−435.

Martin, K.C., Ephrussi, A., 2009. mRNA localization: gene expression in the spatial dimension. Cell 136 (4), 719−730.

Martinez, J., et al., 2002. Single-stranded antisense siRNAs guide target RNA cleavage in RNAi. Cell 110 (5), 563−574.

Mayr, C., Bartel, D.P., 2009. Widespread shortening of 3′UTRs by alternative cleavage and polyadenylation activates oncogenes in cancer cells. Cell 138 (4), 673−684.

Meisenheimer, K.M., Meisenheimer, P.L., Koch, T.H., 2000. Nucleoprotein photo-cross-linking using halopyrimidine-substituted RNAs. Methods Enzymol. 318, 88−104.

Miska, E.A., et al., 2007. Most *Caenorhabditis elegans* microRNAs are individually not essential for development or viability. PLoS Genet. 3 (12), e215.

Moore, M.J., Proudfoot, N.J., 2009. Pre-mRNA processing reaches back to transcription and ahead to translation. Cell 136 (4), 688−700.

Moore, M.J., et al., 2015. miRNA-target chimeras reveal miRNA 3′-end pairing as a major determinant of Argonaute target specificity. Nat. Commun. 6, 8864.

Mukherji, S., et al., 2011. MicroRNAs can generate thresholds in target gene expression. Nat. Genet. 43 (9), 854−859.

Park, C.Y., et al., 2012. A resource for the conditional ablation of microRNAs in the mouse. Cell Rep. 1 (4), 385−391.

Parker, J.S., et al., 2009. Enhancement of the seed-target recognition step in RNA silencing by a PIWI/MID domain protein. Mol. Cell. 33 (2), 204−214.

Pasquinelli, A.E., et al., 2000. Conservation of the sequence and temporal expression of let-7 heterochronic regulatory RNA. Nature 408 (6808), 86−89.

Perkins, K.K., Furneaux, H., Hurwitz, J., 1985. Isolation and characterization of an RNA ligase from HeLa cells. Proc. Natl. Acad. Sci. U. S. A. 82 (3), 684−688.

Popow, J., et al., 2011. HSPC117 is the essential subunit of a human tRNA splicing ligase complex. Science (New York, NY) 331 (6018), 760−764.

Popow, J., et al., 2014. Analysis of orthologous groups reveals archease and DDX1 as tRNA splicing factors. Nature. 1−17.

Rajewsky, N., 2006. microRNA target predictions in animals. Nat. Genet. 38 (Suppl), S8−13.

Rajewsky, N., Socci, N.D., 2004. Computational identification of microRNA targets. Dev. Biol. 267 (2), 529−535.

Reinhart, B.J., et al., 2000. The 21-nucleotide let-7 RNA regulates developmental timing in *Caenorhabditis elegans*. Nature. 403 (6772), 901−906.

Ricci, E.P., et al., 2012. miRNA repression of translation in vitro takes place during 43S ribosomal scanning. Nucleic Acids. Res. 41 (1), 586−598.

Rybak-Wolf, A., et al., 2014. A variety of dicer substrates in human and *C. elegans*. Cell 159 (5), 1153−1167.

Sandberg, R., et al., 2008. Proliferating cells express mRNAs with shortened 3′ untranslated regions and fewer microRNA target sites. Science (New York, NY) 320 (5883), 1643−1647.

Schirle, N.T., Macrae, I.J., 2012. The crystal structure of human Argonaute2. Science (New York, NY) 336 (6084), 1037−1040.

Schmiedel, J.M., et al., 2015. Gene expression. MicroRNA control of protein expression noise. Science (New York, NY) 348 (6230), 128−132.

Selbach, M., et al., 2008. Widespread changes in protein synthesis induced by microRNAs. Nature. 455 (7209), 58−63.

Sethupathy, P., Megraw, M., Hatzigeorgiou, A.G., 2006. A guide through present computational approaches for the identification of mammalian microRNA targets. Nat. Methods. 3 (11), 881−886.

Shin, C., et al., 2010. Expanding the microRNA targeting code: functional sites with centered pairing. Mol. Cell. 38 (6), 789−802.

Siciliano, V., et al., 2013. MiRNAs confer phenotypic robustness to gene networks by suppressing biological noise. Nat Commun. 4, 2364.

Sonenberg, N., Hinnebusch, A.G., 2009. Regulation of translation initiation in eukaryotes: mechanisms and biological targets. Cell 136 (4), 731−745.

Sood, P., et al., 2006. Cell-type-specific signatures of microRNAs on target mRNA expression. Proc. Natl. Acad. Sci. USA 103 (8), 2746−2751.

Su, H., et al., 2009. Essential and overlapping functions for mammalian Argonautes in microRNA silencing. Genes Dev. 23 (3), 304–317.

Taliaferro, J.M., et al., 2016. Distal alternative last exons localize mRNAs to neural projections. Mol. Cell. 61 (6), 821–833.

Vasquez-Rifo, A., et al., 2012. Developmental characterization of the microRNA-specific *C. elegans* Argonautes alg-1 and alg-2. PLoS ONE. 7 (3), e33750.

Vella, M.C., et al., 2004. The *C. elegans* microRNA let-7 binds to imperfect let-7 complementary sites from the lin-41 3′UTR. Genes Dev. 18 (2), 132–137.

Wang, Y., et al., 2010. Mechanism of microRNA-target interaction: molecular dynamics simulations and thermodynamics analysis. PLoS Comput. Biol. 6 (7), e1000866.

Wee, L.M., et al., 2012. Argonaute divides its RNA guide into domains with distinct functions and RNA-binding properties. Cell 151 (5), 1055–1067.

Wightman, B., Ha, I., Ruvkun, G., 1993. Posttranscriptional regulation of the heterochronic gene lin-14 by lin-4 mediates temporal pattern formation in *C. elegans*. Cell 75 (5), 855–862.

Xie, X., et al., 2005. Systematic discovery of regulatory motifs in human promoters and 3′ UTRs by comparison of several mammals. Nature 434 (7031), 338–345.

Yao, S.Y.M., et al., 2002. Functional and molecular characterization of nucleobase transport by recombinant human and rat equilibrative nucleoside transporters 1 and 2. Chimeric constructs reveal a role for the ENT2 helix 5-6 region in nucleobase translocation. J. Biol. Chem. 277 (28), 24938–24948.

Zekri, L., et al., 2009. The silencing domain of GW182 interacts with PABPC1 to promote translational repression and degradation of microRNA targets and is required for target release. Mol. Cell. Biol. 29 (23), 6220–6231.

Zhang, C., Darnell, R.B., 2011. Mapping in vivo protein-RNA interactions at single-nucleotide resolution from HITS-CLIP data. Nat. Biotechnol. 29 (7), 607–614.

Zisoulis, D.G., et al., 2010. Comprehensive discovery of endogenous Argonaute binding sites in *Caenorhabditis elegans*. Nat. Struct. Mol. Biol. 17 (2), 173–179.

COMPUTATIONAL CHALLENGES AND -OMICS APPROACHES FOR THE IDENTIFICATION OF MicroRNAs AND TARGETS

3

Ioannis S. Vlachos[1,2], Georgios Georgakilas[1,2], Spyros Tastsoglou[1,2], Maria D. Paraskevopoulou[1,2], Dimitra Karagkouni[1,2] and Artemis G. Hatzigeorgiou[1,2]

[1]University of Thessaly, Volos, Greece [2]Hellenic Pasteur Institute, Athens, Greece

INTRODUCTION

The noncoding RNA class of microRNAs (miRNAs) is characterized by powerful capabilities for posttranscriptional repression of gene expression. miRNAs identify their target regions (miRNA Recognition Elements, MREs) in 3′UTR (Bartel, 2009), coding sequence (CDS) (Reczko et al., 2012), and more rarely in the 5′UTR of genes (Kloosterman et al., 2004; Lytle et al., 2007) through base complementarity, and induce cleavage, degradation, or translation suppression in the case of protein coding genes (Huntzinger and Izaurralde, 2011).

MiRNAs are being actively researched for more than a decade and in silico tools have been proven from early on to be an invaluable aid to this effort (Enright et al., 2003; Lewis et al., 2003; Kiriakidou et al., 2004). Throughout these years, a significant number of algorithms (Vlachos and Hatzigeorgiou, 2013; Akhtar et al., 2016) have been developed to assist in the elucidation of fundamental research questions related to miRNA biogenesis and function, including miRNA annotation, regulation, and interactions with other RNA species (Fig. 3.1). Advanced applications have been also developed that facilitate complex research inquiries, such as the identification of miRNA biomarkers, the detection of miRNAs with crucial regulatory roles and of miRNA-controlled pathways. A plethora of databases and repositories were also deployed to catalog and organize the large amounts of miRNA-related data in a comprehensible and efficient fashion (Singh, 2016).

A long series of experimental techniques have been devised to seemingly complement the numerous computational algorithms employed in the exploration of miRNA interactomes (Thomson et al., 2011). High-throughput -omics methodologies, ranging from microarrays and RNA sequencing to specialized techniques, such as crosslinking immunoprecipitation sequencing (CLIP-Seq) (Chi et al., 2009), have become central to miRNA research. These methodologies produce large volumes of crucial data, allowing the optimization of existing computational methods and the design of novel algorithms, but also to complement, or even substitute, existing in silico approaches.

Essentials of Noncoding RNA in Neuroscience. DOI: http://dx.doi.org/10.1016/B978-0-12-804402-5.00003-0

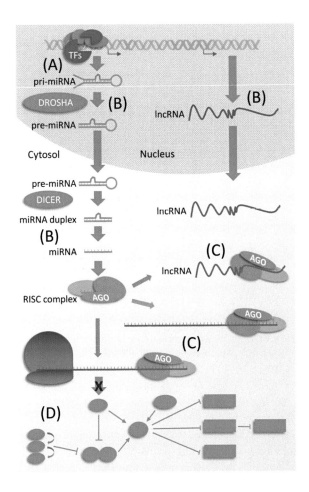

FIGURE 3.1

Schematic overview of points in miRNA biogenesis and function where computational methods facilitate miRNA research. (A) miRNA TSS/regulator databases and prediction tools, (B) sequence and expression annotation databases, (C) miRNA interaction/binding site databases and prediction tools, and (D) miRNA functional analysis/enrichment tools.

In this chapter, we will provide a concise yet comprehensive overview of available computational methods and -omics technologies utilized in miRNA research. Additionally, current limitations and still open challenges will be presented. The computational algorithms, web servers, and databases, as well as -omics technologies, have been divided thematically into the following categories:

- miRNA annotation and regulation
- miRNA—gene interactions
- Functional analysis tools and functional annotation databases

MiRNA ANNOTATION AND REGULATION

MiRNAs are transcribed by Pol II and can originate from intergenic or intronic loci with approximately 1:1 ratio. A pri-miRNA (primary) transcript can host more than one pre-miRNAs (hairpin miRNA precursor), which are then termed miRNA clusters. The usually long primary miRNAs are readily processed by enzyme Drosha within the nucleus, so, as a community, we have initially focused on the much more stable pre-miRNA and mature miRNA forms, which can be more easily identified and quantified with standard methodologies. From very early on, the database *miRBase* (Kozomara and Griffiths-Jones, 2014) has the leading role in cataloging and annotating mature miRNAs and precursors for most species and is now considered the reference online compendium of miRNA sequences and annotation. It is a systematically updated repository allowing miRNA-related studies to utilize a uniform miRNA annotation. Currently, miRBase v21 hosts 28,645 miRNA hairpin precursors, that correspond to more than 35,000 mature miRNA products in 223 species. The numbers of deposited sequences for *Homo sapiens* miRNA precursors and mature sequences have reached 1,881 and 2,588, respectively. It additionally provides levels of confidence to the indexed entries based on analyzed sRNA deep sequencing datasets. miRBase miRNA hairpins are coupled with enriched metadata such as predicted structure, precursor/mature sequences, links to external annotation resources, databases providing miRNA–gene interactions, as well as related bibliography. The nomenclature and naming of miRNAs used within miRBase are also adopted by the community. Currently, a significant effort has started to reexamine identified miRNAs and keep separately those that are annotated with stricter criteria and which are supported by ample experimental methods.

MiRNA-TRANSCRIPTION START SITES (TSSs)

Contrary to the significant progress made in the topic of miRNA target identification, the regulation of miRNA transcription still remains obscure. Such knowledge enables the genome-wide identification of miRNA expression regulators, including transcription factors (TFs), other noncoding RNAs and epigenetic modifiers, as well as shed light to the mechanisms underlying miRNA expression in development and disease.

During the past few years, in silico miRNA promoter recognition methods have emerged, in order to address the increased difficulty of high-throughput miRNA promoter identification. Initial approaches to identify promoters proximal to miRNA loci (Zhou et al., 2007; Saini et al., 2007, 2008) relied on overrepresented k-mers, TF binding motifs, and CpG content, extracted from known promoters of protein-coding genes.

In the light of NGS methodologies, novel miRNA promoter prediction algorithms have been created. Numerous techniques (Barski et al., 2009; Ozsolak et al., 2008; Corcoran et al., 2009; Marson et al., 2008) utilized genome-wide binding sites of known markers of transcription initiation (i.e., H3K4me3 and Pol II), as identified by chromatin immunoprecipitation (ChIP-Seq), to increase prediction accuracy. Alternative approaches include the use of NGS data from transcriptomic analysis experiments (i.e., RNA-Seq and CAGE-Seq) combined with sequence features (Megraw et al., 2009; Marsico et al., 2013), or ChIP-Seq derived binding sites of active transcription marks (Chien et al., 2011) and TF binding sites identified by DNase-Seq (Georgakilas et al., 2014).

A brief overview of the available miRNA promoter identification algorithms and their characteristics is presented below.

CoVote (Zhou et al., 2007): One of the initial attempts for miRNA promoter recognition, CoVote, investigated miRNA gene transcription initiation loci in four species based on sequence motifs found to be overrepresented in promoter regions of annotated protein-coding genes and machine learning. The species-specific models were trained using already annotated core promoter sequences (positives) and 1000 sequences simulated based on the intergenic nucleotide composition of each organism (negatives). Statistically, overrepresented k-mers (5–10 bp) were utilized for training the models. The performance of the algorithm was tested against TSSs derived from experimental methods described in DBTSS (Suzuki et al., 2002).

Saini et al. (2007, 2008): In the following year, two additional studies from the same group were published, attempting to go beyond miRNA TSS identification and facilitate pri-miRNA prediction by utilizing transcriptional features proximal to annotated pre-miRNAs. The $5'$ ends of intergenic pri-miRNAs were predicted by combining Eponine (Down and Hubbard, 2002), CpG islands, and evidence from CAGE data. Likewise, the $3'$ ends were predicted on the basis of poly (A) associated motifs, as well as gene identification signature tags. Expression sequence tags (ESTs), cDNA sequences, and TF binding motif enrichment were also utilized as additional evidences.

S-Peaker (Megraw et al., 2009): In this study, the authors investigated whether solely the presence of TF motifs can facilitate TSS identification. CAGE experimental data were utilized as genomic landmarks for extracting the positive set of promoter sequences. For each positive TSS, a group of 20 randomly selected proximal positions ($+100/-4,000$ bp) formed the negative set along with a random intragenic location. Features were designed to approximate the DNA binding affinities of initially 39 known TFs based on the method of log-likelihood scoring for positional weight matrices. The algorithm computes the enrichment of each TF proximal to the TSSs in the training set and calculates the final set of features by also including the GC percentage of the queried sequence. The performance of this approach was tested on an independent test set derived from a subset of the initial set of CAGE TSSs (positives) and 100,000 randomly selected genomic locations (negatives).

Marson et al. (2008): The algorithm introduced in this study relies on H3K4me3 ChIP-Seq data, CpG island annotation, and ESTs for miRNA promoter identification. Initially, genomic loci enriched in H3K4me3 signals are considered as putative promoters and ranked by an empirically derived scoring system. Enriched sites are positively scored if they are either the start of a known gene or present an overlap with ESTs. Additional positive scores are given to enriched sites in close proximity (5 kb) to miRNAs. Negative scores are assigned on the basis of the number of intervening H3K4me3 sites and in the case where the candidate region could be assigned to a gene or EST that did not overlap the miRNA.

Barski et al. (2009): In the study by Barski et al., putative miRNA TSSs are predicted by scanning the pre-miRNA upstream region for statistically significant ChIP-Seq derived peaks of Pol II, H3K4me3, and H2A.Z. The statistical significance of each peak is assessed by modeling the distribution of tags surrounding each queried nucleotide position using a kernel density estimation function. The resulting profiles keep track of the strongest identified peak for each protein. The algorithm terminates when a peak has been found for all three markers or the search space is exhausted. If the Pol II peak colocalizes with H3K4me3 and H2A.Z, then the Pol II peak is considered as a TSS. If the H3K4me3 peak colocalizes with the H2A.Z peak, then the H3K4me3 peak is selected. If neither of the above conditions holds, or if peaks for all three transcription marks have not been identified, then no prediction is made.

Ozsolak et al. (2008): This is one of the first attempts to incorporate ChIP-derived genomic loci of transcription-related marks into in silico miRNA TSS identification. Genomic regions upstream of miRNAs and enriched in H3K4me3, H3K9/14Ac, RNAPII, and/or RNAPIII experimentally derived signals are treated as promoter loci. In order to pinpoint the actual nucleotide of the transcription initiation event, a scoring function is implemented that combines promoter features, including nucleosome-depleted regions, evolutionary conservation profiles, CpG islands, and TF motif occurrence. The algorithm performance was tested against previously annotated miRNA TSSs, as well as a set of six miRNA TSSs identified with promoter cloning and RT-PCR.

CPPP (Corcoran et al., 2009): Core Promoter Prediction Program (CPPP) attempts to predict miRNA core promoter loci by combining ChIP-Seq data with sequence features and machine learning. The algorithm initially scans the upstream miRNA region and considers the nearest Pol II enriched peak as putative promoter. Subsequently, an SVM model is applied in order to calculate the peak's probability of being an actual core promoter. The model has been trained on previously annotated protein-coding genes (positive set) and random intergenic regions (negative set) based on a set of features that includes 17 TF familial binding profiles, k-mers, and GC content.

miRStart (Chien et al., 2011): miRStart is a computational approach that integrates CAGE with TSS-Seq, and H3K4me3 ChIP-Seq datasets in order to predict miRNA TSSs. The algorithm models the signature profile around the TSS of protein-coding genes by utilizing machine learning (SVM). The model identifies putative promoter regions upstream of mature miRNAs and filters each candidate promoter based on the distance from the corresponding miRNA and the number of overlapping ESTs or protein-coding exons.

PROmiRNA (Marsico et al., 2013): PROmiRNA attempts to tackle the problem of miRNA promoter characterization, by incorporating CAGE data and sequence features into machine learning models. The algorithm considers loci upstream of precursor miRNAs, enriched in CAGE signals, as putative promoters. Each candidate, as well as randomly selected intergenic and intronic regions, serve as positive and negative examples for training an expectation maximization model which additionally incorporates CpG content, conservation, TATA box affinity and mature miRNA proximity. The performance of the algorithm was tested against ChIP-Seq derived Pol II enriched regions, as well as transcriptome data from RNA-Seq and previously validated miRNA TSSs found in the literature.

microTSS (Georgakilas et al., 2014): microTSS is an approach with a focus on intergenic miRNA TSSs. The algorithm integrates RNA sequencing reads to construct transcription "islands" (i.e., regions abundant in RNA-Seq reads) upstream of pre-miRNAs. The key element of microTSS is that each RNA-Seq enriched locus is regarded as a putative TSS, facilitating single nucleotide resolution. Three independent SVM models are utilized to score candidate TSSs and conclude to the final predictions. The models were trained on H3K4me3 and Pol II occupancy around protein-coding TSSs, as well as on the existence of open chromatin domains, as identified by DNase-Seq. Numerous human and mouse TSSs were identified at single nucleotide resolution by applying microTSS on deeply sequenced NGS data derived from the ENCODE consortium. Specifically, 70 intergenic miRNA gene TSSs, corresponding to 118 miRNA precursors in mESCs, 63 TSSs corresponding to 86 pre-miRNAs in hESCs and 50 TSSs associated to 82 precursors in mESCs. The algorithm's performance was evaluated against the only currently available extensive set of experimentally identified miRNA TSSs. They were identified using *Drosha* null/conditional-null mESCs, as well as an alternative set of TSSs detected by deeply sequenced global run-on sequencing (GRO-Seq) data in human IMR90 and ES cells. microTSS was found to significantly outperform all previous implementations and to offer >90% sensitivity and precision against the compendium of experimentally validated promoters.

Databases of miRNA regulation

Most available repositories of miRNA transcription regulation lack accurate TSS predictions. In most cases, heuristic methods are used to cluster pre-miRNAs into transcriptional units on the basis of inter-miRNA distance; an error-prone approach considering the size of pri-miRNAs. Other databases comprise early miRNA TSS identification predictions, many of which are characterized by low resolution since they are derived from wide transcription signals, e.g., histone marks. In this section, the current state-of-the-art is briefly presented (Table 3.1).

ChIPBase (Yang et al., 2013): ChIPBase aims to facilitate the construction of TF—gene and TF—miRNA interaction maps in multiple tissues and cell lines of five species. The database integrates 543 ChIP-Seq libraries against 252 TFs, as well as lncRNA and miRNA-related annotation from external databases and the literature. ChIPBase also integrates miRNA posttranscriptional regulation information from StarBase (Li et al., 2014a) as well as Cytoscape-mediated visualization of TF—miRNA and miRNA-target networks (Shannon et al., 2003). In order to identify the miRNA TSSs, ChIPBase groups pre-miRNAs into transcriptional units and considers the cluster's TSS as the 5′ end of the first cluster member.

CircuitsDB (Friard et al., 2010): CircuitsDB focuses on identifying gene regulatory motifs by supporting TF—miRNA, TF—gene, and miRNA—gene interactions in two species. Pre-miRNAs are clustered into transcriptional units and the 5′-most location of each cluster is considered as a TSS. TF—miRNA and TF—gene interactions are assembled from overlapping TF binding sites that are computationally derived from motif scanning analysis in a limited region (1 kb) surrounding

Table 3.1 Overview of miRNA Transcription Regulation Databases and Their Characteristics

	Expression Info	Network Visualization	miRNA TSS Source	TF: miRNA Source	TF:gene Source	miRNA:gene Source	miRNA: miRNA Source	Functional Annotation
ChIPBase	✓	Cytoscape	Heuristic	ChIP-Seq	ChIP-Seq	starBase	–	GO, KEGG
CircuitsDB	✓	–	Heuristic	Motif scanning	Motif scanning	TargetScan, TargetMiner	–	GO, OMIM, HMDD
TMREC	–	Text	–	Literature	–	Literature	–	–
TransmiR	–	–	–	Literature	–	–	–	–
TSmiR	–	Cytoscape	Landgraf et al., miRStart, CPPP, Ozsolak et al.	ChIP-Seq	BIOGRIDE, GREAT	miRTarBase, miRecords, TargetScan	–	GO, KEGG (Bartel)
miRGen	✓	–	microTSS	DNase-Seq Motif scanning	–	TarBase, LncBase, microT	–	GO, KEGG (miRPath)

miRNA clusters or protein-coding gene TSSs. miRNA−gene interactions are computed using TargetScan (Yassour et al., 2009) and TargetMiner (Bandyopadhyay and Mitra, 2009). The web interface facilitates the investigation of the circuit functional properties by integrating GO ontologies, expression, and disease information.

TMREC and TransmiR (Wang et al., 2010, 2015b): TMREC and TransmiR are the only databases that solely rely on literature text mining for extracting TF−miRNA interactions. Approximately 100 and 5,000 publications, for TMREC and TransmiR, respectively, have been mined in order to associate TF−miRNA interactions with various diseases. TMREC hosts 689 disease and TF/miRNA pairs that involve 21 diseases, 219 TFs, and 85 miRNAs. TransmiR supports 735 interactions that correspond to 201 TFs and 209 miRNAs in 16 organisms. The quality of the hosted information is highly dependent on the curation of the text mining results.

TSmiR (Guo et al., 2014): TSmiR database aims to facilitate the understanding of tissue-specific regulation and function by providing interaction maps between TFs, miRNAs, and their target genes in 12 human tissues for 116 miRNAs. Their TSSs have been obtained from already published in silico and experimental methodologies. ChIP-Seq-derived TF binding sites were derived from the ENCODE project. Approximately 2,347 TF−miRNA interactions were calculated by cataloging TF-binding sites in the miRNA TSS proximal region. TSmiR also hosts experimentally verified miRNA−gene interactions from miRTarBase (Hsu et al., 2011) and miRecords (Xiao et al., 2009), as well as in silico predicted interactions from TargetScan (Grimson et al., 2007) and functional annotation of targeted genes based on DAVID (Huang da et al., 2009).

DIANA-miRGen (Georgakilas et al., 2016): DIANA-miRGen v3.0 is the most recent attempt to elucidate miRNA transcription regulation. To this end, this database provides an accurate genome-wide map of TF−miRNA interactions for multiple tissues and cell-lines in *H. sapiens* and *Mus musculus*. microTSS (Georgakilas et al., 2014), a state-of-the-art computational framework was applied on RNA-, ChIP-, and DNase-Seq deep sequencing data resulting in the identification of 276 tissue/cell-line specific TSSs for 428 miRNA precursors. More than 200 TF position frequency matrices, combined with in-house assembled RNA-Seq expression profiles, created sets of tissue-specific motifs (Mathelier et al., 2014). The database is accessible through a comprehensible and easy-to-use interface incorporating rich meta-data on miRNA and TF function and tissue-specific expression levels, as well as their implication in physiological conditions and diseases. miRGen's interconnection with other DIANA resources facilitates conduction of miRNA pathway analyses with miRPath (Vlachos et al., 2015b), identification of miRNA predicted targets on protein-coding genes with microT (Paraskevopoulou et al., 2013b) and validated targets with TarBase (Vlachos et al., 2015a), or in silico as well as experimentally verified miRNA targets on lncRNAs with LncBase (Paraskevopoulou et al., 2016).

MiRNA−GENE INTERACTIONS

COMPUTATIONAL APPROACHES FOR THE IDENTIFICATION OF miRNA−mRNA PREDICTED INTERACTIONS

MiRNA−gene interactions are considered the backbone of most miRNA functional studies, and significant effort has been devoted to understand the rules underlying this mechanism. Target prediction tools constituted the first in silico approaches in miRNA research. Most of the developed algorithms focused from the very beginning on the prediction of miRNA binding sites solely on the 3′UTR of mRNAs. However, recent advances in high-throughput sequencing revealed a significant portion of target sites in CDS (Hafner et al., 2010). There are numerous widely used and promising applications for de novo identification of miRNA−gene interactions. These in silico implementations often produce radically different outcomes due to the incorporation of diverse features and experimental data utilized for the creation of each model. Therefore, selecting the most appropriate implementation is a common and multifaceted problem.

DIANA-microT-CDS (Paraskevopoulou et al., 2013b; Reczko et al., 2012): DIANA-microT-CDS is a state-of-the-art implementation that identifies canonical miRNA targets both in CDS and 3′UTR regions. It achieves increased performance in terms of sensitivity and precision, due to independent analysis and distinct feature extraction for CDS and 3′UTR regions. Important microT-CDS features are the target site complementarity, A/U upflanking content, accessibility, pairing stability, and conservation of miRNA targeted CDS and 3′UTR regions in 30 and 16 species, respectively. Positive and negative instances are derived from PAR-CLIP data (Hafner et al., 2010). Candidate miRNA binding sites are combined in a general linear model, which is trained on microarray datasets that measure mRNA expression changes after transfection or knockout of a specific miRNA. The potency of each miRNA−gene interaction is described by a combined score that represents the synergistic action of multiple binding sites in the targeted mRNA regions. The overall performance of the algorithm is estimated on quantitative proteomics and HITS-CLIP data (Chi et al., 2009).

TargetScan (Agarwal et al., 2015): TargetScan is an algorithm with high performance in terms of sensitivity and precision. It is a quantitative model that incorporates 14 distinct features, including the target site type, 3′ supplementary pairing, local AU content, 3′UTR binding site abundance, predicted seed-pairing stability, and conservation. It mainly detects canonical (high level of seed complementarity) sites within 3′UTR regions according to a seed-dependent scoring system. Each target site can be evaluated with a cumulative context and/or an aggregated conservation score. The training and testing of the model was performed on microarray datasets with clear sRNA-induced repression using stepwise regression (Garcia et al., 2011).

MIRZA-G (Gumienny and Zavolan, 2015): MIRZA-G is another tool able to predict canonical miRNA binding sites and siRNA off-targets. Decisive features for MIRZA-G include the nucleotide composition around putative targeted regions, the site structural accessibility, the evolutionary conservation, and the location of the site within the 3′UTR region. miRNA binding affinity in mRNA regions is assessed by the MIRZA biophysical model deduced from Argonaute crosslinking and immunoprecipitation (AGO-CLIP) data (Khorshid et al., 2013). The latter implementation assigns base binding energies on the candidate miRNA−mRNA duplexes. The training and testing of the algorithm was performed using a generalized linear model against miRNA/siRNA transfection microarray and proteomics datasets.

miRanda (Betel et al., 2010; John et al., 2004): miRanda is the target prediction model provided by microRNA.org. Its core algorithm identifies putative miRNA−gene interactions which are scored by mirSVR model. mirSVR utilizes a support vector regression approach and is trained on miRNA transfection experiments performed on HeLa cells. The scoring scheme is based on sequence, context, accessibility, conservation, and UTR-relevant features. The performance of miRanda-mirSVR joint usage was assessed on miRNA transfection/inhibition microarray and proteomics datasets.

mirMark (Menor et al., 2014): mirMark uses an extensive list of features, such as complementarity, structural accessibility, composition, and evolutionary conservation, in order to predict putative targets. It incorporates miRanda algorithm for the initial identification of candidate miRNA binding sites. mirMark adopts separate levels of classification with a random forest model for target site and miRNA—UTR interaction evaluation. The training was performed using experimentally verified miRNA-gene interactions derived from miRecords and miRTarBase (Chou et al., 2016), whereas mock miRNA—gene pairs were included as negative target regions. PAR-CLIP data were used for the evaluation of mirMark's performance.

mBSTAR (Bandyopadhyay et al., 2015): mBSTAR constitutes a learning framework designed for predicting specific functional binding sites of miRNAs. It incorporates 40 sequence, structural, and energy features, including nucleotide frequencies, internal loops, bulges, and minimum free energy of the entire flanking region. mBSTAR utilizes a random forest classifier, whereas the training and testing was performed on experimentally validated miRNA-gene interactions derived from miRecords, Tarbase v6.0 (Vergoulis et al., 2012), and StarBase (Li et al., 2014a).

Computational methodologies unambiguously provide the basis for several miRNA-related studies. However, even the most sophisticated implementations still achieve a far from perfect predictive accuracy (Vlachos and Hatzigeorgiou, 2013). The excess of miRNA—gene computationally predicted interactions, along with the increased number of false positives often hinders the conduction of meaningful downstream analyses. To this end, the available algorithms can be further improved, to expand existing knowledge regarding miRNA function, by encompassing the technological advances made in miRNA-related sequencing experiments.

DATABASES CATALOGING EXPERIMENTALLY SUPPORTED miRNA—mRNA—lncRNA INTERACTIONS

Numerous experimental methodologies of varying accuracy have been established, enabling the identification of novel miRNA targets and the validation of predicted interactions. The application of specific and high-throughput methodologies has already partly characterized miRNA interactomes. The method distinction in low/high-throughput techniques is associated with the amount of information each category produces. Low-yield experiments include reporter gene-assays, qPCR, western blotting, and ELISA. Reporter gene-assays focus on specific miRNA binding site validation, whereas qPCR, western blotting or ELISA immunoassays recognize interactions indirectly by identifying events of mRNA or protein concentration reduction; signifying mRNA degradation or translation suppression, respectively. High-throughput methodologies correspond to increased throughput versions of specific techniques with low accuracy, like microarrays and proteomics, as well as miRNA-specific NGS experiments, such as HITS-CLIP, PAR-CLIP, CLASH, and Degradome-Seq (Thomson et al., 2011; Vlachos et al., 2015a). CLIP-Seq experiments combine high-throughput results with increased accuracy and have revolutionized miRNA research. These methodologies usually return miRNA—AGO binding sites on a transcriptome-wide scale. They can be complemented with RNA expression and ribosome profiling experiments. An overview of the experimental methodologies and their intended use is presented in Table 3.2.

Experimentally supported miRNA interactions remain dispersed in numerous publications, supplementary material, and raw NGS datasets. To this end, several repositories have been created aiming to provide a centralized access to miRNA—gene experimentally derived interactions.

Table 3.2 Experimental Methodologies for miRNA−Gene Interaction Identification

Method	Direct Technique	Throughput	Experiment Context
Luciferase reporter gene	✓	Low	Identification of interacting miRNA−gene regions
qPCR, Northern Blots	−	Low	miRNA effect on mRNA levels
Western Blots, ELISA	−	Low	miRNA effect on protein concentration
Microarrays, RNA-Seq	−	High	miRNA effect on mRNA expression
CLIP-Seq/CLASH/CLEAR-CLIP	✓	High	MRE binding site sequencing
3LIFE	✓	High	High-throughput reporter gene assay
RPF	−	High	Ribosome-bound transcript sequencing
Biotin miRNA tagging (Biotin-Seq, Biotin-Microarrays, Biotin-qPCR)	−	High/Low	Biotin-tagged miRNA pull down followed by RNA-Seq/microarrays/qPCR
Quantitative proteomics	−	High	miRNA effect on protein concentration
AGO-IP	−	High	Enriched transcripts in AGO immunoprecipitates
miTRAP	−	High	miRNA trapping by RNA baiting
IMPACT-Seq	−	High	Biotin-tagged miRNA pull down
PARE/Degradome-Seq	−	High	Cleaved mRNA targets
LAMP	−	High	Labeled miRNA pull-down with digoxigenin

(Thomson et al., 2011; Moore et al., 2015; Vlachos et al., 2015a)

DIANA-TarBase was developed in 2006, and since then it is constantly updated (Vlachos et al., 2015a). It is considered a reference database, providing the largest manually curated collection of miRNA−mRNA interactions with experimental support. TarBase v7.0 indexes 29 different experimental methodologies, covering most of the established low-yield and high-throughput techniques. It contains more than half a million entries derived from the analysis of 250 miRNA-related NGS datasets (e.g., 150 CLIP-Seq, CLASH, microarrays, and Degradome-Seq) and approximately 7,500 validated specific miRNA−gene interactions. Indexed entries are coupled with rich metadata on 356 cell types and 59 tissues.

The sixth version of miRTarBase (Chou et al., 2016) includes 366,181 miRNA experimentally supported interactions obtained from low-yield and high-throughput methods. It provides entries comprising ∼22,500 genes and more than 3,500 mature miRNAs for 18 different species. miRTarbase also incorporates miRNA/mRNA expression profiles derived from the Cancer Genome Atlas (TCGA) (Weinstein et al., 2013).

Other databases index a significantly smaller set of interactions, compared to DIANA-TarBase and miRTarBase. *miRecords* (Xiao et al., 2009) comprises 2,705 experimentally validated interactions with 2028 entries derived from low-yield techniques. The supported interactions include 644 miRNAs and 1901 genes from 9 species. *miR2Disease* (Jiang et al., 2009) hosts 3,273 miRNA−mRNA curated interactions for 349 miRNAs coupled with information concerning miRNA deregulation in human diseases. Many interactions are derived from the fifth version of DIANA-TarBase (Papadopoulos et al., 2009b).

There are also repositories that significantly differ from the aforementioned databases, as they are designed to register CLIP-Seq experimental results from different RNA binding proteins (RBP) and to intersect them with target prediction algorithms. *StarBase* (Li et al., 2014a) includes RBP-binding regions from 108 CLIP-Seq datasets and facilitates the online functional analysis of such experimental outcomes. *CLIPZ* (Khorshid et al., 2011) is another open-access repository which enables users to interactively analyze RBP CLIP-Seq experimental data.

Investigating miRNA—lncRNA interactions

Even though miRNAs have been considered mRNA repressors, they were only recently found to target long noncoding RNA (lncRNA) transcripts. There is increasing evidence supporting that lncRNAs are able to sequester miRNAs from their mRNA targets (Paraskevopoulou and Hatzigeorgiou, 2016). This sponge-decoy lncRNA function has been experimentally validated with low-yield, (in)direct techniques in different human cell types, whereas certain interactions are conserved in more than one model organisms (Paraskevopoulou and Hatzigeorgiou, 2016). For instance, CDR1as/ciRS-7 circular antisense transcript harbors multiple miRNA binding sites and linc-MD1 is a muscle specific lncRNA acting as a miRNA decoy in the cytoplasm of both human and mouse species (Cesana et al., 2011; Hansen et al., 2013). However, miRNA—lncRNA endogenous interactions remain widely unexplored and their biological significance is still under investigation. Therefore, NGS experiments can assist the comprehension of miRNA-noncoding interactome.

Currently, there are limited implementations seeking miRNA—lncRNA interactions. *DIANA-LncBase* (Paraskevopoulou et al., 2013a) is a repository dedicated to the indexing of miRNA—lncRNA interactions. Its latest update (Paraskevopoulou et al., 2016) provides a repository of more than 70,000 experimentally supported miRNA—lncRNA interactions. Database entries have been derived from the analysis of numerous high-throughput experiments, as well as a collection of specific techniques in different cell types, tissues, and conditions. Additionally, LncBase caters a distinct module comprising computationally predicted miRNA—lncRNA interactions for an integrative transcript set.

miRcode (Jeggari et al., 2012) accommodates in silico inferred miRNA binding sites for lncRNAs indexed in GENCODE v11 (Harrow et al., 2012). *StarBase* supports miRanda/mirSVR-predicted miRNA binding sites in AGO CLIP-Seq enriched regions localized on lncRNA transcripts. There are also databases such as *LncReg* and *lncRNome* (Bhartiya, Pal et al. 2013; Zhou, Shen et al. 2015) that provide a small number of miRNA—lncRNA interactions complementary to other lncRNA regulatory roles, derived from low/high-throughput published experiments and text mining. *LNCipedia* (Volders et al., 2015) on the other hand, is a repository initially developed for lncRNA cataloging and supports miRNA—lncRNA canonical interactions predicted by MirTarget2 algorithm (Wang and El Naqa, 2008). *miRSponge* (Wang et al., 2015a) is another database that contains more than 500 miRNA sponge-like events and competing endogenous miRNA—lncRNA—mRNA interactions across different species.

The databases indexing miRNA—mRNA—lncRNA endogenous interactions gradually become essential tools. The wealth of information they provide can significantly boost the understanding of miRNA posttranscriptional regulation mechanisms and empower or even, at extensively studied cell types, substitute in silico predicted interactions.

FUNCTIONAL ANALYSIS TOOLS AND FUNCTIONAL ANNOTATION DATABASES

MiRNA FUNCTIONAL ANALYSIS AND FUNCTIONAL ANNOTATION

Despite the evident progress in the field of in silico and experimental miRNA target identification, the elucidation of miRNA function still remains an open challenge. Each miRNA can target dozens to hundreds of genes, whereas numerous miRNAs can be identified as differentially expressed (DE) in a single-NGS experiment. The complexity increases exponentially, when we take into account that miRNAs can collaborate or compete for targets and can also form elaborate motifs by targeting other molecules with important regulatory roles, such as TFs (Vlachos et al., 2016).

Different applications, algorithms, and tools have been implemented to analyze long lists of miRNA interactions and derive biologically meaningful functional insight. Some of these implementations can identify the pathways that one or more miRNAs seem to control, annotate miRNA function using gene ontology (GO) terms, or even define the role of each miRNA within complex regulatory networks. In this section, we will provide an outline of the current status in miRNA functional analysis. The applications have been divided into different categories based on scope and aim.

IDENTIFICATION OF miRNA-CONTROLLED PATHWAYS AND miRNA FUNCTIONS

Applications in this category utilize predetermined miRNA interaction lists to identify pathways under miRNA control or annotate miRNA function using gene ontologies. These tools usually examine the targets of each miRNA and try to identify if the members of a pathway or a functional category are more often targeted than what is expected by chance. The main ingredients that can affect the results of such analyses include, the sources of miRNA targets, pathways and annotation, as well as the employed statistics. The required miRNA—gene interactions are usually derived from in silico prediction algorithms such as TargetScan, DIANA-microT, and miRanda, whereas other implementations also offer the option to complement or substitute them with experimentally supported targets (Vlachos et al., 2012, 2015b). The study of Bleazard et al. has shown that the use of Fisher's exact test for miRNA enrichment analyses introduces bias (Bleazard et al., 2015), especially in the case of GO (Ashburner et al., 2000) enrichment and/or in analyses comprising multiple datasets. The recent introduction of more sophisticated statistics, including empirical distributions and meta-analysis methodologies, bypasses the inherent bias in the structure of annotation datasets and returns the overrepresentation analysis from gene level back to miRNA level (Bleazard et al., 2015; Vlachos et al., 2015b). Importantly, these methodologies have proven to be especially useful when analyzing the function of more than one miRNA. As mentioned above, each miRNA can target hundreds of genes, and therefore the targetome of, e.g., 30 miRNAs could include a large part of all coding genes, diminishing the usefulness of a standard overrepresentation approach. This new generation of analysis methodologies has enabled a more accurate investigation of combined miRNA actions. The field remains actively researched, since it has a high practical usefulness and harnesses any progress achieved in its numerous sources of data and annotation. In the following section, we will briefly present some of the most widely used implementations:

miRSystem (Lu et al., 2012): miRSystem was first implemented in 2012 and aims to combine miRNA−gene interactions derived from multiple target prediction algorithms, as well as TarBase v5.0 (Papadopoulos et al., 2009b) and miRecords (Xiao et al., 2009). miRSystem aims to identify enriched miRNA targets against GO (Ashburner et al., 2000), KEGG (Kanehisa et al., 2016), Biocarta (Nishimura, 2001), Reactome (Croft et al., 2014), and Pathway Interaction Database (Schaefer et al., 2009). It employs standard statistics and empirical distributions. The prediction algorithms can be intersected, whereas around 1500 interactions are derived from the two experimental databases. The tool has been updated to miRBase v21 (Kozomara and Griffiths-Jones, 2014).

miEAA (Backes et al., 2016): miEAA has been recently implemented and enables enrichment analyses of miRNA sets against miRBase, HMDD (Li et al., 2014b), miRWalk (Dweep et al., 2011) and miRTarBase datasets, as well as annotation sets for diseases, age/gender, and immune cells. The application relies on the GeneTrail (Backes et al., 2007) gene set enrichment framework but has been tailored to support miRNAs as input. miEAA utilizes Fisher's exact test and an unweighted variant of gene set enrichment analysis in order to extract the relevant *P* values. It has been first released in 2015 and supports miRBase v21 annotations but also offers a tool for the conversion of miRNAs and precursors into different miRBase versions, as well as a converter between mature miRNA and precursor names. miEAA also permits the optional upload of expressed genes in order to be used as a reference set for the analysis.

miTalos (Preusse et al., 2016): The second version of miTalos was released in 2016 and can perform an overrepresentation analysis using Fisher's exact test against KEGG, Reactome, and Wiki Pathways (Kelder et al., 2011). It comprises baseline gene expression data for 68 human/mouse tissues and cell lines from EBI Expression Atlas (Fonseca et al., 2014), whereas interactions are derived from TargetScan 6.2, StarBase v2, and miRanda. The incorporation of expression data enables users to discard interactions with genes not expressed in a specific tissue or cell type, enabling a more targeted pathway analysis. It currently supports miRBase v21 and 1,621 human and 535 mouse miRNAs.

DIANA-miRPath (Vlachos et al., 2015b): DIANA-miRPath was first deployed in 2009 (Papadopoulos et al., 2009a). The third and most recent version aims to elucidate the combined miRNA functions with experimental support. DIANA-miRPath enables users to incorporate and/or substitute the available predicted miRNA−gene interaction lists (provided by DIANA-microT-CDS and TargetScan) with more than 600,000 experimentally supported interactions derived from TarBase v7.0. The enrichment analysis can be performed using Fisher's exact test and empirical distributions, in order to diminish the relevant bias. miRPath also employs meta-analysis methodologies to elucidate the combined miRNA function. Expressed genes in a specific experiment can be optionally uploaded and utilized as the background set. Users can identify molecular pathways controlled by one or more miRNAs and/or perform functional analyses using ontology terms. The pathways are derived from KEGG, whereas the supported ontologies include GO (and its subcategories), as well as its simplified GOSlim instance. miRPath can also perform sophisticated analyses and visualizations, including miRNA, pathway and ontology dendrograms based on miRNA targeting, as well as miRNA versus pathways and miRNA versus function heat maps. Its reverse search module enables users to identify which miRNAs target a selected pathway or share a specific function. This module can be used to query the extensive experimental miRNA−gene interaction dataset derived from TarBase, practically enabling users to meta-analyze the available literature and hundreds of high-throughput datasets, in order to identify previously unknown miRNA functions, usually hidden in fragmented sources. The current version has been updated in 2015 and supports analyses for *H. sapiens, M. musculus, Rattus norvegicus, Drosophila melanogaster, Caenorhabditis elegans, Gallus gallus,* and *Danio rerio* using miRBase v21.

IDENTIFICATION OF FUNCTIONAL miRNAs USING EXPRESSION DATA

The second category of implementations relies heavily on experimental expression data in order to identify not only which miRNA–mRNA interactions are present within the specific samples but also to infer which miRNAs share crucial regulatory roles. The latter is usually performed by identifying deregulated miRNAs having more DE gene targets than what is expected by chance. Such tools usually focus on analyzing results of differential expression analyses between two conditions and assisting in the detection of miRNAs with potent regulatory activities. These tools have gradually become indispensable, since increased throughput techniques discover dozens of DE miRNAs and hundreds of mRNAs between two conditions. High-throughput methodologies have revolutionized biomedical research by enabling transcriptome-wide hypothesis-free investigations but have also increased exponentially the complexity of downstream data analysis. The most widely used implementations are presented below:

- *Applications using expression for functional miRNA detection*: From very early on, tools such as *Sylamer* (van Dongen et al., 2008) or *DIANA-mirExTra* (Alexiou et al., 2010) performed overrepresentation analyses on combined miRNA/mRNA expression data in order to detect miRNAs having a crucial regulatory role. These tools were based on predicted miRNA–mRNA interactions, which suffered from increased noise, and utilized the miRNA/mRNA differential expression results in order to increase the relevant accuracy but also to uncover miRNAs having the highest functional impact.
- *Implementations elucidating the interplay between miRNAs and TFs using expression data*: These tools extended the scope of such analyses by introducing also TF–mRNA and later on TF–miRNA interactions. TFs and miRNAs can have a significant regulatory impact but their interactions are often complex and their elucidation requires extensive datasets, computational infrastructure, and expertise. miRNAs can target genes, including TFs, whereas TFs can regulate the expression (activate and/or repress) of other genes, as well as miRNAs. These three types of entities (miRNAs, mRNAs, and TFs) usually have intrinsic relationships that create elaborate network motifs. Some of the most important implementations in this field include *MMIA* (Li et al., 2015), *MAGIA2* (Bisognin et al., 2012), *cGRNB* (Xu et al., 2013), *miRConnX* (Huang et al., 2011), *miRTarVis* (Jung et al., 2015), as well as the most recent *TFmiR* (Hamed et al., 2015) and *DIANA-mirExtra v2.0* (Vlachos et al., 2016). These tools can identify miRNAs and TFs with central roles between two conditions, whereas many can also recreate, visualize, and analyze the underlying regulatory networks. These tools can be distinguished by the supporting datasets, the sources of annotation (miRNA and mRNA TSSs), the interaction data (miRNA–mRNA, TF–mRNA, and TF–miRNA), as well as the employed statistics and network visualization/analysis capabilities.

All initial implementations relied solely on miRNA target prediction data in order to identify miRNA–mRNA interactions. Subsequently, implementations such as *miRConnX, MAGIA2,* and

TFmiR also incorporated experimentally supported interactions derived from the then available databases of experimental miRNA targets (TarBase and miRecords). *DIANA-mirExtra v2.0* pushed the envelope further by enabling users to base their analyses solely on extensive datasets of more than 450,000 experimentally supported miRNA–mRNA interactions derived from DIANA-TarBase v7.0.

This shift toward experimental data has also been observed in the other two types of interactions utilized in such applications. Initial implementations, employed position weight matrix-derived TF binding positions, which can also suffer from high false positive numbers. The most recent implementations (*TFmiR* and *mirExTra v2.0*) utilize sources that include experimentally derived interactions, with the latter incorporating more than 2,000,000 experimentally supported TF binding sites (genome-wide) derived from ORegAnno v3.0 (Lesurf et al., 2016) database and from the analysis of DNase-Seq libraries. One of the most important hindrances in relevant research efforts and applications was the lack of experimentally supported TF–miRNA interactions. As mentioned previously, the recent breakthroughs in the annotation of miRNA TSSs have started to revolutionize this field. An extensive compendium of currently available experimentally supported miRNA TSS and TF–miRNA interactions has been incorporated in *DIANA-mirExTra v2.0*, enabling accurate analyses.

Tools belonging to this specific category usually incorporate methodologies to infer and depict the underlying mRNA–miRNA–TF network. For instance, *TFmiR* and *mirExTra v2.0* support the formation of interactive networks, with the former offering online tools for motif and network investigation and the latter export capabilities to Cytoscape desktop for further analysis. Finally, some implementations also support various tools and functionalities, including topological analysis tools (*TFmiR*), extensive NGS libraries comprising more than 400 small RNA-Seq and RNA-Seq samples (*DIANA-mirExTra v2.0*), and a user-friendly interface to perform the differential analysis online using DESeq, Limma, and edgeR (*DIANA-mirExTra v2.0*).

STATE-OF-THE-ART COMPUTATIONAL–EXPERIMENTAL APPROACHES

The multiple available experimental and computational methodologies can be used to create custom-tailored frameworks to explore the role of miRNAs in pathological and physiological conditions. In this last section, we will utilize the classic case-control setup (e.g., healthy vs diseased sample, WT vs KO, treated vs untreated, etc.), in order to show the information gain offered by including different miRNA-related dry and wet lab approaches. It is important to stress here that the optimal setting requires all -omics techniques to be performed in the same samples. Since there can be time and cost constraints inhibiting an all-inclusive approach, we will present also the available alternatives. Certainly, these examples can be extended to almost any other setup (such as exploratory studies, population studies, time series, etc.) following proper adjustments. In the selected case-control design, the main tools to form the required framework can be the following:

1. *Expression*
   a. Small-RNA-Seq/RNA-Seq: In order to identify expressed and DE miRNAs in the two studied conditions. The expressed transcripts are crucial to identify mRNAs, miRNAs, and

TFs present in the samples, to filter the possible interactions, whereas DE miRNAs and mRNAs can be used to detect actual regulatory events in downstream analyses.

   b. Use of expression data present in public repositories and relevant databases: A common alternative, since expression data are usually crucial for most analysis scenarios. There are tools (e.g., DIANA-mirExTra) that offer preanalyzed miRNA/mRNA expression sets that can be used to start or complement relevant analyses.

2. *miRNA−mRNA interactions*
   a. CLIP-Seq or relevant (e.g., CLASH) experiments: These techniques currently form the state-of-the-art for charting the miRNA targetome in specific samples and/or conditions.
   b. miRNA−mRNA interaction databases and in silico tools: They can be used to complement a relevant experiment or support research projects lacking related experimental methods.

3. *TF−mRNA/TF−miRNA interactions*
   a. TF binding NGS techniques (e.g., ChIP-Seq, DNase-Seq, and ATAC-Seq): These methodologies can be used to identify binding and differential binding events in the analyzed samples.
   b. Repositories and tools: They can be used to complement relevant experimental approaches or support research projects lacking related methods.

4. *Downstream functional analyses*
   a. Pathway/Functional in silico tools: These tools can be used to identify targeted pathways and enriched functions. They perform more accurately when combined with expression data.
   b. TF−miRNA−mRNA regulatory analysis tools: They can be used to elucidate the underlying regulatory networks. Their optimal use case is when they are supported by experimental expression and interaction data from the analyzed samples.

CLOSING REMARKS

Despite the evident progress, there are still numerous open challenges in the field of miRNA research. For instance, a major bottleneck is the lack of extensive and accurate annotation. Primary miRNA genes and transcripts are yet to be accurately annotated, whereas the functional characterization of the identified mature transcripts is until now incomplete. Furthermore, TSS positions for a large part of the miRtrons have not been identified, diminishing the accuracy of analyses using TF−miRNA interactions for this specific subcategory of transcripts. Importantly, there is no silver bullet approach for transcriptome-wide miRNA targetome characterization computationally or experimentally. Even the best methodologies in both worlds (dry or wet) still face important shortcomings. A breakthrough in these fields will provide a paradigm shift in the miRNA research cosmos.

Similar challenges are currently also observed in miRNA functional analyses. High quality manually curated pathways include only a small subset of genes, whereas the more inclusive integrated approaches have a significantly higher inherent noise level. Moreover, the novel network and empirical distribution-based methodologies have increased the accuracy of the provided results but also require dramatically longer computation times, rendering their incorporation in online web servers quite tedious and complex. Furthermore, we are still lacking implementations that could directly analyze expression datasets along with the appropriately matched interaction datasets (e.g., CLIP-Seq and ChIP-Seq or DNase-Seq) and perform targeted investigations. The challenge becomes even greater if they are provided in raw format. Finally, there are no applications that could efficiently extract the most important network motifs or identify the changes and/or the culprit regulators (e.g., TFs or miRNAs) responsible for the observed perturbations between two regulatory network states. Nevertheless, this remains a fascinating field, which advances as we learn more about miRNAs, genes, and their functions, as well as through algorithmic and computational breakthroughs.

REFERENCES

Agarwal, V., Bell, G.W., Nam, J.W., et al., 2015. Predicting effective microRNA target sites in mammalian mRNAs. eLife 4, e05005.

Akhtar, M.M., Micolucci, L., Islam, M.S., et al., 2016. Bioinformatic tools for microRNA dissection. Nucleic Acids Res. 44, 24−44.

Alexiou, P., Maragkakis, M., Papadopoulos, G.L., et al., 2010. The DIANA-mirExTra web server: from gene expression data to MicroRNA function. PLoS ONE. 5, e9171.

Ashburner, M., Ball, C.A., Blake, J.A., et al., 2000. Gene ontology: tool for the unification of biology. The gene ontology consortium. Nat. Genet. 25, 25−29.

Backes, C., Keller, A., Kuentzer, J., et al., 2007. GeneTrail—advanced gene set enrichment analysis. Nucleic Acids Res. 35, W186−W192.

Backes, C., Khaleeq, Q.T., Meese, E., et al., 2016. miEAA: microRNA enrichment analysis and annotation. Nucleic Acids Res. 44, W110−W116.

Bandyopadhyay, S., Mitra, R., 2009. TargetMiner: microRNA target prediction with systematic identification of tissue-specific negative examples, Bioinformatics, 25, pp. 2625−2631.

Bandyopadhyay, S., Ghosh, D., Mitra, R., et al., 2015. MBSTAR: multiple instance learning for predicting specific functional binding sites in microRNA targets. Sci. Rep. 5, 8004.

Barski, A., Jothi, R., Cuddapah, S., et al., 2009. Chromatin poises miRNA- and protein-coding genes for expression. Genome Res. 19, 1742−1751.

Bartel, D.P., 2009. MicroRNAs: target recognition and regulatory functions. Cell 136, 215−233.

Betel, D., Koppal, A., Agius, P., et al., 2010. Comprehensive modeling of microRNA targets predicts functional non-conserved and non-canonical sites. Genome. Biol. 11, R90.

Bhartiya, D., Pal, K., Ghosh, S., et al., 2013. lncRNome: a comprehensive knowledgebase of human long non-coding RNAs. Database (Oxford) bat034.

Bisognin, A., Sales, G., Coppe, A., et al., 2012. MAGIA2: from miRNA and genes expression data integrative analysis to microRNA−transcription factor mixed regulatory circuits (2012 update). Nucleic Acids Res. 40, W13−W21.

Bleazard, T., Lamb, J.A., Griffiths-Jones, S., 2015. Bias in microRNA functional enrichment analysis. Bioinformatics. 31, 1592−1598.

Cesana, M., Cacchiarelli, D., Legnini, I., et al., 2011. A long noncoding RNA controls muscle differentiation by functioning as a competing endogenous RNA. Cell 147, 358−369.

Chi, S.W., Zang, J.B., Mele, A., et al., 2009. Argonaute HITS-CLIP decodes microRNA-mRNA interaction maps. Nature. 460, 479−486.

Chien, C.H., Sun, Y.M., Chang, W.C., et al., 2011. Identifying transcriptional start sites of human microRNAs based on high-throughput sequencing data. Nucleic Acids Res. 39, 9345−9356.

Chou, C.H., Chang, N.W., Shrestha, S., et al., 2016. miRTarBase 2016: updates to the experimentally validated miRNA-target interactions database. Nucleic Acids Res. 44, D239–D247.

Corcoran, D.L., Pandit, K.V., Gordon, B., et al., 2009. Features of mammalian microRNA promoters emerge from polymerase II chromatin immunoprecipitation data. PLoS ONE. 4, e5279.

Croft, D., Mundo, A.F., Haw, R., et al., 2014. The reactome pathway knowledgebase. Nucleic Acids Res. 42, D472–D477.

Down, T.A., Hubbard, T.J., 2002. Computational detection and location of transcription start sites in mammalian genomic DNA. Genome Res. 12, 458–461.

Dweep, H., Sticht, C., Pandey, P., et al., 2011. miRWalk—database: prediction of possible miRNA binding sites by "walking" the genes of three genomes. J. Biomed. Inform. 44, 839–847.

Enright, A.J., John, B., Gaul, U., et al., 2003. MicroRNA targets in Drosophila. Genome Biol. 5, R1.

Fonseca, N.A., Marioni, J., Brazma, A., 2014. RNA-Seq gene profiling—a systematic empirical comparison. PLoS ONE. 9, e107026.

Friard, O., Re, A., Taverna, D., et al., 2010. CircuitsDB: a database of mixed microRNA/transcription factor feed-forward regulatory circuits in human and mouse. BMC Bioinform. 11, 435.

Garcia, D.M., Baek, D., Shin, C., et al., 2011. Weak seed-pairing stability and high target-site abundance decrease the proficiency of lsy-6 and other microRNAs. Nat. Struct. Mol. Biol. 18, 1139–1146.

Georgakilas, G., Vlachos, I.S., Paraskevopoulou, M.D., et al., 2014. microTSS: accurate microRNA transcription start site identification reveals a significant number of divergent pri-miRNAs. Nat. Commun. 5, 5700.

Georgakilas, G., Vlachos, I.S., Zagganas, K., et al., 2016. DIANA-miRGen v3.0: accurate characterization of microRNA promoters and their regulators. Nucleic Acids Res. 44, D190–D195.

Grimson, A., Farh, K.K., Johnston, W.K., et al., 2007. MicroRNA targeting specificity in mammals: determinants beyond seed pairing. Mol. Cell. 27, 91–105.

Gumienny, R., Zavolan, M., 2015. Accurate transcriptome-wide prediction of microRNA targets and small interfering RNA off-targets with MIRZA-G. Nucleic Acids Res. 43, 9095.

Guo, Z., Maki, M., Ding, R., et al., 2014. Genome-wide survey of tissue-specific microRNA and transcription factor regulatory networks in 12 tissues. Sci. Rep. 4, 5150.

Hafner, M., Landthaler, M., Burger, L., et al., 2010. Transcriptome-wide identification of RNA-binding protein and microRNA target sites by PAR-CLIP. Cell 141, 129–141.

Hamed, M., Spaniol, C., Nazarieh, M., et al., 2015. TFmiR: a web server for constructing and analyzing disease-specific transcription factor and miRNA co-regulatory networks. Nucleic Acids Res. 43, W283–W288.

Hansen, T.B., Jensen, T.I., Clausen, B.H., et al., 2013. Natural RNA circles function as efficient microRNA sponges. Nature. 495, 384–388.

Harrow, J., Frankish, A., Gonzalez, J.M., et al., 2012. GENCODE: the reference human genome annotation for The ENCODE Project. Genome Res. 22, 1760–1774.

Hsu, S.D., Lin, F.M., Wu, W.Y., et al., 2011. miRTarBase: a database curates experimentally validated microRNA-target interactions. Nucleic Acids Res. 39, D163–D169.

Huang da, W., Sherman, B.T., Lempicki, R.A., 2009. Bioinformatics enrichment tools: paths toward the comprehensive functional analysis of large gene lists. Nucleic Acids Res. 37, 1–13.

Huang, G.T., Athanassiou, C., Benos, P.V., 2011. mirConnX: condition-specific mRNA-microRNA network integrator. Nucleic Acids Res. 39, W416–W423.

Huntzinger, E., Izaurralde, E., 2011. Gene silencing by microRNAs: contributions of translational repression and mRNA decay. Nat. Rev. Genet. 12, 99–110.

Jeggari, A., Marks, D.S., Larsson, E., 2012. miRcode: a map of putative microRNA target sites in the long non-coding transcriptome. Bioinformatics. 28, 2062–2063.

Jiang, Q., Wang, Y., Hao, Y., et al., 2009. miR2Disease: a manually curated database for microRNA deregulation in human disease. Nucleic Acids Res. 37, D98–D104.

John, B., Enright, A.J., Aravin, A., et al., 2004. Human MicroRNA targets. PLoS Biol. 2, e363.

Jung, D., Kim, B., Freishtat, R.J., et al., 2015. miRTarVis: an interactive visual analysis tool for microRNA−mRNA expression profile data. BMC Proc. 9, S2.

Kanehisa, M., Sato, Y., Kawashima, M., et al., 2016. KEGG as a reference resource for gene and protein annotation. Nucleic Acids Res. 44, D457−D462.

Kelder, T., van Iersel, M.P., Hanspers, K., et al., 2011. WikiPathways: building research communities on biological pathways. Nucleic Acids Res. 40, D1301−D1307.

Khorshid, M., Rodak, C., Zavolan, M., 2011. CLIPZ: a database and analysis environment for experimentally determined binding sites of RNA-binding proteins. Nucleic Acids Res. 39, D245−D252.

Khorshid, M., Hausser, J., Zavolan, M., et al., 2013. A biophysical miRNA−mRNA interaction model infers canonical and noncanonical targets. Nat. Methods. 10, 253−255.

Kiriakidou, M., Nelson, P.T., Kouranov, A., et al., 2004. A combined computational-experimental approach predicts human microRNA targets. Genes Dev. 18, 1165−1178.

Kloosterman, W.P., Wienholds, E., Ketting, R.F., et al., 2004. Substrate requirements for let-7 function in the developing zebrafish embryo. Nucleic Acids Res. 32, 6284−6291.

Kozomara, A., Griffiths-Jones, S., 2014. miRBase: annotating high confidence microRNAs using deep sequencing data. Nucleic Acids Res. 42, D68−D73.

Lesurf, R., Cotto, K.C., Wang, G., et al., 2016. ORegAnno 3.0: a community-driven resource for curated regulatory annotation. Nucleic Acids Res. 44, D126−D132.

Lewis, B.P., Shih, I.H., Jones-Rhoades, M.W., et al., 2003. Prediction of mammalian microRNA targets. Cell 115, 787−798.

Li, J.-H., Liu, S., Zhou, H., et al., 2014a. starBase v2.0: decoding miRNA−ceRNA, miRNA−ncRNA and protein−RNA interaction networks from large-scale CLIP-Seq data. Nucleic Acids Res. 42, D92−D97.

Li, Y., Qiu, C., Tu, J., et al., 2014b. HMDD v2.0: a database for experimentally supported human microRNA and disease associations. Nucleic Acids Res. 42, D1070−D1074.

Li, Z., Wang, L., Xu, J., et al., 2015. MiRNA expression profile and miRNA−mRNA integrated analysis (MMIA) during podocyte differentiation. Mol. Genet. Genomics. 290, 863−875.

Lu, T.-P., Lee, C.-Y., Tsai, M.-H., et al., 2012. miRSystem: an integrated system for characterizing enriched functions and pathways of microRNA targets. PLoS ONE. 7, e42390.

Lytle, J.R., Yario, T.A., Steitz, J.A., 2007. Target mRNAs are repressed as efficiently by microRNA-binding sites in the 5′ UTR as in the 3′ UTR, Proc. Natl. Acad. Sci. USA 104, pp. 9667−9672.

Marsico, A., Huska, M.R., Lasserre, J., et al., 2013. PROmiRNA: a new miRNA promoter recognition method uncovers the complex regulation of intronic miRNAs. Genome Biol. 14, R84.

Marson, A., Levine, S.S., Cole, M.F., et al., 2008. Connecting microRNA genes to the core transcriptional regulatory circuitry of embryonic stem cells. Cell 134, 521−533.

Mathelier, A., Zhao, X., Zhang, A.W., et al., 2014. JASPAR 2014: an extensively expanded and updated open-access database of transcription factor binding profiles. Nucleic Acids Res. 42, D142−D147.

Megraw, M., Pereira, F., Jensen, S.T., et al., 2009. A transcription factor affinity-based code for mammalian transcription initiation. Genome Res. 19, 644−656.

Menor, M., Ching, T., Zhu, X., et al., 2014. mirMark: a site-level and UTR-level classifier for miRNA target prediction. Genome Biol. 15, 500.

Moore, M.J., Scheel, T.K., Luna, J.M., et al., 2015. miRNA-target chimeras reveal miRNA 3′-end pairing as a major determinant of Argonaute target specificity. Nat. Commun. 6, 8864.

Nishimura, D., 2001. BioCarta. Biotech. Softw. Internet Rep. 2, 117−120.

Ozsolak, F., Poling, L.L., Wang, Z., et al., 2008. Chromatin structure analyses identify miRNA promoters. Genes Dev. 22, 3172−3183.

Papadopoulos, G.L., Alexiou, P., Maragkakis, M., et al., 2009a. DIANA-mirPath: integrating human and mouse microRNAs in pathways. Bioinformatics. 25, 1991−1993.

Papadopoulos, G.L., Reczko, M., Simossis, V.A., et al., 2009b. The database of experimentally supported targets: a functional update of TarBase. Nucleic Acids Res. 37, D155−D158.

Paraskevopoulou, M.D., Georgakilas, G., Kostoulas, N., et al., 2013a. DIANA-LncBase: experimentally verified and computationally predicted microRNA targets on long non-coding RNAs. Nucleic Acids Res. 41, D239−D245.

Paraskevopoulou, M.D., Georgakilas, G., Kostoulas, N., et al., 2013b. DIANA-microT web server v5.0: service integration into miRNA functional analysis workflows. Nucleic Acids Res. 41, W169−W173.

Paraskevopoulou, M.D., Hatzigeorgiou, A.G., 2016. Analyzing miRNA−lncRNA interactions. Methods Mol. Biol. 1402, 271−286.

Paraskevopoulou, M.D., Vlachos, I.S., Karagkouni, D., et al., 2016. DIANA-LncBase v2: indexing microRNA targets on non-coding transcripts. Nucleic Acids Res. 44, D231−238.

Preusse, M., Theis, F.J., Mueller, N.S., 2016. miTALOS v2: analyzing tissue specific microRNA function. PLoS ONE. 11, e0151771.

Reczko, M., Maragkakis, M., Alexiou, P., et al., 2012. Functional microRNA targets in protein coding sequences. Bioinformatics. 28, 771−776.

Saini, H.K., Enright, A.J., Griffiths-Jones, S., 2008. Annotation of mammalian primary microRNAs. BMC Genomics. 9, 564.

Saini, H.K., Griffiths-Jones, S., Enright, A.J., 2007. Genomic analysis of human microRNA transcripts. Proc. Natl. Acad. Sci. USA 104, 17719−17724.

Schaefer, C.F., Anthony, K., Krupa, S., et al., 2009. PID: the pathway interaction database. Nucleic Acids Res. 37, D674−D679.

Shannon, P., Markiel, A., Ozier, O., et al., 2003. Cytoscape: a software environment for integrated models of biomolecular interaction networks. Genome Res. 13, 2498−2504.

Singh, N.K., 2016. microRNAs databases: developmental methodologies, structural and functional annotations. Interdiscip. Sci. 2016, 1−21.

Suzuki, Y., Yamashita, R., Nakai, K., et al., 2002. DBTSS: DataBase of human transcriptional start sites and full-length cDNAs. Nucleic Acids Res. 30, 328−331.

Thomson, D.W., Bracken, C.P., Goodall, G.J., 2011. Experimental strategies for microRNA target identification. Nucleic Acids Res. 39, 6845−6853.

van Dongen, S., Abreu-Goodger, C., Enright, A.J., 2008. Detecting microRNA binding and siRNA off-target effects from expression data. Nat. Methods. 5, 1023−1025.

Vergoulis, T., Vlachos, I.S., Alexiou, P., et al., 2012. TarBase 6.0: capturing the exponential growth of miRNA targets with experimental support. Nucleic Acids Res. 40, D222−D229.

Vlachos, I.S., Hatzigeorgiou, A.G., 2013. Online resources for miRNA analysis. Clin. Biochem. 46, 879−900.

Vlachos, I.S., Kostoulas, N., Vergoulis, T., et al., 2012. DIANA miRPath v.2.0: investigating the combinatorial effect of microRNAs in pathways. Nucleic Acids Res. 40, W498−W504.

Vlachos, I.S., Paraskevopoulou, M.D., Karagkouni, D., et al., 2015a. DIANA-TarBase v7.0: indexing more than half a million experimentally supported miRNA:mRNA interactions. Nucleic Acids Res. 43, D153−D159.

Vlachos, I.S., Zagganas, K., Paraskevopoulou, M.D., et al., 2015b. DIANA-miRPath v3.0: deciphering microRNA function with experimental support. Nucleic Acids Res. 43, W460−W466.

Vlachos, I.S., Vergoulis, T., Paraskevopoulou, M.D., et al., 2016. DIANA-mirExTra v2.0: uncovering microRNAs and transcription factors with crucial roles in NGS expression data. Nucleic Acids Res. 44, W128−W134.

Volders, P.J., Verheggen, K., Menschaert, G., et al., 2015. An update on LNCipedia: a database for annotated human lncRNA sequences. Nucleic Acids Res. 43, 4363−4364.

Wang, J., Lu, M., Qiu, C., et al., 2010. TransmiR: a transcription factor-microRNA regulation database. Nucleic Acids Res. 38, D119−D122.

Wang, P., Zhi, H., Zhang, Y., et al., 2015a. miRSponge: a manually curated database for experimentally supported miRNA sponges and ceRNAs. Database (Oxford). 2015, bav098.

Wang, S., Li, W., Lian, B., et al., 2015b. TMREC: a database of transcription factor and MiRNA regulatory cascades in human diseases. PLoS ONE. 10, e0125222.

Wang, X., El Naqa, I.M., 2008. Prediction of both conserved and nonconserved microRNA targets in animals. Bioinformatics. 24, 325−332.

Weinstein, J.N., Collisson, E.A., Mills, G.B., et al., 2013. The cancer genome atlas pan-cancer analysis project. Nat. Genet. 45, 1113−1120.

Xiao, F., Zuo, Z., Cai, G., et al., 2009. miRecords: an integrated resource for microRNA-target interactions. Nucleic Acids Res. 37, D105−D110.

Xu, H., Yu, H., Tu, K., et al., 2013. cGRNB: a web server for building combinatorial gene regulatory networks through integrated engineering of seed-matching sequence information and gene expression datasets. BMC. Syst. Biol. 7 (Suppl 2), S7.

Yang, J.H., Li, J.H., Jiang, S., et al., 2013. ChIPBase: a database for decoding the transcriptional regulation of long non-coding RNA and microRNA genes from ChIP-Seq data. Nucleic Acids Res. 41, D177−D187.

Yassour, M., Kaplan, T., Fraser, H.B., et al., 2009. Ab initio construction of a eukaryotic transcriptome by massively parallel mRNA sequencing. Proc. Natl. Acad. Sci. USA 106, 3264−3269.

Zhou, X., Ruan, J., Wang, G., et al., 2007. Characterization and identification of microRNA core promoters in four model species. PLoS Comput. Biol. 3, e37.

Zhou, Z., Shen, Y., Khan, M.R., et al., 2015. LncReg: a reference resource for lncRNA-associated regulatory networks. Database (Oxford). 2015, bav083.

METHODOLOGICAL CHALLENGES IN FUNCTIONAL INVESTIGATION AND THERAPEUTIC USE OF MicroRNAs

Federica Marinaro*, Meritxell Pons-Espinal* and Davide De Pietri Tonelli

Istituto Italiano di Tecnologia, Genoa, Italy

RNA ISOLATION METHODS

The choice for the best RNA extraction method is the first step of any experimental pipeline aimed at the identification and expression analysis of microRNAs (miRNAs) and their target mRNAs. This choice is mainly driven by the RNA concentration/purity needed for the type of downstream analyses. For example, if the purpose of the investigation is the expression analysis of both miRNAs and their potential mRNA targets, researchers would definitively need to use an extraction method suitable for the isolation of both small and long RNAs.

Several protocols are available to obtain total RNA extracts from various sources. Most of protocols are characterized by the absence of an initial size-selection step, allowing the extraction of any RNA type from the sample, regardless their sequence, structure and size. An important point to take into account when isolating RNA is that all extraction methods can introduce biases, such as enrichment or elimination of RNA molecules of a given size, sequence, or posttranscriptional modification. This problem becomes particularly relevant when studying low-abundant miRNAs, such as those isolated from exosomes or from few cells such as those obtained by laser capture microdissection. Kim et al. (2012) observed that miRNAs with low GC content are selectively lost during RNA extraction, indicating that the measured expression level of certain miRNAs might vary depending from the number of cells from which RNA is isolated, rather than their actual expression in the samples. Size bias can be also introduced by sample preparation and storage, such as formalin-fixed, paraffin-embedded (FFPE) tissues. FFPE has been in use for nearly a century and there are millions of FFPE bio-banks worldwide, many of them decades old, but extremely valuable as they are often associated with clinical data. Remarkably, in FFPE tissue miRNAs are generally less degraded than mRNAs, perhaps due to their small size/high copy numbers. Thus analysis of miRNA isolated from FFPE tissues could provide novel biomarkers for several human pathologies. However at the present time, it remains challenging to perform reliable expression analysis from

---

*Equal contribution.

Essentials of Noncoding RNA in Neuroscience. DOI: http://dx.doi.org/10.1016/B978-0-12-804402-5.00004-2

these samples. The above-mentioned examples emphasize how crucial is the choice of the most appropriate RNA extraction method for the specimen subject of the investigation. Nowadays, total RNA can be extracted from a wide range of specimens (e.g., tissue, cells, body fluids). In the case of brain the typical sources from which miRNA are isolated are fresh specimens (e.g., whole tissue, subregions, cells, cerebrospinal fluid, etc., that are manually dissected/collected, obtained by enzymatic dissociation or laser capture microdissection) and cell cultures. Intermediate steps can be used to decrease cell-type heterogeneity in the starting sample, such as cell sorting (e.g., FACS-; or microfluidics based, such as the one from Fludigm inc) or to enrich for specific cell type in vitro (e.g., neurons, glia and neural stem/progenitor cells), see for review (Poulin et al., 2016; Pritchard et al., 2012).

Most of the methods used for RNA extraction from cells and tissues are evolutions of the Chomczynski and Sacchi protocol, in which RNA is separated from DNA and proteins in consecutive incubation steps in guanidinium thiocyanate, sodium acetate, phenol, and chloroform, ending with RNA precipitation in isopropanol (Chomczynski and Sacchi, 1987). This extraction method is mostly known with commercial names such as TRIzol (Thermofisher), Qiazol (Qiagen), or TRI Reagent (Sigma-Aldrich). MiRNAs and mRNAs can be directly analyzed from total RNA isolated with this method, nonetheless several RNA purification kits were developed in the last 30 years by combining this extraction method with downstream size-purification steps, such as solid-phase binding and elution of RNA of specific sizes on silica columns. This combination of extraction and purification methods yields highly pure RNAs, with relatively minor loss of samples (almost negligible when large amounts of starting material are available). If miRNAs enrichment is required, polyacrylamide gel electrophoresis or immunoprecipitation of proteins bound to miRNAs (such as Ago) can be performed (for more detail on RNA IP see, Chapter 2, Essentials of MiRNA-Dependent Control of mRNA Translation and Decay, miRNA Targeting Principles and Methods for Target Identification, of this book). The choice of this additional purification step is dependent on the downstream analysis that has to be performed.

A last important step of RNA isolation workflow is the assessment of RNA quality and yield after extraction: this aspect is crucial if downstream analysis such as miRNAs and target mRNAs profiling require high accuracy and reproducibility (Pritchard et al., 2012). The most widely used methods are based on spectrophotometry, which can suffer of too low RNA yields. Alternatively, on chip capillary electrophoresis is a sensitive technique that gives clear information regarding RNA integrity (Fleige and Pfaffl, 2006).

ANALYSIS OF miRNAs EXPRESSION

The main challenges that accompany the quantification of the expression level of miRNAs in a given biological sample, hereby referred as "profiling," are related to the intrinsic properties of miRNAs. The first challenge is due to their small size: mature miRNAs are ≈ 22 nt long molecules and for this reason it is difficult to target them with primers/probes. In such short RNA molecules the GC content is particularly relevant. The GC content is a crucial parameter to predict the melting temperature (T_m) of successful primers/probes. GC content in annotated human miRNAs, varies from 16% to 91% (miRBase; Griffiths-Jones et al., 2008), thus it is extremely difficult to design primers/probes with somehow equal T_m, allowing both a highly sensitive binding and discriminative profiling for all the miRNA. The second challenge is due to the absence of a common sequence, such as the poly (A) tail, to be used as an annealing site for universal primer-based assays (for example, during reverse transcription). A third challenge is represented by the lack of the sequence differences between certain mature miRNAs such as those belonging to the same family or between isomiRs, which can differ for few (if not only 1) nucleotide (Lee et al., 2010). These challenges and limitations need to be taken in account when performing miRNA profiling.

The three most widely used methods for miRNA profiling are quantitative real time polymerase chain reaction (qRT-PCR), hybridization-based methods (such as microarrays, etc.) and next-generation sequencing (RNA-Seq). Considerations that need to be taken in account when determining the most suitable profiling approach for a specific biological question are sensitivity; the need for quantitative data; the effective costs (ranging from few dollars for qRT-PCR to hundreds for RNA-seq, per sample) and the throughput time (spanning from few hours for qRT-PCR; days for microarrays, to weeks in the case of RNA-Seq) (Baker, 2010; Pritchard et al., 2012).

Because of its sensitivity, qRT-PCR is particularly indicated to perform profiling of miRNAs from low-abundant starting materials (e.g., exosomes, single cells). This approach starts with reverse transcription of miRNAs. Reverse transcription methods include primers binding to stem loop (Chen et al., 2005), oligo-dT primers annealing to an exogenously added poly (A) tail (e.g., those developed by Qiagen and Stratagene). Primers might also include modified nucleotides such as locked nucleic acids (LNAs), to increase their binding specificity (Raymond et al., 2005). After reverse transcription, DNA fragments are amplified by TaqMan (Applied Biosystems) or SYBR® (Invitrogen, Thermo Fisher Scientific inc.) green-based qRT-PCR (Chen et al., 2011). Although qRT-PCR is very powerful, it is often misused and for the reasons discussed above it does not easily allow absolute quantifications of different miRNAs in a given sample (without proper calibration curve). An evolution of qRT-PCR is droplet-digital-PCR (dd-PCR, Vogelstein and Kinzler, 1999). This method overcomes limitations of standard qRT-PCR techniques by massively partitioning one sample in thousands of droplets. By analyzing tens of thousands of PCR reactions simultaneously for each point, dd-PCR enormously empowers statistical analysis, thus allowing reliable quantitative measurements.

Historically, microarray-based technology has been the first high-throughput profiling method for miRNAs (Krichevsky et al., 2003). MiRNA-microarray technology mostly utilizes DNA-probes that are spotted in an array on a solid substrate on which miRNAs are later hybridized. Different miRNA microarrays variants were developed in the last years and currently several of them are commercially available, allowing to profile simultaneously all the miRNAs or subgroups related by pathway or disease (Liu et al., 2008; Pritchard et al., 2012). miRNA-microarray is particularly useful as it allows easy customization/profile of specific miRNA panels related to a certain experimental applications. However, given that microarray technology relies on probes, parameters such as hybridization temperature, washing steps, etc., need to be optimized. Thus data obtained with microarrays are not easily comparable between labs, especially if acquired with different platforms.

RNA-Seq (also known as "Next-generation," or "Deep-sequencing") represents the most recent approach for profiling that derives from the advent of "next-generation" sequencing platforms. Since its introduction, RNA-Seq rapidly supplanted microarrays as the "gold standard" for profiling, as it enables massive parallel sequencing of millions of RNA molecules from a cDNA library and simultaneously provides both quantitative and qualitative information of known and novel

RNAs (reviewed in Pritchard et al., 2012). At the time of the present writing the three most-widely used RNA-Seq platforms are the 454 Pyrosequencing (Roche Diagnostics), the Illumina sequencing and the SOLiD sequencing (Life Technologies). RNA-Seq is an extremely fast-developing technology, since new platform are being released while writing this chapter we do not describe it in details here (for review see Shah et al., 2010).

When analyzing miRNA expression, especially in heterogeneous tissues such as brain, it is important to relate its expression with the cell-specific context in which the miRNA of interest is acting. The "golden standard" technique used for spatial and quantitative detection of miRNAs at the tissue and cellular level is the in situ hybridization (ISH) (Wienholds et al., 2005). As this technique suffers of poor cellular resolution, we have implemented ISH protocols enabling the simultaneous detection of the expression of miRNAs by fluorescence ISH and specific markers immuno-fluorescence staining within the same optical section, thus achieving a much greater cellular resolution, compared to the conventional bright field-based ISH. The use of a laser scanning confocal microscope (excitation 633 nm; emission between 707.3 nm and 795 nm) allows the detection of a fluorescent ISH signal from the conventional alkaline phosphate-NBT (nitro-blue tetrazolium chloride) /BCIP (5-bromo-4-chloro-3′-indolyphosphate p-toluidine salt) chromogenic staining (which precipitates upon reaction with alkaline phosphatase) together with specific cell-markers (De Pietri Tonelli et al., 2014). This represents a great advantage allowing investigating the cellular context in which miRNAs are expressed.

Another approach that has been widely used for the investigation of spatial-temporal distribution of miRNA expression is the "Negative Sensor." This assay is based on the intrinsic ability of miRNAs to inactivate genes by RNA interference (RNAi), when the entire miRNA binds to a target sequence. Typically the negative sensor is a transcript encoding for a reporter (e.g., GFP or LacZ) protein, fused with miRNA binding sequences in its 3′UTR, triggering degradation of the sensor transcript in presence of the miRNA (Brennecke et al., 2003; De Pietri Tonelli et al., 2006). In contrast with all the profiling techniques described above, which are direct methods for miRNA quantification and/or visualization, the negative sensor approach is an indirect method. However, the latter allows to visualize/measure miRNA expression and also to monitor miRNA activity over time, in vivo. Further details about the "Negative Sensor" technique are provided in the "Indirect validation" subsection, below (and Table 4.1).

FUNCTIONAL INVESTIGATION OF miRNAs

To validate a functional miRNA:mRNA interaction is necessary to demonstrate a direct association of a specific miRNA to its target mRNA and to examine the biological significance of such regulation. Therefore, altering the stoichiometry of miRNAs to their target mRNAs by loss-or gain-of-function experiments has been extensively used to examine the physiological role of miRNAs.

APPROACHES FOR LOSS OF miRNA FUNCTION

MiRNA antagonists (hereby referred as "antagonists") are modified antisense oligonucleotides harboring the full or partial complementary reverse sequence of a mature miRNA, which interfere with the ability of an endogenous miRNA to bind to its target mRNAs (Table 4.1) .

Table 4.1 Current Methodologies for the Functional Investigation of miRNAs

Application	Method	Advantages	Disadvantages	References
miRNA inhibition	miRNA inhibitor	Costs	Transient effect	Hutvágner et al. (2004)
	Antagomir	Specificity, stability, improved cell penetration	Transient effect	Krützfeldt et al. (2005)
	LNA	Specificity, stability	Transient effect	Elmén et al. (2008)
	Tiny LNA	Inhibition of miRNAs families with common seed region	Transient effect	Obad et al. (2011)
	Phosphorothioate	Nuclease resistance	Transient effect, short half-life in vivo, low specificity	Krützfeldt et al. (2005)
	PNA	Specificity, stability, increased delivery without transfection reagents	Transient effect	Nielsen et al. (1991)
	Morpholino	Specificity, stability, nontoxic	Mainly used in Zebrafish and *Xenopus laevis*	Flynt et al. (2007)
	Target protector	Specific binding to selected mRNA target	Few reports in mammalian systems	Staton and Giraldez (2011)
	Sponge /TuD decoy	Stable inhibition, cost-effective	Difficult to assess the degree of inhibition of targeted miRNAs of the same family	Ebert et al. (2007)
	siRNA	Costs	Possible off-target effects	Vasudevan et al. (2007)
	Genetic ablation	In vivo analysis of miRNA functions	Generation of mouse models	De Pietri Tonelli et al. (2008), Park et al. (2012)
	CRISPR-CAS	Stable, cost-effective	Few reports in mammalian cells	Chang et al. (2016), Zhao et al. (2014)
miRNA overexpression	Mimic	Acute	Target delivery, transient effect, supraphysiological levels, possible off-target effects	Chiu and Rana (2003)
	Plasmid	Stable	Target delivery, supraphysiological levels	Schratt et al. (2006)
	Viral vector	Stable	Supraphysiological levels	Cheng et al. (2009)
Target validation (indirect)	Luciferase assay	Costs; in vitro and in vivo target validation	Lack of information regarding direct mRNA: miRNA interaction	Clovis et al. (2012)

(Continued)

Table 4.1 Current Methodologies for the Functional Investigation of miRNAs *Continued*

Application	Method	Advantages	Disadvantages	References
Target validation (direct)	Negative sensor	Spatial and temporal resolution; in vitro and in vivo target validation		Brennecke et al.(2003), De Pietri Tonelli et al. (2006)
	ISH/IHC	Detection of miRNAs expression in fixed tissues; spatial resolution	Lack of temporal resolution; not self-standing technique for the validation of mRNA: miRNA interaction	Clovis et al. (2012), De Pietri Tonelli et al. (2014)
	Ago-CLIP	Direct functional mRNA:miRNA target	Only for stable mRNA: miRNA interactions after coimmunoprecipitation	Easow et al. (2007)
	HITS-CLIP	Protection from RNAses, identification miRNA binding sites, useful for in vivo experiments	Low resolution of 30−60 nt	Licatalosi et al. (2008)
	PAR-CLIP	Increased RNA recovery, high resolution of miRNA binding sites	Requires preincubation of cytotoxic nucleosides, limited to in vitro cell culture experiments, variable nucleoside uptake	Hafner et al. (2010)
	CLASH	Not restricted to Ago2 interactions	The tag and overexpression of proteins may influence the binding pattern of protein and RNA, low efficiency of RNA−RNA ligation	Helwak et al. (2013)
	iCLIP	Captures truncated cDNA produced, individual nucleotide resolution	Circularization may introduce biases	König et al. (2010)
	Biotin	Specificity of single miRNA pull-down	Requires introduction of miRNA duplexes (overexpression)	Orom and Lund (2007)
	TAP-Tar	Reduces number of false positives	Requires introduction of miRNA duplexes (overexpression)	Nonne et al. (2010)

The success of the antagonist strategy requires optimization of the binding affinity of the antagonist for the miRNA of interest, efficiency of its intracellular delivery and stability (especially in vivo). Several chemical modifications have been introduced in the antagonists to enhance their efficacy. Typical modifications of antagonists include, but are not limited to, modifications of the sugar, such as the nucleobase or the internucleotide linkage and modifications of the 2′-OH in the ribose (easily attacked by nucleases). The most popular sugar modification are LNAs, in which the ribose moiety of a nucleotide is modified with an $2'-4'$ bridge "locking" the ribose (Elmén et al., 2008). Introduction of LNA-modified nucleotides in antagonists has been shown to increase affinity for the miRNA improving the efficiency of the antagonists. For example, introduction of LNAs was used in "tiny LNAs." These 8-mer antagonists designed to target the seed region of miRNAs are very powerful despite their small size as they can simultaneously silence miRNAs families sharing identical seed regions (Obad et al., 2011). Common $2'$-OH modifications are $2'$-O-methyl-group (OMe), $2'$-O-methoxyethyl and $2'$-fluoro. $2'$-OMe modification has been widely used to modify siRNAs, and shown to increase in vivo stability of antagonists (Krützfeldt et al., 2005).

In addition to these modifications, morpholino- and PNA-modified oligomers have been also used to increase stability of miRNA inhibition. Morpholinos are uncharged, sequence specific, nontoxic and potent inhibitors of both pri-miRNA and mature miRNA activity, but mainly used in lower vertebrates (Flynt et al., 2007). PNA are uncharged oligonucleotides analogues in which the sugar-phosphodiester backbone of DNA/RNA has been replaced by N-(2-aminoethyl)-glycine units (Nielsen et al., 1991). Given the lack of charge, PNAs do not require transfection reagents for intracellular delivery and since are not easily recognized by either nucleases or proteases, PNAs are resistant to degradation. Finally, conjugation of antagonists with cholesterol or cell-penetrating peptides (e.g., rabies virus glycoprotein (RGV)) has been shown to improve their intracellular delivery. Since the initial studies demonstrating in vivo efficacy of the antagonist strategy (Krützfeldt et al., 2005), this approach has raised increasing interest for therapeutic applications (see "Delivery of MiRNA Antagonists or Mimics" section).

Although synthetic miRNA antagonists are powerful molecules and their use straightforward to perform loss of miRNA function, this approach is not suitable for long-term studies or on proliferating cells, such as neural stem cells since they rapidly dilute them. To overcome this limitation the approach of miRNA "sponges" or "decoys" was developed. A typical sponge or decoy consists of a RNA transcript (typically encoded by a plasmid) bearing multiple binding sites for the miRNA of interest (Ebert et al., 2007). Improved RNA decoys (single strand RNAs with one antisense miRNA binding domain (Decoy) or a stabilized stem-loop with two miRNA binding domains (Tough decoys, TuD)) can achieve efficient and long-term suppression of specific miRNAs for over 1 month in mammalian cells (Haraguchi et al., 2009). However, it is extremely difficult to optimize sponges for the simultaneous inhibition of different miRNAs. Loss of function of a miRNA can also be accomplished by knocking-down its precursor by means of RNAi (Vasudevan et al., 2007). This approach however is not very popular as it is difficult to optimize, and the mechanism of RNAi on miRNAs is not entirely known.

A completely different approach to achieve loss-of-function of miRNAs is the "target protector." Instead of inhibiting the miRNA, target protectors interfere with a target of a selected miRNA. They typically consist of morpholino-based oligonucleotides that interfere with a single miRNA:mRNA pair through specifically binding to the mRNA $3'$UTR region (Staton and Giraldez, 2011). Target protectors are mainly used in Zebrafish, but their efficiency has also been reported in mice (Goljanek-Whysall et al., 2011). Recently, plasmid-based protectors were also used in neural progenitors of the developing mouse cortex, both in vitro and in vivo (Knauss et al., 2013).

The genetic knockout of a single miRNA approach has also been used to investigate the consequences of miRNAs depletion in rodents (Rodriguez et al., 2007; Thai et al., 2007; Zhao et al., 2007) and a repository of mice bearing floxed-miRNAs genes is now available (Park et al., 2012). However, most individual miRNA mutants show only subtle phenotypes (Smibert and Lai, 2008).

Indeed, this is not surprising as each miRNA can target directly between hundreds or thousands mRNAs (Bartel, 2009) and accumulating evidence demonstrate that miRNAs act in cooperation, a phenomenon that we named as "miRNA Convergence" (Barca-Mayo and De Pietri Tonelli, 2014). These evidence, therefore complicate the use of miRNA-specific knockouts to infer phenotypically relevant functional aspects of miRNA biology. Alternatively, several studies used genetically ablation of key components of the miRNA biogenesis machinery, such as Dicer, Drosha, and Dgcr8 to achieve global miRNA depletion in vivo (Barca-Mayo and De Pietri Tonelli, 2014). However, although the main cause of phenotypes is the miRNA loss, global miRNA depletion approach complicates the identification of specific contribution of each miRNA on the phenotype observed. Moreover, it is also important to take into account the existence of alternative miRNA independent functions of these proteins and the existence of alternative miRNA pathways independent of these proteins (Yang and Lai, 2011; Marinaro et al., 2017).

More recently, the use of CRISPR/Cas9 genome-editing technology has been also used for miRNA inhibition in Zebrafish (Xiao et al., 2013) and mammalian cells (Zhao et al., 2014) with long-term stability of miRNA knockdown phenotype up to 30 days both in vitro and in xenograft transplants in vivo in mice (Chang et al., 2016). Since the discovery of Cas9, the applications of this enzyme are exponentially growing. At present most scientists believe that "*CRISPR's powerful possibilities are almost infinite*," it is therefore very plausible to foresee that in the near future CRISPR/Cas9-based methods will become the new standard for functional investigation of miRNAs (Table 4.1).

APPROACHES FOR GAIN-OF miRNA FUNCTION

MiRNA "mimics" have been widely used in vitro to perform gain of miRNA functions studies. Mimics are synthetic RNA duplexes designed to mimic the sequence of endogenous pre-miRNAs (stem loop) or diced double-strand immature miRNAs (Table 4.1). The main limitation of miRNA mimics is their high susceptibility to nuclease degradation upon in vivo delivery, and the induction of immuno-toxicity response through Toll-like receptors (Peacock et al., 2011). Some chemical modifications in the "passenger strand" have been introduced to avoid nuclease degradation, such as $2'$-fluoro (Chiu and Rana, 2003), or cholesterol linking to improve cellular uptake. However, because the "guide strand" must function as an endogenous recognizable miRNA, possibilities for chemical modifications are very limited. The main limitation of this approach is that it only leads to transient overexpression of miRNAs, and since it is exogenously provided, it suffers from dilution problems in fast proliferating cells. To improve such limitations (especially for long-term and in vivo studies), stable expression vectors (plasmids or viral) encoding stem-loop miRNA precursor transcripts have been developed (Cheng et al., 2009; Schratt et al., 2006).

Although the potential therapeutic use of mimics to restore the level of down-regulated miRNAs is promising, there are some limitations to be considered. First the re-expression of miRNAs should not exceed supraphysiological levels and second, the absence of cell-target specific sequence could result in unspecific/off-target effects due to the introduction of miRNAs in a tissue/cell that do not normally express the miRNA. To overcome this limitation, it is necessary to develop better vectors enclosing tissue/cell-specific promoters to confine expression of the miRNA in a stable manner.

DELIVERY OF miRNA ANTAGONISTS OR MIMICS

One of the main issues in RNA-based therapies is the delivery method used for in vivo purposes. Beside the discussed modifications of oligonucleotides to improve their intracellular delivery, binding efficiency and stability; intracerebral delivery is complicated by the presence of the blood-brain barrier (BBB, see also "Future Perspectives") impermeable to RNA.

Currently, there are two main delivery strategies: local and systemic. Effective gene silencing or induction has been successfully performed through intracerebral delivery ("behind BBB" approach) of antagomirs or mimics to treat brain tumors, such as glioblastoma multiforme (Christensen et al., 2010; Krützfeldt et al., 2005; Møller et al., 2013). The advantage of local delivery is the confined administration at the injection site, avoiding undesired uptake by nearby tissues or healthy regions. This approach also allows a better control of the concentration of the oligonucleotide delivered and it is less affected by nuclease degradation in blood, liver, or other organs. However, in brain it requires stereotactic surgery, thus limiting their use for human therapy. Conversely, systemic delivery ("across BBB") although less effective and costly, it is becoming attractive for human therapy. The physical properties of the synthetic oligonucleotides as polyanionic molecules confer them high water solubility, being suitable for intravenous and subcutaneous administration. However, they need permeation enhancers to be absorbed by the tissues, and even with these modifications, oligonucleotides size and charge limit their absorption. Moreover, oligonucleotide levels in brain upon transvascular delivery are modest and their plasma levels drop quickly, while their accumulation in the liver and kidney last up to 24 hours after injection (Wang et al., 2012). This evidence clearly indicates that new technological development aimed to improve pharmacodynamics of synthetic RNA-based drugs, is needed.

As mentioned above, chemical modifications increase cellular uptake, stability and nuclease resistance. However, although in vitro they are very useful, only few promising results have been obtained in vivo, emphasizing the need of novel strategies.

VIRAL VECTORS

Virally encoded antagonists or mimics confer valuable deliver properties. First, the miRNA of interest can be continuously expressed due to the replication capacity of the virus and, in some cases, the integration into the genome. Second, the targeting moieties for a specific cell type or region of interest can be included in the viral capsid, increasing their target specificity. Nowadays, many viral vectors are available with different properties.

Adenoviruses and adeno-associated viruses (AAV) are nonenveloped viruses from the Adenoviridae and Parvoviridae family, respectively. They infect both dividing and nondividing cells, they replicate in the nucleus of mammalian cells, without integrating their DNA into the host cell genome, making them promising delivery vectors to avoid unspecific effects of mutation insertions. The main advantage of using AAV compared to adenovirus is that they poorly transduce antigen-presenting cells, thereby reducing dramatically the immunogenic response. Moreover, they are not pathogenic making them safe for human therapy. Using recombinant AAV, Schratt group demonstrated that miR-134 impaired dendritogenesis in the mouse brain in vivo (Christensen et al., 2010).

Conversely, *retroviruses* and *lentiviruses* are enveloped viruses from the Retroviridae family that integrate into the host genome allowing long-term expression, but with the inconvenient of undesired insertional mutagenesis. Retroviruses can only infect dividing cells, making them valuable for targeting neural stem cells without compromising nondividing cells, such as neurons. Retroviruses and lentiviruses for miR-124 have been used to demonstrate a role on neuronal fate determination in the adult brain (Åkerblom et al., 2012; Cheng et al., 2009).

Although viruses have the potential to specifically target the central nervous system (CNS), there are several challenges that are still unresolved. In fact, although systemic administration of AAV viruses with CNS tropism has been successfully used, the liver is the major target for AAV vectors, and transduction to other tissues has also been shown (Zincarelli et al., 2008). Such off-targets effects increase probable toxicities into other tissues and reduce therapeutic efficacy. Historically, cell/tissue specific promoters have been introduced in vectors to improve the specificity of viral-mediated expression. Another promising approach named "de-targeting" used insertion of endogenous miRNA target sequences into the viral transcript, to suppress transgene expression outside the CNS (Brown et al., 2006).

Nonviral carriers

Despite viruses are attractive delivery agents due to the high infection efficiency, variability available and duration; unspecific off-targets effects due to integration properties of retroviruses and immune response to adenovirus have been observed (Anson, 2004). Moreover, their efficacy in brain is limited by the lack of transport across the BBB following intravenous administration. Therefore, nonviral carriers have been developed to allow systemic delivery of synthetic or virally encoded RNAs into the CNS in vivo.

Several nonviral carriers are cationic molecules that condense and neutralize the negatively charged polynucleic acids through electrostatic interactions. Although cationic carriers display lower delivery efficiency compared to viral vectors, their molecular composition can be easily manufactured, they tolerate different cargo sizes and are poorly immunogenic. Moreover, small ligands targeting diseased cell surface specific receptors have been used to increase their intracellular delivery efficiency. Some examples include, but are not limited to cationic polymers, cell-penetrating peptides, cationic lipids, dendrimers, surface modified carbon nanotubes, and a multitude of nanoparticles.

Liposome-like lipid-based carriers encapsulate RNA cargoes to protect them from degradation and increase the stability in blood circulation, thus improving systemic delivery of RNAs. Diverse formulations for lipid carriers include cationic or positively charged lipids that enhance the uptake of the incorporated oligonucleotides. However, liposomes use is limited by their toxicity due to their strong cationic charge. To overcome this problem, neutral lipid emulsion was developed and successfully used for delivery of miR-7 and miR-34a into lung cancer cells (Trang et al., 2011). Limitations of this approach are the short half-life of liposomes, thus requiring a continuous infusion or frequent administration. To solve these problems, synthetic polymers have been developed.

The "Trojan Horse Liposome" system presents a promising solution for the transvascular gene therapy in brain. It encapsulates synthetic oligonucleotides within the liposome with polyethylene glycol (PEG) to stabilize the liposome. The PEG can be covered with monoclonal antibodies targeting specific endogenous receptors located in the brain, such as insulin receptor. Moreover, the expression of the transgene can be restricted using tissue/cell-specific gene promoters (Boado and

Pardridge, 2011). Recently, a new class of lipid-based vectors has been successfully applied for in vivo delivery of anti-miR-21 to suppress glioblastoma tumor. Costa et al. demonstrated that intravenously administered chlorotoxin (CTX)-coupled (targeted) stable nucleic acid lipid particle-formulated anti-miR-21 oligonucleotides accumulate preferentially within brain tumors and promote efficient miR-21 silencing, decreasing tumor size. Therefore the use of nanoparticle-coupled ligands toward overexpressed tumor receptors such as CTX, a scorpion-derived peptide that was reported as a reliable and specific marker for gliomas enables tumor-specific delivery, while minimizing side effects to normal cells (Costa et al., 2015). "Bubble liposomes" are made of PEG modified with echo-contrast gas, which can function as plasmid DNA and siRNA delivery tool when used with ultrasound exposure in vitro and in vivo. Researches successfully delivered miR-126 via intravascular injection, to treat hindlimb ischemia (Endo-Takahashi et al., 2014). A modified version of these liposomes is now emerging for brain-targeted delivery and ultrasound imaging via systemic administration.

Inorganic nanoparticles composed by gold, carbon, or silica with their unique physicochemical properties such as shape surface area and functionalization offer the potential of a biocompatible, nonimmunogenic and nontoxic delivery system for oligonucleotides. Gold nanoparticles showed efficient gene silencing in cancer cells (Ghosh et al., 2013). Systemic administration of silica nanoparticles with antibody modifications, specifically delivers miR-34a in neuroblastoma (Tivnan et al., 2012). However, the lack of cargo protection and inefficient endosome escape limits their use.

Polymers such as poly(lactic-glycolic) acid (PLGA) and polyethyleneimine (PEI) are good candidates for miRNA delivery due to stable and high electrostatic interactions with miRNA cargoes, allowing the release of miRNA over time. Recently, PLGA-based nanoparticles were used to deliver antagonists for miR-21 to gliobastoma cells resulting in antiproliferative and apoptotic effect (Ananta et al., 2015); or for miR-155 to malignant pre-B lymphoma cells in mouse models with good results (Babar et al., 2012). Intracranial delivery of miR-145 mimics coupled to a polyurethane-short branch PEI (PU-PEI) significantly reduced glioblastoma tumorigenesis (Yang et al., 2012). More recently, PEI nanoparticles bound to RGV was shown to target specifically nicotinic acetylcholine receptors on neuronal cells, achieving efficient delivery of miR-124a mimics and promoting neurogenesis. Although this technique shows promising results for brain accumulation, it is limited by the use of permeabilizing agents, such as mannitol (Hwang et al., 2011). Similarly, Pillai group delivered a RVG peptide with an additional non-D-arginine stretch for better interaction with the LNA-modified-miR-29 mimics by a single tail vein injection, observing specific brain targeting and behavioral effects in mice (Kumar et al., 2007).

Dendrimers are potent delivery agents due to their defined architecture and high ratio of surface moieties to molecular volume. Delivery of miR-21 antagonists with a poly(amidoamine) dendrimer into glioblastoma cells reduced significantly migration of tumor cells (Ren et al., 2010). Although this study was conducted in vitro, dendrimers capacity to entry into cells avoiding endosomal and lysosomal compartments make them attractive for in vivo purposes (Hong et al., 2004).

Cell-based carriers have recently become attractive method for delivering mimics and inhibitors for therapies, especially when treating gliomas, particularly through stem cells due to their tropism to glioblastoma cells (Bexell et al., 2013). Neural stem cells and mesenchymal stem cells (MSCs) are currently in trials as drug delivery method for gliomas. Indeed, MSCs were successfully used to deliver miR-9 antagonists, through secreted exosomes, in glioblastoma cells (Munoz et al., 2013) and miR-124 and miR-145 mimics into glioma xenografts, when MSCs were administered intracranially, reducing the migration and self-renewal of transplanted cells in vivo (Lee et al., 2013).

MiRNA:TARGET VALIDATION
INDIRECT VALIDATION

Manipulation of miRNA levels determines a change in the expression of target mRNAs and proteins. In order to analyze the functional consequences of miRNA manipulation, *in silico* analyses of miRNA:target interactions is a first necessary step that needs to be followed by experimental validation (for more detail, see Chapter 3, Computational Challenges and -Omics Approaches for the Identification of MicroRNAs and Targets).

The experimental evaluation of changes in mRNA/protein expression upon miRNA manipulation can be accomplished by several molecular and biochemical assays. For simplicity, we categorize these approaches as "indirect" or "direct" validation assays. Indirect assays allow obtaining indications of the effect of a given miRNA on their target mRNAs (and in turn, proteins), but do not provide information regarding the direct interaction (binding) between miRNA and its target mRNAs. In contrast, direct assays prove the interaction of miRNA with its targets (Table 4.1).

The most widely used indirect miRNA:mRNA functional validation assay is based on luciferase, considered as the "gold-standard" procedure to assess the miRNA-dependent control of mRNA target expression. This assay starts with the cloning of the 3'UTR sequence of the target mRNA of interest downstream to the protein coding sequence of a luciferase (at present a genome wide repository of luciferase 3'UTR plasmids is available at OriGene Technologies, Inc). This plasmid is transiently transfected into a host cell in which the endogenously expressed miRNA is inhibited or exogenously administered and luciferase activity is measured after 24−72 hours. Typical controls in luciferase assays are parallel experiments involving plasmids that do not contain the 3'UTR of interest, and/or plasmids in which the 3'UTR is cloned in an inverse orientation and/or the predicted miRNA binding site in the 3'UTR is mutated (if the miRNA is endogenously expressed). Luciferase assays also require co-transfection of a second plasmid encoding for a different luciferase that is used for normalization of transfection efficiency. In addition, scrambled double-stranded RNA control molecules, unable to target any mRNA, can be used as negative controls. Although luciferase assay is a powerful method for the functional investigation of miRNA:mRNA interaction, its approach is often misused, thus leading to potentially misleading results. Indeed this assay is often performed in vitro, sometimes in cell cultures that derive from tissue or even organism different from the biological context in which the initial observations about miRNA effects were observed. Moreover, several labs perform luciferase assays on minimal miRNA responsive elements instead of testing the entire 3'UTR of the target mRNA, thus leading to results that are potentially due to a different RNA structural context. To overcome these limitations, we introduced in vivo luciferase approach. In particular, by the use of in utero electroporation, we investigated the effect of miRNA manipulation (overexpression or inhibition) on the stability of the 3'UTR sequence of specific target mRNAs in embryonic mouse neocortex (Clovis et al., 2012; Nigro et al., 2012).

As above explained, the Negative Sensor assay has been generally used to detect or monitor expression of a given miRNA with high specificity and single-cell resolution. However, since the negative sensor assay takes advantage of the natural capability of miRNAs to silence a target mRNA, it can be also used as miRNA:target functional validation assay. In vivo, this assay has been mostly implemented with transgenic animals expressing the sensor for the miRNA of interest, thus making this approach not suitable for rapid validation of different miRNA:targets interactions, in parallel (Brennecke et al., 2003; Mansfield et al., 2004). To overcome this limitation, we modified the negative sensor technique by developing a Dual Fluorescence Reporter Sensor plasmid that can be transiently administered in mice in in vivo (De Pietri Tonelli et al., 2006).

DIRECT VALIDATION

Although the use of chimeric reporter assays to analyze the miRNA translational repressive effect is a widely used for both miRNA visualization and target validation, it should be considered as an indirect method that only takes into account the putative interaction of a given miRNA with a specific target. Moreover, the transfection of supraphysiological concentrations of a specific miRNA, as well as the introduction of a chimeric plasmid inside the cell, could lead to misleading information by creating a nonphysiological situation. To avoid these problems, direct interacting studies should be performed using "pull-down" assays where the miRNA:mRNA target pairs coprecipitate with RNA Induced Silencing Complex (RISC) proteins. Since Rajewsky and colleagues have extensively discussed this topic in Chapter 2, Essentials of miRNA-dependent control of mRNA translation and decay, and miRNA targeting principles, including target recognition, and methods for target identification, of this book, we do not enter into details here. However, we compare different methodologies and highlight weaknesses and advantages of the different methods.

RNA and protein can be coimmunoprecipitated (RIP). However, to avoid unspecific RNA-protein binding during sample processing, crosslinking and immunoprecipitation (CLIP) method has been introduced. This approach stabilizes protein−RNA interactions with UV light, prior to RNA:proteins extraction. Some variants of CLIP experiments include high-throughput sequencing of crosslinking immunoprecipitation (HITS-CLIP), CLIP-seq, or CLIP-Chip (Licatalosi et al., 2008) in which crosslinking is followed by deep-sequencing, or microarray. This approach allows profiling of RISC-bound RNAs, and the identification of miRNA binding sites location on target mRNA region at a resolution of ∼30−60 nucleotides. Photoactivatable-ribonucleoside-enhanced crosslinking and immunoprecipitation (PAR-CLIP) incorporates photoreactive ribonucleoside analogs (e.g., 4-thiouridine or 6-thioguanosine) into nascent RNA transcripts leading to higher efficient crosslinking efficiency upon UV irradiation (Hafner et al., 2010). The conversion of thioribonucleosides (T to C or G to A) allow a precise identification of RNA binding sites. However, it is limited by the efficiency of the nucleoside uptake and the potential cytotoxicity (Burger et al., 2013). The iCLIP method (individual nucleotide resolution CLIP) achieves nucleotide resolution through replacing one of the intermolecular RNA ligation steps with an intramolecular cDNA circularization step to capture the truncated cDNAs that are produced lacking the 5′adaptor region during standard CLIP library preparation (König et al., 2010).

Currently, most of the CLIP protocols are restricted to immunoprecipitation of Ago2, which is the only mammalian Argonaute protein with slicer activity, but not the only member of the Argonaute family that binds to miRNAs. To overcome this problem, Crosslinking Ligation and Sequencing of Hybrids (CLASH) approaches have been developed to map RNA–RNA interactions by ligating the miRNA to the mRNA fragments prior to sequencing (Helwak et al., 2013). This technique revealed that miRNA associate in comparable distributions with the four mammalian AGO homologs. Moreover, although the seed region was the most frequent miRNA interaction site with a target, noncanonical miRNA–mRNA targeting were substantially represented in this study.

The use of biotinylated tagged miRNA duplexes and obtainment of the miRNA–mRNA complex with streptavidin beads followed by analysis with deep-sequencing allows the specific pulldown of a single miRNA (Orom and Lund, 2007). However, it has the disadvantage of introducing exogenous modified miRNAs.

Finally, TAP-Tar procedure (Tandem affinity purification of miRNA target mRNAs) decreases the number of false-positive miRNA:mRNA interactions, due to the sequential pull-down of mRNA/miRNA complexes. In this approach a specific biotinylated miRNA mimic is transfected, and the ribonucleoprotein complexes are selected by immunoprecipitation with an anti-AGO antibody first, and then by pull-down with streptavidin beads (Nonne et al., 2010).

FUTURE PERSPECTIVES

Billions of neurons and glial cells compose the mammalian brain. Even considering neurons only, their cellular specialization is particularly prominent. There are several unresolved problems to consider for brain therapy: brain diseases often need a long time to manifest and when symptoms appear is often too late (e.g., neurodegenerative diseases); therapy is not available or has poor efficacy and when brain tissue is damaged, it cannot be easily repaired or replaced. Nonetheless, the recent years have witnessed a terrific technological advance, which has certainly opened exciting scientific questions, as well as innovative therapeutic possibilities in the broad field of brain science.

With rapid advances in single-cell isolation and profiling technologies it is now possible to undertake the immense task of unraveling brain cell heterogeneity. High-throughput single-cell RNA sequencing and multiplexed qRT-PCR have recently become accessible to many labs. These technologies enable now the systematic categorization of individual brain cells, including stem cells (Poulin et al., 2016). Given that most of the human genome transcribes noncoding RNAs, it is highly likely that future studies aimed at the systematic profiling of single brain cells will certainly reveal novel important functions for miRNAs and other noncoding RNAs. These studies will certainly lead us to a better understanding of molecular mechanisms underlying brain physiology and pathology.

With an ageing population in developed countries, brain health is becoming an urgent challenge. Hence, there is an utmost need for innovative approaches for early diagnosis, novel therapies and innovative strategies to repair the damaged brain. Indeed, miRNAs are exciting candidates as biomarkers for diseases, as novel drugs/drug targets and to improve cell reprogramming protocols. As we are writing this chapter several miRNA-based drugs are in pipeline or clinical trials. "Miravirsen" (SPC3649) is a LNA-modified oligonucleotide antagonist of miR-122 (MiRNA Therapeutics, http://www.mirnarx.com) that completed two Phase-I clinical trials, and is currently in Phase-II to treat hepatitis C virus chronic infection (Janssen et al., 2013; Lanford et al., 2010). Another miRNA-based drug is "MRX34" (MiRNA therapeutics), a miR-34a mimic encapsulated in a liposomal nanoparticle formulation, currently in clinical Phase-I stage with tumor suppressor properties (Liu et al., 2011). By the end of 2016 MiRNA therapeutics plans to advance MRX34 into two Phase-II clinical trials. Regulus therapeutics (http://regulusrx.com) has in pipeline several miRNA-based drugs, such as RG-101 to treat HCV; RG-012 for kidney diseases and RG-125 for type-2 diabetes. Finally, MiRagen therapeutics (http://miragentherapeutics.com) is currently developing MRG-201, MRG-106, and MRG-107, miRNA-based drugs for fibrosis, cancer and amyotrophic lateral sclerosis, respectively.

However, to further exploit miRNAs for diagnosis and therapy in brain, key problems remain, such as improving sensitivity of isolation/detection of circulating miRNAs, improving strategies to deliver miRNAs across the BBB, and identifying crucial miRNA species that specify brain cell identities. Development of miRNA-based technologies will certainly propel innovative solutions to these challenges. Accordingly, we foresee that miRNAs have the power to become the next-generation of diagnostics and therapeutics.

ACKNOWLEDGMENTS

This study was supported by Fondazione Istituto Italiano di Tecnologia (IIT) and by grant from Fondazione CARIPLO (no. 2015-0590) to DDPT.

REFERENCES

Åkerblom, M., Sachdeva, R., Barde, I., Verp, S., Gentner, B., Trono, D., et al., 2012. MicroRNA-124 is a subventricular zone neuronal fate determinant. J. Neurosci. 32, 8879−8889.

Ananta, J.S., Paulmurugan, R., Massoud, T.F., 2015. Nanoparticle-delivered antisense microRNA-21 enhances the effects of temozolomide on glioblastoma cells. Mol. Pharm. 12, 4509−4517.

Anson, D.S., 2004. The use of retroviral vectors for gene therapy-what are the risks? A review of retroviral pathogenesis and its relevance to retroviral vector-mediated gene delivery. Genet. Vaccines Ther. 2, 9.

Babar, I.A., Cheng, C.J., Booth, C.J., Liang, X., Weidhaas, J.B., Saltzman, W.M., et al., 2012. Nanoparticle-based therapy in an in vivo microRNA-155 (miR-155)-dependent mouse model of lymphoma. Proc. Natl. Acad. Sci. USA 109, E1695−E1704.

Baker, M., 2010. MicroRNA profiling: separating signal from noise. Nat. Methods 7, 687−692.

Barca-Mayo, O., De Pietri Tonelli, D., 2014. Convergent microRNA actions coordinate neocortical development. Cell. Mol. Life Sci. CMLS 71, 2975−2995.

Bartel, D.P., 2009. MicroRNAs: target recognition and regulatory functions. Cell 136, 215−233.

Bexell, D., Svensson, A., Bengzon, J., 2013. Stem cell-based therapy for malignant glioma. Cancer Treat. Rev. 39, 358−365.

Boado, R.J., Pardridge, W.M., 2011. The Trojan horse liposome technology for nonviral gene transfer across the blood−brain barrier. J. Drug Deliv. 2011, 296151.

Brennecke, J., Hipfner, D.R., Stark, A., Russell, R.B., Cohen, S.M., 2003. bantam encodes a developmentally regulated microRNA that controls cell proliferation and regulates the proapoptotic gene hid in Drosophila. Cell 113, 25−36.

Brown, B.D., Venneri, M.A., Zingale, A., Sergi Sergi, L., Naldini, L., 2006. Endogenous microRNA regulation suppresses transgene expression in hematopoietic lineages and enables stable gene transfer. Nat. Med. 12, 585–591.

Burger, K., Mühl, B., Kellner, M., Rohrmoser, M., Gruber-Eber, A., Windhager, L., et al., 2013. 4-Thiouridine inhibits rRNA synthesis and causes a nucleolar stress response. RNA Biol. 10, 1623–1630.

Chang, H., Yi, B., Ma, R., Zhang, X., Zhao, H., Xi, Y., 2016. CRISPR/cas9, a novel genomic tool to knock down microRNA in vitro and in vivo. Sci. Rep. 6, 22312.

Chen, C., Ridzon, D.A., Broomer, A.J., Zhou, Z., Lee, D.H., Nguyen, J.T., et al., 2005. Real-time quantification of microRNAs by stem-loop RT-PCR. Nucleic Acids Res. 33, e179–e179.

Chen, C., Tan, R., Wong, L., Fekete, R., Halsey, J., 2011. Quantitation of microRNAs by real-time RT-qPCR. Methods Mol Biol. 687, 113–134.

Cheng, L.-C., Pastrana, E., Tavazoie, M., Doetsch, F., 2009. miR-124 regulates adult neurogenesis in the subventricular zone stem cell niche. Nat. Neurosci. 12, 399–408.

Chiu, Y.-L., Rana, T.M., 2003. siRNA function in RNAi: a chemical modification analysis. RNA NYN 9, 1034–1048.

Chomczynski, P., Sacchi, N., 1987. Single-step method of RNA isolation by acid guanidinium thiocyanate-phenol-chloroform extraction. Anal. Biochem. 162, 156–159.

Christensen, M., Larsen, L.A., Kauppinen, S., Schratt, G., 2010. Recombinant adeno-associated virus-mediated microRNA delivery into the postnatal mouse brain reveals a role for miR-134 in dendritogenesis in vivo. Front. Neural Circuits 3, 16.

Clovis, Y.M., Enard, W., Marinaro, F., Huttner, W.B., De Pietri Tonelli, D., 2012. Convergent repression of Foxp2 3'UTR by miR-9 and miR-132 in embryonic mouse neocortex: implications for radial migration of neurons. Dev. Camb. Engl. 139, 3332–3342.

Costa, P.M., Cardoso, A.L., Custódia, C., Cunha, P., Pereira de Almeida, L., Pedroso de Lima, M.C., 2015. miRNA-21 silencing mediated by tumor-targeted nanoparticles combined with sunitinib: a new multimodal gene therapy approach for glioblastoma. J. Control. Release Off. J. Control. Release Soc. 207, 31–39.

De Pietri Tonelli, D., Calegari, F., Fei, J.-F., Nomura, T., Osumi, N., Heisenberg, C.-P., et al., 2006. Single-cell detection of microRNAs in developing vertebrate embryos after acute administration of a dual-fluorescence reporter/sensor plasmid. BioTechniques 41, 727–732.

De Pietri Tonelli, D., Pulvers, J.N., Haffner, C., Murchison, E.P., Hannon, G.J., Huttner, W.B., 2008. miRNAs are essential for survival and differentiation of newborn neurons but not for expansion of neural progenitors during early neurogenesis in the mouse embryonic neocortex. Dev. Camb. Engl. 135, 3911–3921.

De Pietri Tonelli, D., Clovis, Y.M., Huttner, W.B., 2014. Detection and monitoring of microRNA expression in developing mouse brain and fixed brain cryosections. Methods Mol. Biol. Clifton NJ 1092, 31–42.

Easow, G., Teleman, A.A., Cohen, S.M., 2007. Isolation of microRNA targets by miRNP immunopurification. RNA NYN 13, 1198–1204.

Ebert, M.S., Neilson, J.R., Sharp, P.A., 2007. MicroRNA sponges: competitive inhibitors of small RNAs in mammalian cells. Nat. Methods 4, 721–726.

Elmén, J., Lindow, M., Schütz, S., Lawrence, M., Petri, A., Obad, S., et al., 2008. LNA-mediated microRNA silencing in non-human primates. Nature 452, 896–899.

Endo-Takahashi, Y., Negishi, Y., Nakamura, A., Ukai, S., Ooaku, K., Oda, Y., et al., 2014. Systemic delivery of miR-126 by miRNA-loaded Bubble liposomes for the treatment of hindlimb ischemia. Sci. Rep. 4, 3883.

Fleige, S., Pfaffl, M.W., 2006. RNA integrity and the effect on the real-time qRT-PCR performance. Mol. Aspects Med. 27, 126–139.

Flynt, A.S., Li, N., Thatcher, E.J., Solnica-Krezel, L., Patton, J.G., 2007. Zebrafish miR-214 modulates Hedgehog signaling to specify muscle cell fate. Nat. Genet. 39, 259–263.

Ghosh, R., Singh, L.C., Shohet, J.M., Gunaratne, P.H., 2013. A gold nanoparticle platform for the delivery of functional microRNAs into cancer cells. Biomaterials 34, 807−816.

Goljanek-Whysall, K., Sweetman, D., Abu-Elmagd, M., Chapnik, E., Dalmay, T., Hornstein, E., et al., 2011. MicroRNA regulation of the paired-box transcription factor Pax3 confers robustness to developmental timing of myogenesis. Proc. Natl. Acad. Sci. USA 108, 11936−11941.

Griffiths-Jones, S., Saini, H.K., van Dongen, S., Enright, A.J., 2008. miRBase: tools for microRNA genomics. Nucleic Acids Res. 36, D154−D158.

Hafner, M., Landthaler, M., Burger, L., Khorshid, M., Hausser, J., Berninger, P., et al., 2010. Transcriptome-wide identification of RNA-binding protein and microRNA target sites by PAR-CLIP. Cell 141, 129−141.

Haraguchi, T., Ozaki, Y., Iba, H., 2009. Vectors expressing efficient RNA decoys achieve the long-term suppression of specific microRNA activity in mammalian cells. Nucleic Acids Res. 37, e43.

Helwak, A., Kudla, G., Dudnakova, T., Tollervey, D., 2013. Mapping the human miRNA interactome by CLASH reveals frequent noncanonical binding. Cell 153, 654−665.

Hong, S., Bielinska, A.U., Mecke, A., Keszler, B., Beals, J.L., Shi, X., et al., 2004. Interaction of poly(amidoamine) dendrimers with supported lipid bilayers and cells: hole formation and the relation to transport. Bioconjug. Chem. 15, 774−782.

Hutvágner, G., Simard, M.J., Mello, C.C., Zamore, P.D., 2004. Sequence-specific inhibition of small RNA function. PLoS Biol. 2, E98.

Hwang, D.W., Son, S., Jang, J., Youn, H., Lee, S., Lee, D., et al., 2011. A brain-targeted rabies virus glycoprotein-disulfide linked PEI nanocarrier for delivery of neurogenic microRNA. Biomaterials 32, 4968−4975.

Janssen, H.L.A., Reesink, H.W., Lawitz, E.J., Zeuzem, S., Rodriguez-Torres, M., Patel, K., et al., 2013. Treatment of HCV infection by targeting microRNA. N. Engl. J. Med. 368, 1685−1694.

Kim, Y.-K., Yeo, J., Kim, B., Ha, M., Kim, V.N., 2012. Short structured RNAs with low GC content are selectively lost during extraction from a small number of cells. Mol. Cell 46, 893−895.

Knauss, J.L., Bian, S., Sun, T., 2013. Plasmid-based target protectors allow specific blockade of miRNA silencing activity in mammalian developmental systems. Front. Cell. Neurosci. 7, 163.

König, J., Zarnack, K., Rot, G., Curk, T., Kayikci, M., Zupan, B., et al., 2010. iCLIP reveals the function of hnRNP particles in splicing at individual nucleotide resolution. Nat. Struct. Mol. Biol. 17, 909−915.

Krichevsky, A.M., King, K.S., Donahue, C.P., Khrapko, K., Kosik, K.S., 2003. A microRNA array reveals extensive regulation of microRNAs during brain development. RNA 9 (10), 1274−1281.

Krützfeldt, J., Rajewsky, N., Braich, R., Rajeev, K.G., Tuschl, T., Manoharan, M., et al., 2005. Silencing of microRNAs in vivo with "antagomirs". Nature 438, 685−689.

Kumar, P., Wu, H., McBride, J.L., Jung, K.-E., Kim, M.H., Davidson, B.L., et al., 2007. Transvascular delivery of small interfering {RNA} to the central nervous system. Nature 448, 39−43.

Lanford, R.E., Hildebrandt-Eriksen, E.S., Petri, A., Persson, R., Lindow, M., Munk, M.E., et al., 2010. Therapeutic silencing of microRNA-122 in primates with chronic hepatitis C virus infection. Science 327, 198−201.

Lee, H.-K., Finniss, S., Cazacu, S., Bucris, E., Ziv-Av, A., Xiang, C., et al., 2013. Mesenchymal stem cells deliver synthetic microRNA mimics to glioma cells and glioma stem cells and inhibit their cell migration and self-renewal. Oncotarget 4, 346−361.

Lee, L.W., Zhang, S., Etheridge, A., Ma, L., Martin, D., Galas, D., et al., 2010. Complexity of the microRNA repertoire revealed by next-generation sequencing. RNA NYN 16, 2170−2180.

Licatalosi, D.D., Mele, A., Fak, J.J., Ule, J., Kayikci, M., Chi, S.W., et al., 2008. HITS-CLIP yields genome-wide insights into brain alternative RNA processing. Nature 456, 464−469.

Liu, C., Kelnar, K., Liu, B., Chen, X., Calhoun-Davis, T., Li, H., et al., 2011. The microRNA miR-34a inhibits prostate cancer stem cells and metastasis by directly repressing CD44. Nat. Med. 17, 211−215.

Liu, C.-G., Calin, G.A., Volinia, S., Croce, C.M., 2008. MicroRNA expression profiling using microarrays. Nat. Protoc. 3, 563−578.

Mansfield, J.H., Harfe, B.D., Nissen, R., Obenauer, J., Srineel, J., Chaudhuri, A., et al., 2004. MicroRNA-responsive "sensor" transgenes uncover Hox-like and other developmentally regulated patterns of vertebrate microRNA expression. Nat. Genet. 36, 1079−1083.

Marinaro, F., Marzi, M.J., Hoffmann, N., Amin, H., Pelizzoli, R., Niola, F., et al., 2017. MicroRNA-independent functions of DGCR8 are essential for neocortical development and TBR1 expression. EMBO Rep. pii: e201642800.

Møller, H.G., Rasmussen, A.P., Andersen, H.H., Johnsen, K.B., Henriksen, M., Duroux, M., 2013. A systematic review of microRNA in glioblastoma multiforme: micro-modulators in the mesenchymal mode of migration and invasion. Mol. Neurobiol. 47, 131−144.

Munoz, J.L., Bliss, S.A., Greco, S.J., Ramkissoon, S.H., Ligon, K.L., Rameshwar, P., 2013. Delivery of functional anti-miR-9 by mesenchymal stem cell-derived exosomes to glioblastoma multiforme cells conferred chemosensitivity. Mol. Ther. Nucleic Acids 2, e126.

Nielsen, P.E., Egholm, M., Berg, R.H., Buchardt, O., 1991. Sequence-selective recognition of {DNA} by strand displacement with a thymine-substituted polyamide. Science 254, 1497.

Nigro, A., Menon, R., Bergamaschi, A., Clovis, Y.M., Baldi, A., Ehrmann, M., et al., 2012. MiR-30e and miR-181d control radial glia cell proliferation via HtrA1 modulation. Cell Death Dis. 3, e360.

Nonne, N., Ameyar-Zazoua, M., Souidi, M., Harel-Bellan, A., 2010. Tandem affinity purification of miRNA target mRNAs (TAP-Tar). Nucleic Acids Res. 38, e20.

Obad, S., dos Santos, C.O., Petri, A., Heidenblad, M., Broom, O., Ruse, C., et al., 2011. Silencing of microRNA families by seed-targeting tiny LNAs. Nat. Genet. 43, 371−378.

Orom, U.A., Lund, A.H., 2007. Isolation of microRNA targets using biotinylated synthetic microRNAs. Methods San Diego Calif. 43, 162−165.

Park, C.Y., Jeker, L.T., Carver-Moore, K., Oh, A., Liu, H.J., Cameron, R., et al., 2012. A resource for the conditional ablation of microRNAs in the mouse. Cell Rep. 1, 385−391.

Peacock, H., Fucini, R.V., Jayalath, P., Ibarra-Soza, J.M., Haringsma, H.J., Flanagan, W.M., et al., 2011. Nucleobase and ribose modifications control immunostimulation by a microRNA-122-mimetic RNA. J. Am. Chem. Soc. 133, 9200−9203.

Poulin, J.-F., Tasic, B., Hjerling-Leffler, J., Trimarchi, J.M., Awatramani, R., 2016. Disentangling neural cell diversity using single-cell transcriptomics. Nat Neurosci. 19, 1131−1141.

Pritchard, C.C., Cheng, H.H., Tewari, M., 2012. MicroRNA profiling: approaches and considerations. Nat. Rev. Genet. 13, 358−369.

Raymond, C.K., Roberts, B.S., Garrett-Engele, P., Lim, L.P., Johnson, J.M., 2005. Simple, quantitative primer-extension PCR assay for direct monitoring of microRNAs and short-interfering RNAs. RNA 11, 1737−1744.

Ren, Y., Zhou, X., Mei, M., Yuan, X.-B., Han, L., Wang, G.-X., et al., 2010. MicroRNA-21 inhibitor sensitizes human glioblastoma cells U251 (PTEN-mutant) and LN229 (PTEN-wild type) to taxol. BMC Cancer 10, 27.

Rodriguez, A., Vigorito, E., Clare, S., Warren, M.V., Couttet, P., Soond, D.R., et al., 2007. Requirement of bic/microRNA-155 for normal immune function. Science 316, 608−611.

Schratt, G.M., Tuebing, F., Nigh, E.A., Kane, C.G., Sabatini, M.E., Kiebler, M., et al., 2006. A brain-specific microRNA regulates dendritic spine development. Nature 439, 283−289.

Shah, A.A., Meese, E., Blin, N., 2010. Profiling of regulatory microRNA transcriptomes in various biological processes: a review. J. Appl. Genet. 51, 501−507.

Smibert, P., Lai, E.C., 2008. Lessons from microRNA mutants in worms, flies and mice. Cell Cycle Georget. Tex 7, 2500−2508.

Staton, A.A., Giraldez, A.J., 2011. Use of target protector morpholinos to analyze the physiological roles of specific {miRNA}−{mRNA} pairs in vivo. Nat. Protoc. 6, 2035−2049.

Thai, T.-H., Calado, D.P., Casola, S., Ansel, K.M., Xiao, C., Xue, Y., et al., 2007. Regulation of the germinal center response by microRNA-155. Science 316, 604−608.

Tivnan, A., Orr, W.S., Gubala, V., Nooney, R., Williams, D.E., McDonagh, C., et al., 2012. Inhibition of neuroblastoma tumor growth by targeted delivery of microRNA-34a using anti-disialoganglioside GD2 coated nanoparticles. PloS One 7, e38129.

Trang, P., Wiggins, J.F., Daige, C.L., Cho, C., Omotola, M., Brown, D., et al., 2011. Systemic delivery of tumor suppressor microRNA mimics using a neutral lipid emulsion inhibits lung tumors in mice. Mol. Ther. J. Am. Soc. Gene Ther. 19, 1116−1122.

Vasudevan, S., Tong, Y., Steitz, J.A., 2007. Switching from repression to activation: microRNAs can up-regulate translation. Science 318, 1931−1934.

Vogelstein, B., Kinzler, K.W., 1999. Digital PCR. Proc. Natl. Acad. Sci. USA 96, 9236−9241.

Wang, H., Chiu, M., Xie, Z., Chiu, M., Liu, Z., Chen, P., et al., 2012. Synthetic microRNA cassette dosing: pharmacokinetics, Tissue Distribution and Bioactivity. Mol. Pharm. 9, 1638−1644.

Wienholds, E., Kloosterman, W.P., Miska, E., Alvarez-Saavedra, E., Berezikov, E., de Bruijn, E., et al., 2005. MicroRNA expression in zebrafish embryonic development. Science 309, 310−311.

Xiao, A., Wang, Z., Hu, Y., Wu, Y., Luo, Z., Yang, Z., et al., 2013. Chromosomal deletions and inversions mediated by TALENs and CRISPR/Cas in zebrafish. Nucleic Acids Res. 41, e141.

Yang, J.-S., Lai, E.C., 2011. Alternative miRNA biogenesis pathways and the interpretation of core miRNA pathway mutants. Mol. Cell 43, 892−903.

Yang, Y.-P., Chien, Y., Chiou, G.-Y., Cherng, J.-Y., Wang, M.-L., Lo, W.-L., et al., 2012. Inhibition of cancer stem cell-like properties and reduced chemoradioresistance of glioblastoma using microRNA145 with cationic polyurethane-short branch PEI. Biomaterials 33, 1462−1476.

Zhao, Y., Ransom, J.F., Li, A., Vedantham, V., Drehle, M., von, Muth, A.N., et al., 2007. Dysregulation of cardiogenesis, cardiac conduction, and cell cycle in mice lacking miRNA-1-2. Cell 129, 303−317.

Zhao, Y., Dai, Z., Liang, Y., Yin, M., Ma, K., He, M., et al., 2014. Sequence-specific inhibition of microRNA via CRISPR/CRISPRi system. Sci. Rep. 4, 3943.

Zincarelli, C., Soltys, S., Rengo, G., Rabinowitz, J.E., 2008. Analysis of AAV serotypes 1-9 mediated gene expression and tropism in mice after systemic injection. Mol. Ther. J. Am. Soc. Gene Ther. 16, 1073−1080.

FURTHER READING

Choi, Y., Hwang, D. won, Kim, M.Y., Kim, J.Y., Sun, W., Lee, D.S., 2016. Transgenic mouse expressing optical microRNA reporter for monitoring microRNA-124 action during development. Front. Mol. Neurosci. 9, 52.

Mishima, Y., Abreu-Goodger, C., Staton, A.A., Stahlhut, C., Shou, C., Cheng, C., et al., 2009. Zebrafish miR-1 and miR-133 shape muscle gene expression and regulate sarcomeric actin organization. Genes Dev. 23, 619−632.

Pathania, M., Torres-Reveron, J., Yan, L., Kimura, T., Lin, T.V., Gordon, V., et al., 2012. miR-132 enhances dendritic morphogenesis, spine density, synaptic integration, and survival of newborn olfactory bulb neurons. PloS One 7, e38174.

Pena, J.T.G., Sohn-Lee, C., Rouhanifard, S.H., Ludwig, J., Hafner, M., Mihailovic, A., et al., 2009. miRNA in situ hybridization in formaldehyde and EDC-fixed tissues. Nat. Methods 6, 139−141.

Xie, J., Xie, Q., Zhang, H., Ameres, S.L., Hung, J.-H., Su, Q., et al., 2011. MicroRNA-regulated, systemically delivered rAAV9: a step closer to CNS-restricted transgene expression. Mol. Ther. 19, 526−535.

THE CELL BIOLOGY OF NEURAL STEM AND PROGENITOR CELLS AND NEOCORTEX EXPANSION IN DEVELOPMENT AND EVOLUTION

Stefania Tavano and Wieland B. Huttner

Max Planck Institute of Molecular Cell Biology and Genetics, Dresden, Germany

The neocortex, the evolutionarily youngest part of the brain, is a hallmark of mammals and the seat of higher cognitive functions. The adult neocortex is characterized by six distinct layers of neurons. The most abundant class of cortical neurons are the pyramidal neurons (Lodato and Arlotta, 2015). They are generated during cortical neurogenesis from neural progenitor cells (NPCs) that reside in the neocortex itself. A concerted program of NPC proliferation and the production of postmitotic neurons is a crucial determinant for neocortex development. Understanding how this program is regulated and how it has changed during neocortex evolution has been an important goal of neuroscientists and developmental biologists.

The first part of this chapter describes the cytoarchitecture of the developing neocortex, the different NPC types, and their cell biological characteristics, in a paradigm of neurogenesis during neocortical development—the mouse. The second part focuses on recent advances in understanding the differences between this canonical lissencephalic species and selected gyrencephalic species, notably primates, which occurred during neocortex evolution.

THE DEVELOPMENT OF THE MOUSE NEOCORTEX FROM THE ONSET OF NEUROGENESIS TO THE FORMATION OF NEOCORTICAL ZONES

In mouse the process of cortical neurogenesis starts between 10 and 11 days after conception, and the neurogenic period lasts about 7 days. Before the start of neuron production, the neocortex is a pseudostratified neuroepithelium composed of a single layer of one type of highly proliferative progenitor, the neuroepithelial cells (NECs) (Fig. 5.1A and B) (Götz and Huttner, 2005). NECs undergo mostly symmetric proliferative divisions, giving rise to two daughter cells with the same characteristics as the mother cell. This proliferative phase drives an increase in the initial pool of NPCs and consequently an expansion of the cortical wall in the lateral dimension. As other epithelia, the neuroepithelium exhibits an apical and a basal side that correspond to the ventricular surface and the basal lamina, respectively. This apical–basal polarity of the cortical wall is maintained after the onset of neurogenesis, when NECs gradually lose some of their epithelial

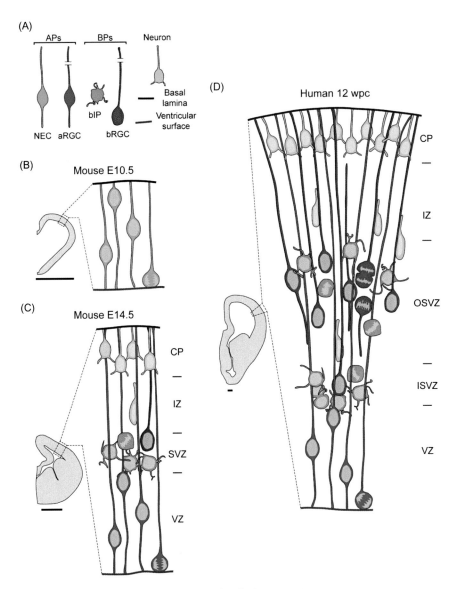

FIGURE 5.1 Schematic representations of the mouse developing neocortex

. (A) Cell types in the developing neocortex. Two main groups of NPCs are present in the mouse developing neocortex: APs, which includes NEC (light blue) and aRGC (blue), and BPs, consisting of bIP (orange) and bRGC (dark red). Neurons are represented in green. The ventricular surface and the basal lamina are in red and black, respectively. (B) Schematic representation of the mouse neocortical wall prior onset of neurogenesis, at embryonic day (E) 10.5. (C) Schematic representation of the mouse neocortical wall at midneurogenesis, E14.5. The different neocortical zones (VZ, SVZ, IZ, CP) are labeled on the right. (D) Schematic representation of the human neocortical wall 12-week post conception (WPC). The different neocortical zones (VZ, ISVZ, OSVZ, IZ, CP) are labeled on the right. All scale bars are 500 μm.

features and start to express glial markers, becoming a different type of NPC—the apical radial glial cell (aRGC). Both NECs and aRGCs belong to one of the two classes of NPCs, the apical progenitors (APs) (Fig. 5.1A). The second class of NPCs are the basal progenitors (BPs) (Fig. 5.1A), which arise from asymmetric cell divisions of aRGCs (Haubensak et al., 2004; Miyata et al., 2004; Noctor et al., 2004; Attardo et al., 2008). In the mammalian neocortex, only a minority of neurons are born form APs and the main source of neurons is the pool of BPs.

With the appearance of new types of NPCs and the production of neurons, the cytoarchitecture of the neocortical wall changes drastically: new neocortical zones start to form basal to the initial single layer of cells (Fig. 5.1C). At the peak of neuron production, four different zones are distinguishable in the developing mouse neocortex:

- The ventricular zone (VZ), the most apical zone, which is reminiscent of the initial neuroepithelium and harbors the nuclei of the aRGCs;
- The subventricular zone (SVZ), located basal to VZ, which is the destination of most of the newly born migrating BPs;
- The intermediate zone (IZ), which is largely composed of migrating neurons;
- The cortical plate (CP), the final destination of the neurons and "ancestor" of the adult six-layered neocortex.

In all mammals the CP forms in an inside-out fashion: following layer I, the first pyramidal neurons will form layer VI, and the next-born neurons will migrate basally to form layers V-II (i.e., deep-layer neuron generation followed by upper-layer neuron generation).

NEURAL PROGENITOR CELLS IN THE EMBRYONIC MOUSE NEOCORTEX

The location of the cell body at mitosis divides the NPC pool in the aforementioned two classes. APs undergo mitosis at, or in the proximity of, the ventricular surface (Fig. 5.1B and C). In contrast, BPs undergo mitosis at a basal, abventricular location, predominantly within the SVZ (Fig. 5.1C). Each NPC class comprises different types of NPCs according to their morphology at mitosis and the expression of known markers. In the developing mouse neocortex, these features are generally sufficient to deduce the proliferative potential of the respective NPC type.

APICAL PROGENITORS

In the developing mouse neocortex, the two main types of APs are NECs and aRGCs (Fig. 5.1A) (Taverna et al., 2014).

Neuroepithelial cells

NECs are the most proliferative NPCs present in the developing mouse neocortex. Similar to other epithelial cells, NECs exhibit apical−basal cell polarity. Each NEC contacts both, the ventricle and the basal lamina, through its apical surface and basal endfoot, respectively (Fig. 5.1B). At the apical-most region of the basolateral membrane, there are tight junctions and adherens junctions (AJs) that form the border to the apical membrane and primary cilium (Taverna et al., 2014). The presence of the primary cilium at the apical side tethers the centrosome, the mother centriole of which constitutes the basal body of the cilium, at the ventricle until the start of mitosis. While the centrosome is thus fixed in position during interphase, the nuclei of NECs migrate along the apical−basal axis of the cortical wall in concert with the cell cycle: during the G1 phase the nuclei move toward the basal side, undergo S phase near the basal lamina, before migrating apically during the G2 phase to undergo mitosis at the ventricular surface (Fig. 5.1B) (Taverna and Huttner, 2010; Kosodo, 2012). This process is called interkinetic nuclear migration (INM) and causes the pseudostratified appearance of the neuroepithelium (Taverna and Huttner, 2010; Kosodo, 2012).

Apical radial glial cells

With the onset of neurogenesis, NECs transform into aRGCs, which share many features with the former: aRGCs exhibit apical−basal polarity and contact both the basal lamina and the ventricle (Fig. 5.1C); their nuclei undergo INM and divide at the ventricular surface (Fig. 5.1C) where the centrosome and primary cilia are located in interphase (Fig. 5.2B.1) (Taverna and Huttner, 2010; Kosodo, 2012); they are integrated into an apical AJ belt that separates their apical and basolateral membrane (Fig. 5.2B.1) (Taverna et al., 2014); they express NEC markers, e.g., nestin. However, aRGCs also display their own characteristics: they express several glial markers, such as GFAP and BLBP; in aRGCs, the INM is confined to the VZ and, consequently, their nuclei never occupy the basal process (Fig. 5.1C); aRGCs lack tight junctions, even though some tight junction proteins (e.g., ZO1) are present at their AJs; their basal process elongates considerably with the growth of the neocortex and serves as guide for migrating neurons (Fig. 5.1C) (Götz and Huttner, 2005; Taverna et al., 2014).

aRGCs divide both symmetrically and asymmetrically (Taverna et al., 2014). During the early stages of cortical neurogenesis, the symmetric aRGC divisions are usually proliferative, generating two aRGCs, whereas during the late stages, they are consumptive, generating two BPs. The asymmetric aRGC divisions generate an aRGC and a BP (Fig. 5.2), or more rarely an aRGC and a neuron (Götz and Huttner, 2005; Taverna et al., 2014). aRGCs express several transcription factors important for their proliferative potential, the best-known among them being Pax6 (Englund et al., 2005).

APICAL−BASAL POLARITY OF APICAL PROGENITORS

The apical−basal polarity of aRGCs is of fundamental importance for their proliferative potential and cell fate. As in other epithelial cells, this polarity underlies the compartmentalization of subcellular structures and functions. Impairing either apical or basal polarity results in dramatic changes of aRGC behavior (Taverna et al., 2014).

Apical plasma membrane and primary cilium

The apical plasma membrane and the primary cilium represent the only contact of aRGCs with the ventricle and the cerebrospinal fluid. Hence, they are the only mediators of the signaling cues present in the cerebrospinal fluid, such as fibroblast growth factors (FGFs) and sonic hedgehog (Shh) (Johansson, 2014). Decreased exposure to these signaling molecules represents one of the possible reasons of the decreased proliferative potential of BPs (Taverna et al., 2014). The primary cilium is mainly composed of microtubules (MT) and ciliary membrane and acts as a platform for different signaling cascades via several membrane receptors. The ciliary MTs are nucleated by the mother

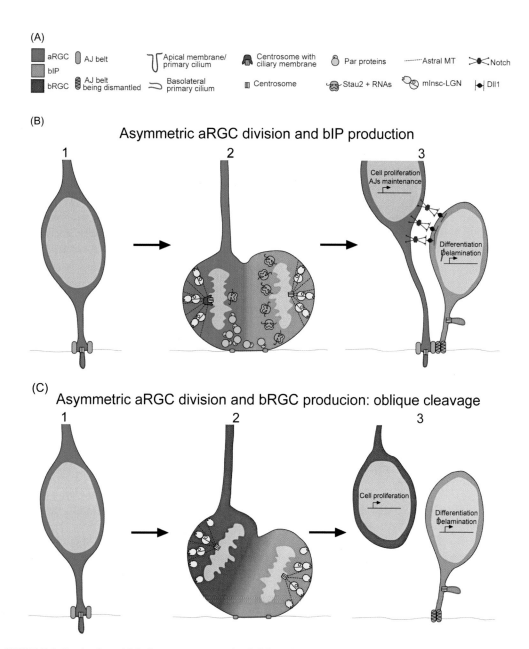

FIGURE 5.2 Production of BPs from an asymmetric division of an aRGC

(A) Key to the cartoons in (B) and (C). (B) bIP-producing asymmetric cell division. (1) An interphase aRGC characterized by AJs, apical primary cilium and centrosome at the ventricular surface. (2) An aRGC dividing asymmetrically. Basal process, Par proteins and centrosome with ciliary membrane are asymmetrically inherited

(Continued)

◀ by the future daughter cell fated to remain an aRGC (blue side); the complex Stau2-RNAs are enriched in the future bIP (orange side). (3) Sibling cells after asymmetric division. The newborn aRGC (blue) reestablishes an apical primary cilium and expresses genes that sustain cell proliferation and AJs. The newborn bIP (orange) has a basolateral cilium and its gene expression program promotes differentiation and degradation of the AJs. (C) bRGC-producing asymmetric cell division. (1) An interphase aRGC characterized by AJs, apical primary cilium and centrosome at the ventricular surface. (2) An aRGC division with an oblique cleavage plane. The future daughter cell retaining the basal process will become a bRGC (dark red side), whereas the other sibling cell will keep the apical contact (orange side). (3) Sibling cells after asymmetric division. The newborn bRGC maintains a basal process and proliferative capacity. The newborn bIP (orange) has a basolateral cilium and its gene expression program promotes differentiation and degradation of the AJs.

centriole (basal body) (Taverna et al., 2014). As in all eukaryotic cells, the centrosome and the peri-centriolar material form the main microtubule-organizing center (MTOC). In the specific case of aRGCs, the centrosome and the MTOC are docked to the apical membrane and are released only upon mitosis or delamination (Taverna et al., 2014).

Adherens junctions

The AJ belt constitutes the major cell-to-cell interaction between aRGCs, is located just beneath the apical membrane, and is a fundamental cell biological feature of aRGCs. AJs are multiprotein complexes containing cell adhesion molecules (CAMs), cytoskeletal structures consisting mainly of actomyosin bundles and MTs, along with several proteins, such as α-catenins, β-catenins, and afadin, which directly or indirectly bind to the cytoplasmic domain of the CAMs. The interaction between CAMs and AJ-interacting proteins ensures the binding to the cytoskeleton and controls the stability of the AJs themselves (Paridaen and Huttner, 2014; Taverna et al., 2014; Stocker and Chenn, 2015). The AJs are important for maintenance of aRGC polarity and for regulation of their proliferation and differentiation. The depletion of N-cadherin or αE-catenin from aRGCs causes misregulation of their proliferation and disruption of the cytoarchitecture of the cortical wall (Gil-Sanz et al., 2014; Paridaen and Huttner, 2014; Taverna et al., 2014; Stocker and Chenn, 2015). The effects on proliferation and differentiation are in part explained by the existing crosstalk between AJs and the Wnt signaling pathway through β-catenin (Taverna et al., 2014).

Among the proteins interacting with AJs are also several small GTPases, and their function is critical to control the stability of AJs and the proliferation and cell fate of aRGCs (Cappello et al., 2012a; Azzarelli et al., 2014; Lian and Sheen, 2015). A neocortex-specific deletion of *cdc42* causes the loss of both AJs and Par proteins from aRGCs (Cappello et al., 2006), whereas the lack of RhoA causes not only the disruption of AJs but also a transient increase in proliferation (Katayama et al., 2011; Cappello et al., 2012a).

Basal process

With the thickening of the neocortical wall and the appearance of new cell types, the basolateral plasma membrane of aRGCs, in particular that of the basal process, greatly expands. The basal process has been known to serve as the rail to the CP for the migrating neurons (Kosodo and Huttner, 2009; Taverna et al., 2014). Studies during the last decade, however, have pointed to an additional, prominent role of the basal process as a platform for the interaction between an aRGC and a BP and as a hub for receptors of several signaling pathways, such as integrins (Kosodo and Huttner, 2009; Taverna et al., 2014). Moreover, the contact of the basal endfoot with the basal lamina is not only important for the proper migration of immature neurons, but also controls aRGC proliferation (Kosodo and Huttner, 2009; Bae et al., 2014).

ASYMMETRIC CELL DIVISION OF APICAL PROGENITORS

The production of newborn BPs in mouse is linked to the occurrence of asymmetric division of aRGCs. Several studies in the last decade identified proteins involved in the regulation of these asymmetric divisions, as well as components that are asymmetrically inherited by one of the daughter cells.

Mitotic spindle orientation

The regulation of mitotic spindle orientation plays an important role in the switch from symmetric to asymmetric aRGC divisions, and consequently for the fate of the daughter cells (Mora-Bermudez and Huttner, 2015). The orientation of the mitotic spindle depends on a complex machinery that includes the centrosomes, astral MTs and MT-binding proteins present at the cell cortex (Fig. 5.2B.2) (Mora-Bermudez and Huttner, 2015). Recent studies showed that specific manipulation of this machinery is sufficient to induce production of BPs. Down-regulation of apical−basal astral MTs (Mora-Bermudez et al., 2014), ablation of the cell cortex protein LGN (Konno et al., 2008; Shitamukai et al., 2011), or perturbation of the complex Ndel1/Lis1 (Yingling et al., 2008; Pawlisz et al., 2008; Xie et al., 2013) cause an increase in aRGCs that divide with a random orientation of the mitotic spindle. This results in an increase in asymmetric aRGC divisions and in the production of BPs, at the expense of aRGC proliferation. Moreover, many genes mutated in microcephaly and lissencephaly in humans encode proteins that regulate the orientation of the mitotic spindle (Lancaster and Knoblich, 2012).

Inheritance of cell fate determinants

In asymmetric aRGC division, different cellular components are inherited asymmetrically between the two daughter cells. One example is the basal process. Previous studies showed that the daughter cell retaining the basal process (Fig. 5.2B.2), and the *Ccnd2* transcript within it, is more likely to become an aRGC (Shitamukai and Matsuzaki, 2012; Tsunekawa et al., 2012). Similarly, Par proteins and other components of the AJs are preferentially inherited by the daughter cell that remains an aRGC (Fig. 5.2B.2) (Kosodo et al., 2004; Bultje et al., 2009; Marthiens and Ffrench-Constant, 2009). In contrast, the RNA binding protein Stau2, associated with different transcripts, is preferentially inherited by the daughter cell that becomes a BP (Fig. 5.2B.2) (Kusek et al., 2012; Vessey et al., 2012). The mother centriole and the ciliary membrane are usually inherited by the future aRGC (Wang et al., 2009; Paridaen et al., 2013). Moreover, the association of the ciliary membrane with the mother centriole allows a faster formation of a new primary cilium in the respective daughter cell (Fig. 5.2B.2) (Paridaen et al., 2013). Such rapid formation of the primary cilium is thought to facilitate the exposure to signaling from the cerebrospinal fluid and thereby promote aRGC fate (Paridaen et al., 2013). In agreement with this, another study showed that newborn BPs establish a basolateral rather than apical primary cilium, and this process precedes their delamination from the apical AJ belt (Fig. 5.2B.3) (Wilsch-Bräuninger et al., 2012).

BASAL PROGENITORS

In contrast to aRGCs, BPs undergo abventricular mitoses, mostly within the SVZ. In the developing mouse neocortex, asymmetric division of aRGCs is the main source of BPs (Haubensak et al., 2004; Noctor et al., 2004; Miyata et al., 2004). We can recognize two main types of BPs based on their morphology: basal intermediate progenitors (bIPs) and basal radial glial cells (bRGCs) (Fig. 5.1A) (Florio and Huttner, 2014; Taverna et al., 2014). In mouse, both BP types have a limited proliferative potential and express the transcription factor Tbr2 (Englund et al., 2005; Florio et al., 2015); however, bRGCs also maintain the expression of Pax6 (Florio and Huttner, 2014).

Basal intermediate progenitors

bIPs are by far the most abundant type of BPs in the mouse neocortex. Unlike aRGCs, they exhibit neither apical−basal polarity nor contact with the basal lamina and the ventricle (Fig. 5.1C). In the mouse, most bIPs are neurogenic, dividing once to produce two neurons.

Basal radial glial cells

bRGCs (also called outer radial glia) constitute only a small minority of the BPs in mouse, where they are mostly neurogenic (Wang et al., 2011b). They are closely related to aRGCs with which they share not only expression of Pax6, but also of glial markers. Moreover, they exhibit a monopolar morphology throughout the cell cycle, with a basal process contacting the basal lamina (Fig. 5.1C). Their soma undergoes a rapid movement toward the basal process shortly before mitosis (mitotic somal translocation) (Florio and Huttner, 2014). This recently characterized type of BP was found to be particularly abundant in gyrencephalic species (Fietz et al., 2010; Hansen et al., 2010), but recent studies showed that bRGCs are also abundant in the near-lissencephalic marmoset (Kelava et al., 2012; Garcia-Moreno et al., 2012). In gyrencephalic primates, bRGCs show high levels of heterogeneity in their morphology, proliferative capacity, and fate of their progeny (Betizeau et al., 2013), as will be discussed in greater detail below.

BASAL PROGENITOR PRODUCTION

Signaling pathways involved in basal progenitor fate

Signaling pathways have an important role both in maintaining the self-renewal of aRGCs and in inducing BP fate (Paridaen and Huttner, 2014).

The activation of the Notch receptor in NPCs causes the release of the Notch intracellular domain, which induces the expression of *Hes* genes, including *Hes1*. Hes1 inhibits BP identity through the repression of proneural basic helix-loop-helix (bHLH) TFs. Moreover, the Notch ligand Dll1 is a direct target of the BP transcription factor Ngn2 (Castro et al., 2006). The expression of *Dll1* in the newborn BP is thought to be important to sustain the self-renewal capacity of the sibling cell in an asymmetric aRGC division (Fig. 5.2B.3) (Shimojo et al., 2008; Ochiai et al., 2009).

There is a close link between AJs and the Wnt signaling pathway. The activation of the Wnt receptors Frizzled/LRP inhibits β-catenin degradation, allowing translocation of β-catenin to the nucleus and expression of LEF/TCF target genes (Harrison-Uy and Pleasure, 2012). During early neocortical development, the activation of the Wnt/β-catenin pathway leads to proliferation of aRGCs (Wrobel et al., 2007). At later stages, Wnt signaling induces production of BPs through the upregulation of N-myc (Kuwahara et al., 2010; Munji et al., 2011).

Various studies showed that the Shh signaling pathway is important for proper development of the neocortex. Lack of the ligand Shh or of the transmembrane protein Smo decreases the proliferative potential and cell survival of NPCs, whereas ectopic expression of Shh increases the pool of proliferative NPCs (Komada et al., 2008). However, the production of BPs requires a decrease of Shh signaling and the consequent increase of the Gli3 repressor (Wang et al., 2011a). In the presence of low Shh, the transcription factor Gli3 undergoes proteolytic cleavage of its transcriptional activator domain and becomes a repressor of Shh target genes (Varjosalo and Taipale, 2008).

From epigenetic to posttranscriptional regulation of basal progenitor fate

Several intrinsic and extrinsic signals that regulate the fate of NPCs ultimately affect gene expression programs. In this context, the epigenetic control of gene expression has emerged as a major regulatory mechanism.

DNA methylation and histone modifications are known to play important roles in regulating the proliferative capacity and fate of stem and progenitors cells in several model systems (Vastenhouw and Schier, 2012). In the lineage from a stem cell to a terminally differentiated cell, the epigenetic signature at the promoter of relevant genes changes in order to promote or repress gene expression. For example, the promoters of genes important for stemness maintenance are enriched in the permissive trimethylation of histone H3 at lysine 4 (H3K4me3) in stem cells and in the repressive trimethylation at lysine 27 (H3k27me3) in differentiated cells (Hirabayashi and Gotoh, 2010). The promoters of certain genes are characterized by either histone modification and are therefore called bivalent (Vastenhouw and Schier, 2012). Such genes are usually important for the developmental program of an organ; they are repressed in stem cells but poised for activation once exposed to appropriate developmental signals (Voigt et al., 2013). In NPCs, the promoter of the aRGC gene *Pax6*, as well as those of the BP identity genes *Ngn1* and *Ngn2*, carry a bivalent signature (Mikkelsen et al., 2007). In agreement with the importance of epigenetic control during neurogenesis, the knockout or knockdown of enzymes necessary for DNA methylation (e.g., Tet2 and Tet3) or of histone modifiers (e.g., Ezh2) induces premature production of BPs and neurons, and impair neocortical development (Pereira et al., 2010; Hahn et al., 2013).

Moreover, the production and specification of newborn BPs are controlled by several transcription factors. Among them, there are Ngn1 and Ngn2. They are part of the family of proneural bHLH TFs, and their expression is sufficient to induce neuronal differentiation (Bertrand et al., 2002; Wilkinson et al., 2013). In particular, Ngn2 is expressed in nascent BPs a few hours after aRGC cytokinesis (Ochiai et al., 2009). Its expression depends on the oscillation of the TF *Hes1*. *Hes1* and *Ngn2* expression oscillate in a counter-phase fashion in dividing NPCs, with *Ngn2* being sustained only when *Hes1* repression is maintained (Shimojo et al., 2008). One of the direct targets of Ngn2 is *Eomes* (Ochiai et al., 2009). *Eomes* encodes Tbr2, the best-known marker of BPs. Tbr2 is expressed by newborn BPs within the VZ shortly after Ngn2, but unlike Ngn2 its expression is sustained in BPs within the SVZ (Englund et al., 2005; Ochiai et al., 2009). Tbr2 is important to specify BP identity.

Its absence induces neocortical malformation, causes a drastic decrease of abventricular mitoses, and impairs the expression of BP-characteristic and neuronal genes (Arnold et al., 2008; Sessa et al., 2008). Another transcription factor involved in the production of BPs is Insm1 (Duggan et al., 2008; Farkas et al., 2008). Perturbations of *Insm1* expression strongly affect the pool of BPs in the mouse neocortex. The forced expression of *Insm1* at the onset of neurogenesis leads to an increase in proliferative BPs together with a reduction in the pool of APs, whereas *Insm1* knockout causes an increase in apical mitoses at the expense of the basal mitoses (Farkas et al., 2008).

Two other important processes that fine-control gene expression and affect the fate of the daughter cells are alternative splicing and other posttranscriptional control processes. It has recently been shown that the expression of alternative isoforms of key genes can dictate the fate of the daughter cells, and impairing the machinery controlling alternative splicing in APs affects neurogenesis (Paridaen and Huttner, 2014). For example, the knockdown of the REST splicing regulator nSR100 promotes AP fate at the expense of neuronal differentiation (Raj et al., 2011).

Similarly, a proper progression from APs to neurons depends on the posttranscriptional control exerted by miRNAs, as shown by several studies on the conditional knockout of DICER (Barca-Mayo and De Pietri Tonelli, 2014). To ensure proper neuronal differentiation, the expression of N-cadherin is fine-tuned by three miRNAs belonging to the miR379—410 cluster. On the contrary, miR-92 has recently been linked to the repression of BP fate through the binding to the 3′-UTR of *Eomes* mRNA and consequent inhibition of Tbr2 translation (Nowakowski et al., 2013; Bian et al., 2013). Interestingly, miR-92 also controls the expression of the prodifferentiative gene *Tis21* through two binding sites within the 3′-UTR (Fei et al., 2014). How miR-92-mediated silencing of *Tis21* affects neocortical development will be discussed in the final section of this chapter.

Delamination from the adherens junction belt

In order to migrate toward the SVZ, the newborn BP needs to disengage from the apical AJ belt (Fig. 5.2B.3). The stability and turnover of the AJ components have been studied in different epithelial cells and organisms. How this process is regulated in the developing neocortex is still not completely clear. Recent studies showed that the SNAG proteins Scrt1 and Scrt2 repress the expression of *Cdh1* (the gene encoding E-cadherin) (Itoh et al., 2013), whereas FoxP2 and FoxP4 suppress *Cdh2* (the gene encoding N-cadherin) (Rousso et al., 2012). The transcription factors Ngn2, Ascl1, and Tbr2 also directly or indirectly suppress the expression of AJ proteins (Sessa et al., 2008; Singh and Solecki, 2015).

Even less is known about the machinery that actively controls AJ disassembly. As mentioned previously, the action of Rho GTPases is crucial for the maintenance of the AJ belt in APs (Cappello et al., 2012b; Azzarelli et al., 2014; Lian and Sheen, 2015). Moreover, the presence of the protein kinase αPKCλ (Imai et al., 2006) and of the nectin-binding protein afadin (Yamamoto et al., 2015) is necessary for the integrity of the AJ belt.

DIFFERENCES BETWEEN LISSENCEPHALIC AND GYRENCEPHALIC DEVELOPING NEOCORTEX

One of most striking differences between lissencephalic and gyrencephalic mammals is the absence vs presence of cortical folds. This gyrification presumably reflects the dramatic increase in the number of cortical neurons and the requirement of a larger cortical sheet to accommodate all of them (Sun and Hevner, 2014). Not only is the number of neurons increased in the human, but also the proportion between the different layers is altered, with a more pronounced expansion of the upper-layer neurons (Defelipe et al., 2002). Major parameters that determine neuron output are NPC pool size and the length of the neurogenic period. In the past decade, major efforts have been undertaken in order to understand how these two parameters changed during evolution, and which molecular and cell biological determinants underlie those changes.

At present, most of our knowledge pertains to the pool size of progenitors and how it differs between the lissencephalic mouse and gyrencephalic species such as human, macaque and ferret. The first study describing a major difference in NPCs between gyrencephalic species and mouse was that of Smart et al. (2002). The authors described, at the histological level, the splitting of the SVZ into two separate zones in the fetal macaque neocortex: an inner SVZ (ISVZ) and an outer SVZ (OSVZ) (Fig. 5.1D) (Smart et al., 2002). In general, in primates, even though the OSVZ typi-cally is the last of the germinal zones to appear during cortical development, it is usually the most enlarged one. Moreover, in contrast to the SVZ in rodents and the ISVZ in gyrencephalic species, the nuclei of OSVZ NPCs are radially organized (Fig. 5.1D). In the discussion of their study, Smart et al. (2002) suggested that, during macaque neocortical development, the neuron output is achieved "by allocating radial glial cells to the OSVZ and these cells acquiring a proliferate mode that does not necessitate interkinetic migration, thereby avoiding nuclear congestion at the ventricu-lar surface". The presence of NPCs with radial glia-like features was subsequently demonstrated by three independent studies. In those studies, the authors showed that the OSVZ of human and ferret fetal neocortex is characterized by a new type of BP, called bRGC (or outer radial glia) (Fig. 5.1D) (Fietz et al., 2010; Hansen et al., 2010; Reillo and Borrell, 2012). Although the presence of bRGCs and an OSVZ alone may not be sufficient to induce gyrification (Kelava et al., 2012; Garcia-Moreno et al., 2012; Hevner and Haydar, 2012), it is generally thought that an increase in the abun-dance of bRGCs and in the proliferative potential of BPs is likely to underlie neocortex expansion and gyrification (Borrell and Götz, 2014; Florio and Huttner, 2014; Dehay et al., 2015). The remaining sections of this chapter focus on the molecular and cell biological features that contribute to the diversity and proliferative potential of BPs during neocortical development in gyrencephalic species.

FROM APICAL TO BASAL RADIAL GLIAL CELLS

Various studies have sought to dissect the cell biological and molecular mechanisms that cause the shift of radial glia from the VZ to the OSVZ in gyrencephalic species. From a morphological point of view, two important features of bRGCs are the retention of the basal process and the loss of api-cal contact. Considering what is known about asymmetric division of APs (Asymmetric cell divi-sion of apical progenitors section), one (though not the only) cell biological mechanism is that the cleavage plane of an aRGC during a bRGC-producing cell division is oriented such that one of the daughter cells inherits the basal process but none of the apical components. Indeed, it has recently been observed that at the peak of neuron production during human neocortical development, there is an increase of aRGC divisions with oblique cleavages, and such divisions result in the production of bRGCs (Fig. 5.2C) (Lamonica et al., 2013). The cues that induce horizontal cleavages in human aRGCs are largely unknown, although it is likely that they involve the machinery that controls the orientation of the mitotic spindle. Indeed, a selective, drug-induced decrease of a specific astral MT subpopulation, the apical—basal astral MT, has been shown to be sufficient to tilt the spindle from vertical to oblique in mouse proliferative aRGCs (Fig. 5.2C.2) (Mora-Bermudez et al., 2014). Furthermore, two studies from Matsuzaki and colleagues have shown that a decrease of LGN or the overexpression of the cell polarity protein mInsc induce oblique aRGC cleavages and an increase in bRGCs (Fig. 5.2C.2). In addition, these studies have suggested that the basal process may be crucial to maintain the progenitor fate, as the daughter cell inheriting only the apical complex was found to be more likely to become a neuron (Konno et al., 2008; Shitamukai et al., 2011).

Is the OSVZ always the final destination of bRGCs born from aRGC divisions? Interestingly, very recent work from Borrell and colleagues shows that in embryonic ferret neocortex, aRGCs produce bRGCs that will then reside within the OSVZ, only for a short window of time (from E34 to E38). Moreover, the beginning and end of this critical phase are regulated by the expression of *Cdh1* and *Trnp1*, respectively (Martinez-Martinez et al., 2016). It will be interesting to study if other genes are also important to determine that time window, and if this finding holds true also for other gyrencephalic species.

BASAL RADIAL GLIA PROCESSES

As aforementioned (Basal radial glia section), in mouse, bRGCs are morphologically distinguishable from bIPs by virtue of their basal process. However, in gyrencephalic species, the situation is more complex: bRGCs can have a basal process only, an apically directed process only, or both, an apically directed process and a basal process, in a transient or stable fashion (Fig. 5.1D) (Betizeau et al., 2013; Martinez-Martinez et al., 2016). Moreover, upon a self-amplifying division of a bRGC, the daughter cell initially born without a process may regrow it soon after cytokinesis (Betizeau et al., 2013; Martinez-Martinez et al., 2016). How bRGCs control the regrowth of their processes and their direction is not known, nor is the functional significance of bearing one or two processes. Regarding the latter issue, Dehay and colleagues have proposed that bRGC processes may function as an intra-tissue antenna for proliferative signals (Betizeau et al., 2013).

BASAL PROGENITOR PROLIFERATION

A key difference in BP behavior between gyrencephalic species and mouse is the massive BP proliferation throughout the neurogenic period in the former but not the latter. While BP generation from aRGCs is the primary step to establish a BP pool, it is evident that the expansion of the pool of BPs depends on their proliferative capacity. Of note, as underscored by the very recent findings of Borrell and colleagues (Martinez-Martinez et al., 2016), the dependence on proliferation is especially true for the OSVZ, where the contribution from aRGCs to the pool of OSVZ-residing bRGCs is limited.

Extracellular matrix and growth factors

It is generally known that the microenvironment has an important role for stem cell niches (Li and Xie, 2005). It would therefore not be surprising if the microenvironment were found to be a major factor for maintaining the proliferative potential of BPs in gyrencephalic species. Indeed, a first clue that this is the case came from the differential transcriptome analysis of neocortical germinal zones in human and mouse, where it was found that extracellular matrix (ECM) components and their receptors are highly expressed in all three human germinal zones and in the mouse VZ but not SVZ, reflecting the differential proliferative potential of these germinal zones in the two species (Fietz et al., 2012). Similar results have been obtained in other transcriptome analyses (Lui et al., 2014; Pollen et al., 2014; Johnson et al., 2015). Moreover, activation of the ECM receptor integrin $\alpha v\beta 3$ has been shown to suffice to increase BP cell cycle reentry in embryonic mouse neocortex (Stenzel et al., 2014).

Similarly, an important role is played by growth factors. Thus, platelet-derived growth factor D (PDGFD) and its receptor PDGFRB are highly expressed by radial glia in human, whereas the corresponding genes *Pdgfd* and *Pdgfrb* are not, or only lowly, expressed in mouse (Lui et al., 2014). Moreover, impairing PDGFD-PDGFRβ signaling in human neocortical slice cultures is sufficient to decrease NPC proliferation. Conversely, the activation of the same signaling pathway in mouse causes an increase in proliferating NPCs both in the VZ and SVZ (Lui et al., 2014).

Intracellular regulators of BP proliferation

The above-mentioned transcriptome analyses also revealed striking differences between mouse and human BPs with regard to the expression of cell cycle regulators (Fietz et al., 2012; Lui et al., 2014; Florio et al., 2015). In fetal human neocortex, BPs maintain the expression of several genes that promote cell cycle reentry. Forced expression of some of these in mouse BPs results in an expansion of the BP pool and, in some cases, promotes the appearance of gyrus-like structures (Nonaka-Kinoshita et al., 2013; Stahl et al., 2013; Florio et al., 2015; Wong et al., 2015). Moreover, with regard to the evolutionary differences that set humans apart from other primates, a recent study identified the first human-specific gene, *ARHGAP11B*, that upon expression in embryonic mouse neocortex promotes the generation and proliferation of BPs and is capable of inducing cortical folding (Florio et al., 2015).

miRNAs

With the exception of *ARHGAP11B*, the genes so far implicated in the expansion of a gyrencephalic neocortex are present in the mouse genome but either are not expressed, or their level and pattern of expression are different, in developing mouse neocortex. These gene expression differences reflect, on the one hand, human-specific differences in promoters and enhancers (Somel et al., 2013; Silbereis et al., 2016). On the other hand, miRNAs are well known to play important roles regulating gene expression (Fabian et al., 2010). A first line of evidence that miRNAs can control brain size was reported by Fei et al. (2014). Specifically, deletion of the 3′-UTR of the prodifferentiative gene *Tis21* in mouse caused an increase in *Tis21* mRNA and protein and a shift of neurogenic aRGCs to consumptive cell divisions. These APs would normally divide asymmetrically to produce an aRGC and a bIP. However, the lack of the *Tis21* 3′-UTR led to the production of two bIPs and, as a consequence, to the depletion of the aRGC pool, resulting in a mild microcephaly. Further investigation as to which miRNA is responsible for *Tis21* repression identified miR-92, a miRNA already known for regulating *Eomes* expression (Bian et al., 2013; Nowakowski et al., 2013; Fei et al., 2014).

Of note, a recent study indicates that the emergence of new miRNAs is another possible cause of evolutionary changes in gene expression in developing neocortex (Arcila et al., 2014). Analysis of miRNA expression in the VZ, OSVZ, and CP of two separate areas in the developing macaque neocortex revealed that more than 300 miRNAs are primate specific. Moreover, several of these are differentially expressed between the various neocortical zones (Arcila et al., 2014). Finding the targets of these new miRNAs and understanding their function in neocortical development clearly are one of the future challenges to understand the evolutionary expansion of the human neocortex (see also see Chapter 16: MicroRNA and Neocortical Evolution.

REFERENCES

Arcila, M.L., Betizeau, M., Cambronne, X.A., Guzman, E., Doerflinger, N., Bouhallier, F., et al., 2014. Novel primate miRNAs coevolved with ancient target genes in germinal zone-specific expression patterns. Neuron 81, 1255−1262.

Arnold, S.J., Huang, G.J., Cheung, A.F., Era, T., Nishikawa, S., Bikoff, E.K., et al., 2008. The T-box transcription factor Eomes/Tbr2 regulates neurogenesis in the cortical subventricular zone. Genes Dev. 22, 2479−2484.

Attardo, A., Calegari, F., Haubensak, W., Wilsch-Bräuninger, M., Huttner, W.B., 2008. Live imaging at the onset of cortical neurogenesis reveals differential appearance of the neuronal phenotype in apical versus basal progenitor progeny. PLoS ONE 3, e2388.

Azzarelli, R., Kerloch, T., Pacary, E., 2014. Regulation of cerebral cortex development by Rho GTPases: insights from in vivo studies. Front. Cell Neurosci. 8, 445.

Bae, B.I., Tietjen, I., Atabay, K.D., Evrony, G.D., Johnson, M.B., Asare, E., et al., 2014. Evolutionarily dynamic alternative splicing of GPR56 regulates regional cerebral cortical patterning. Science 343, 764−768.

Barca-Mayo, O., De Pietri Tonelli, D., 2014. Convergent microRNA actions coordinate neocortical development. Cell Mol. Life Sci. 71, 2975−2995.

Bertrand, N., Castro, D.S., Guillemot, F., 2002. Proneural genes and the specification of neural cell types. Nat. Rev. Neurosci. 3, 517−530.

Betizeau, M., Cortay, V., Patti, D., Pfister, S., Gautier, E., Bellemin-Ménard, A., et al., 2013. Precursor diversity and complexity of lineage relationships in the outer subventricular zone of the primate. Neuron 80, 442−457.

Bian, S., Hong, J., Li, Q., Schebelle, L., Pollock, A., Knauss, J.L., et al., 2013. MicroRNA cluster miR-17-92 regulates neural stem cell expansion and transition to intermediate progenitors in the developing mouse neocortex. Cell Rep 3, 1398−1406.

Borrell, V., Götz, M., 2014. Role of radial glial cells in cerebral cortex folding. Curr. Opin. Neurobiol. 27, 39−46.

Bultje, R.S., Castaneda-Castellanos, D.R., Jan, L.Y., Jan, Y.N., Kriegstein, A.R., Shi, S.H., 2009. Mammalian Par3 regulates progenitor cell asymmetric division via notch signaling in the developing neocortex. Neuron 63, 189−202.

Cappello, S., Attardo, A., Wu, X., Iwasato, T., Itohara, S., Wilsch-Brauninger, M., et al., 2006. The Rho-GTPase cdc42 regulates neural progenitor fate at the apical surface. Nat. Neurosci. 9, 1099−1107.

Cappello, S., Bohringer, C.R., Bergami, M., Conzelmann, K.K., Ghanem, A., Tomassy, G.S., et al., 2012a. A radial glia-specific role of RhoA in double cortex formation. Neuron 73, 911−924.

Cappello, S., Genovese, M., Della Torre, C., Crisari, A., Hassanshahian, M., Santisi, S., et al., 2012b. Effect of bioemulsificant exopolysaccharide (EPS(2)(0)(0)(3)) on microbial community dynamics during assays of oil spill bioremediation: a microcosm study. Marine Pollu. Bull. 64, 2820−2828.

Castro, D.S., Skowronska-Krawczyk, D., Armant, O., Donaldson, I.J., Parras, C., Hunt, C., et al., 2006. Proneural bHLH and Brn proteins coregulate a neurogenic program through cooperative binding to a conserved DNA motif. Dev. Cell 11, 831−844.

Defelipe, J., Alonso-Nanclares, L., Arellano, J.I., 2002. Microstructure of the neocortex: comparative aspects. J. Neurocytol. 31, 299−316.

Dehay, C., Kennedy, H., Kosik, K.S., 2015. The outer subventricular zone and primate-specific cortical complexification. Neuron 85, 683−694.

Duggan, A., Madathany, T., De Castro, S.C., Gerrelli, D., Guddati, K., Garcia-Anoveros, J., 2008. Transient expression of the conserved zinc finger gene INSM1 in progenitors and nascent neurons throughout embryonic and adult neurogenesis. J. Comp. Neurol. 507, 1497−1520.

Englund, C., Fink, A., Lau, C., Pham, D., Daza, R.A., Bulfone, A., et al., 2005. Pax6, Tbr2, and Tbr1 are expressed sequentially by radial glia, intermediate progenitor cells, and postmitotic neurons in developing neocortex. J. Neurosci. 25, 247−251.

Fabian, M.R., Sonenberg, N., Filipowicz, W., 2010. Regulation of mRNA translation and stability by microRNAs. Annu. Rev. Biochem. 79, 351−379.

Farkas, L.M., Haffner, C., Giger, T., Khaitovich, P., Nowick, K., Birchmeier, C., et al., 2008. Insulinoma-associated 1 has a panneurogenic role and promotes the generation and expansion of basal progenitors in the developing mouse neocortex. Neuron 60, 40−55.

Fei, J.F., Haffner, C., Huttner, W.B., 2014. 3′ UTR-dependent, miR-92-mediated restriction of Tis21 expression maintains asymmetric neural stem cell division to ensure proper neocortex size. Cell Rep. 7, 398−411.

Fietz, S.A., Kelava, I., Vogt, J., Wilsch-Brauninger, M., Stenzel, D., Fish, J.L., et al., 2010. OSVZ progenitors of human and ferret neocortex are epithelial-like and expand by integrin signaling. Nat. Neurosci. 13, 690−699.

Fietz, S.A., Lachmann, R., Brandl, H., Kircher, M., Samusik, N., Schroder, R., et al., 2012. Transcriptomes of germinal zones of human and mouse fetal neocortex suggest a role of extracellular matrix in progenitor self-renewal. Proc. Natl. Acad. Sci. USA 109, 11836−11841.

Florio, M., Huttner, W.B., 2014. Neural progenitors, neurogenesis and the evolution of the neocortex. Development 141, 2182−2194.

Florio, M., Albert, M., Taverna, E., Namba, T., Brandl, H., Lewitus, E., et al., 2015. Human-specific gene ARHGAP11B promotes basal progenitor amplification and neocortex expansion. Science 347, 1465−1470.

Garcia-Moreno, F., Vasistha, N.A., Trevia, N., Bourne, J.A., Molnar, Z., 2012. Compartmentalization of cerebral cortical germinal zones in a lissencephalic primate and gyrencephalic rodent. Cereb. Cortex 22, 482−492.

Gil-Sanz, C., Landeira, B., Ramos, C., Costa, M.R., Muller, U., 2014. Proliferative defects and formation of a double cortex in mice lacking Mltt4 and Cdh2 in the dorsal telencephalon. J. Neurosci. 34, 10475−10487.

Götz, M., Huttner, W.B., 2005. The cell biology of neurogenesis. Nat. Rev. Mol. Cell Biol. 6, 777−788.

Hahn, M.A., Qiu, R., Wu, X., Li, A.X., Zhang, H., Wang, J., et al., 2013. Dynamics of 5-hydroxymethylcytosine and chromatin marks in Mammalian neurogenesis. Cell Rep. 3, 291−300.

Hansen, D.V., Lui, J.H., Parker, P.R., Kriegstein, A.R., 2010. Neurogenic radial glia in the outer subventricular zone of human neocortex. Nature 464, 554−561.

Harrison-Uy, S.J., Pleasure, S.J., 2012. Wnt signaling and forebrain development. Cold Spring Harb. Perspect. Biol. 4, a008094.

Haubensak, W., Attardo, A., Denk, W., Huttner, W.B., 2004. Neurons arise in the basal neuroepithelium of the early mammalian telencephalon: A major site of neurogenesis. Proc. Natl. Acad. Sci. USA 101, 3196−3201.

Hevner, R.F., Haydar, T.F., 2012. The (not necessarily) convoluted role of basal radial glia in cortical neurogenesis. Cereb. Cortex 22, 465−468.

Hirabayashi, Y., Gotoh, Y., 2010. Epigenetic control of neural precursor cell fate during development. Nat. Rev. Neurosci. 11, 377−388.

Imai, F., Hirai, S., Akimoto, K., Koyama, H., Miyata, T., Ogawa, M., et al., 2006. Inactivation of aPKClambda results in the loss of adherens junctions in neuroepithelial cells without affecting neurogenesis in mouse neocortex. Development 133, 1735−1744.

Itoh, Y., Moriyama, Y., Hasegawa, T., Endo, T.A., Toyoda, T., Gotoh, Y., 2013. Scratch regulates neuronal migration onset via an epithelial-mesenchymal transition-like mechanism. Nat. Neurosci. 16, 416−425.

Johansson, P.A., 2014. The choroid plexuses and their impact on developmental neurogenesis. Front. Neurosci. 8, 340.

Johnson, M.B., Wang, P.P., Atabay, K.D., Murphy, E.A., Doan, R.N., Hecht, J.L., et al., 2015. Single-cell analysis reveals transcriptional heterogeneity of neural progenitors in human cortex. Nat. Neurosci. 18, 637−646.

Katayama, K., Melendez, J., Baumann, J.M., Leslie, J.R., Chauhan, B.K., Nemkul, N., et al., 2011. Loss of RhoA in neural progenitor cells causes the disruption of adherens junctions and hyperproliferation. Proc. Natl. Acad. Sci. USA 108, 7607−7612.

Kelava, I., Reillo, I., Murayama, A.Y., Kalinka, A.T., Stenzel, D., Tomancak, P., et al., 2012. Abundant occurrence of basal radial glia in the subventricular zone of embryonic neocortex of a lissencephalic primate, the common marmoset *Callithrix jacchus*. Cereb. Cortex 22, 469−481.

Komada, M., Saitsu, H., Kinboshi, M., Miura, T., Shiota, K., Ishibashi, M., 2008. Hedgehog signaling is involved in development of the neocortex. Development 135, 2717−2727.

Konno, D., Shioi, G., Shitamukai, A., Mori, A., Kiyonari, H., Miyata, T., et al., 2008. Neuroepithelial progenitors undergo LGN-dependent planar divisions to maintain self-renewability during mammalian neurogenesis. Nat. Cell Biol. 10, 93−101.

Kosodo, Y., Huttner, W.B., 2009. Basal process and cell divisions of neural progenitors in the developing brain. Dev. Growth Differ. 51, 251−261.

Kosodo, Y., 2012. Interkinetic nuclear migration: beyond a hallmark of neurogenesis. Cell Mol. Life Sci. 69, 2727−2738.

Kosodo, Y., Röper, K., Haubensak, W., Marzesco, A.-M., Corbeil, D., Huttner, W.B., 2004. Asymmetric distribution of the apical plasma membrane during neurogenic divisions of mammalian neuroepithelial cells. EMBO J. 23, 2314−2324.

Kusek, G., Campbell, M., Doyle, F., Tenenbaum, S.A., Kiebler, M., Temple, S., 2012. Asymmetric segregation of the double-stranded RNA binding protein Staufen2 during mammalian neural stem cell divisions promotes lineage progression. Cell Stem Cell 11, 505−516.

Kuwahara, A., Hirabayashi, Y., Knoepfler, P.S., Taketo, M.M., Sakai, J., Kodama, T., et al., 2010. Wnt signaling and its downstream target N-myc regulate basal progenitors in the developing neocortex. Development 137, 1035−1044.

Lamonica, B.E., Lui, J.H., Hansen, D.V., Kriegstein, A.R., 2013. Mitotic spindle orientation predicts outer radial glial cell generation in human neocortex. Nat. Commun. 4, 1665.

Lancaster, M.A., Knoblich, J.A., 2012. Spindle orientation in mammalian cerebral cortical development. Curr. Opin. Neurobiol. 22, 737−746.

Li, L., Xie, T., 2005. Stem cell niche: structure and function. Annu. Rev. Cell Dev. Biol. 21, 605−631.

Lian, G., Sheen, V.L., 2015. Cytoskeletal proteins in cortical development and disease: actin associated proteins in periventricular heterotopia. Front. Cell Neurosci. 9, 99.

Lodato, S., Arlotta, P., 2015. Generating neuronal diversity in the mammalian cerebral cortex. Annu. Rev. Cell Dev. Biol. 31, 699−720.

Lui, J.H., Nowakowski, T.J., Pollen, A.A., Javaherian, A., Kriegstein, A.R., Oldham, M.C., 2014. Radial glia require PDGFD-PDGFRbeta signalling in human but not mouse neocortex. Nature 515, 264−268.

Marthiens, V., Ffrench-Constant, C., 2009. Adherens junction domains are split by asymmetric division of embryonic neural stem cells. EMBO Rep. 10, 515−520.

Martinez-Martinez, M.A., De Juan Romero, C., Fernandez, V., Cardenas, A., Gotz, M., Borrell, V., 2016. A restricted period for formation of outer subventricular zone defined by Cdh1 and Trnp1 levels. Nat. Commun. 7, 11812.

Mikkelsen, T.S., Ku, M., Jaffe, D.B., Issac, B., Lieberman, E., Giannoukos, G., et al., 2007. Genome-wide maps of chromatin state in pluripotent and lineage-committed cells. Nature 448, 553−560.

Miyata, T., Kawaguchi, A., Saito, K., Kawano, M., Muto, T., Ogawa, M., 2004. Asymmetric production of surface-dividing and non-surface-dividing cortical progenitor cells. Development 131, 3133−3145.

Mora-Bermudez, F., Huttner, W.B., 2015. Novel insights into mammalian embryonic neural stem cell division: focus on microtubules. Mol. Biol. Cell 26, 4302−4306.

Mora-Bermudez, F., Matsuzaki, F., Huttner, W.B., 2014. Specific polar subpopulations of astral microtubules control spindle orientation and symmetric neural stem cell division. eLife 3, e02875.

Munji, R.N., Choe, Y., Li, G., Siegenthaler, J.A., Pleasure, S.J., 2011. Wnt signaling regulates neuronal differentiation of cortical intermediate progenitors. J. Neurosci. 31, 1676−1687.

Noctor, S.C., Martinez-Cerdeno, V., Ivic, L., Kriegstein, A.R., 2004. Cortical neurons arise in symmetric and asymmetric division zones and migrate through specific phases. Nat. Neurosci. 7, 136−144.

Nonaka-Kinoshita, M., Reillo, I., Artegiani, B., Martinez-Martinez, M.A., Nelson, M., Borrell, V., et al., 2013. Regulation of cerebral cortex size and folding by expansion of basal progenitors. EMBO J. 32, 1817−1828.

Nowakowski, T.J., Fotaki, V., Pollock, A., Sun, T., Pratt, T., Price, D.J., 2013. MicroRNA-92b regulates the development of intermediate cortical progenitors in embryonic mouse brain. Proc. Natl. Acad. Sci. USA 110, 7056−7061.

Ochiai, W., Nakatani, S., Takahara, T., Kainuma, M., Masaoka, M., Minobe, S., et al., 2009. Periventricular notch activation and asymmetric Ngn2 and Tbr2 expression in pair-generated neocortical daughter cells. Mol. Cell Neurosci. 40, 225−233.

Paridaen, J.T., Huttner, W.B., 2014. Neurogenesis during development of the vertebrate central nervous system. EMBO Rep. 15, 351−364.

Paridaen, J.T., Wilsch-Brauninger, M., Huttner, W.B., 2013. Asymmetric inheritance of centrosome-associated primary cilium membrane directs ciliogenesis after cell division. Cell 155, 333−344.

Pawlisz, A.S., Mutch, C., Wynshaw-Boris, A., Chenn, A., Walsh, C.A., Feng, Y., 2008. Lis1-Nde1-dependent neuronal fate control determines cerebral cortical size and lamination. Hum. Mol. Genet. 17, 2441−2455.

Pereira, J.D., Sansom, S.N., Smith, J., Dobenecker, M.W., Tarakhovsky, A., Livesey, F.J., 2010. Ezh2, the histone methyltransferase of PRC2, regulates the balance between self-renewal and differentiation in the cerebral cortex. Proc. Natl. Acad. Sci. USA 107, 15957−15962.

Pollen, A.A., Nowakowski, T.J., Shuga, J., Wang, X., Leyrat, A.A., et al., 2014. Low-coverage single-cell mRNA sequencing reveals cellular heterogeneity and activated signaling pathways in developing cerebral cortex. Nat. Biotechnol. 32, 1053−1058.

Raj, B., O'hanlon, D., Vessey, J.P., Pan, Q., Ray, D., Buckley, N.J., et al., 2011. Cross-regulation between an alternative splicing activator and a transcription repressor controls neurogenesis. Mol. Cell 43, 843−850.

Rcillo, I., Borrell, V., 2012. Germinal zones in the developing cerebral cortex of ferret: ontogeny, cell cycle kinetics, and diversity of progenitors. Cereb. Cortex 22, 2039−2054.

Rousso, D.L., Pearson, C.A., Gaber, Z.B., Miquelajauregui, A., Li, S., Portera-Cailliau, C., et al., 2012. Foxp-mediated suppression of N-cadherin regulates neuroepithelial character and progenitor maintenance in the CNS. Neuron 74, 314−330.

Sessa, A., Mao, C.A., Hadjantonakis, A.K., Klein, W.H., Broccoli, V., 2008. Tbr2 directs conversion of radial glia into basal precursors and guides neuronal amplification by indirect neurogenesis in the developing neocortex. Neuron 60, 56−69.

Shimojo, H., Ohtsuka, T., Kageyama, R., 2008. Oscillations in notch signaling regulate maintenance of neural progenitors. Neuron 58, 52−64.

Shitamukai, A., Matsuzaki, F., 2012. Control of asymmetric cell division of mammalian neural progenitors. Dev. Growth Differ. 54, 277−286.

Shitamukai, A., Konno, D., Matsuzaki, F., 2011. Oblique radial glial divisions in the developing mouse neocortex induce self-renewing progenitors outside the germinal zone that resemble primate outer subventricular zone progenitors. J. Neurosci. 31, 3683−3695.

Silbereis, J.C., Pochareddy, S., Zhu, Y., Li, M., Sestan, N., 2016. The cellular and molecular landscapes of the developing human central nervous system. Neuron 89, 248−268.

Singh, S., Solecki, D.J., 2015. Polarity transitions during neurogenesis and germinal zone exit in the developing central nervous system. Front. Cell Neurosci. 9, 62.

Smart, I.H., Dehay, C., Giroud, P., Berland, M., Kennedy, H., 2002. Unique morphological features of the proliferative zones and postmitotic compartments of the neural epithelium giving rise to striate and extrastriate cortex in the monkey. Cereb. Cortex 12, 37−53.

Somel, M., Liu, X., Khaitovich, P., 2013. Human brain evolution: transcripts, metabolites and their regulators. Nat. Rev. Neurosci. 14, 112−127.

Stahl, R., Walcher, T., De Juan Romero, C., Pilz, G.A., Cappello, S., Irmler, M., et al., 2013. Trnp1 regulates expansion and folding of the Mammalian cerebral cortex by control of radial glial fate. Cell 153, 535−549.

Stenzel, D., Wilsch-Brauninger, M., Wong, F.K., Heuer, H., Huttner, W.B., 2014. Integrin alphavbeta3 and thyroid hormones promote expansion of progenitors in embryonic neocortex. Development 141, 795−806.

Stocker, A.M., Chenn, A., 2015. The role of adherens junctions in the developing neocortex. Cell Adh. Migr. 9, 167−174.

Sun, T., Hevner, R.F., 2014. Growth and folding of the mammalian cerebral cortex: from molecules to malformations. Nat. Rev. Neurosci. 15, 217−232.

Taverna, E., Huttner, W.B., 2010. Neural progenitor nuclei IN Motion. Neuron 67, 906−914.

Taverna, E., Götz, M., Huttner, W.B., 2014. The cell biology of neurogenesis: toward an understanding of the development and evolution of the neocortex. Annu. Rev. Cell Dev. Biol. 30, 465−502.

Tsunekawa, Y., Britto, J.M., Takahashi, M., Polleux, F., Tan, S.S., Osumi, N., 2012. Cyclin D2 in the basal process of neural progenitors is linked to non-equivalent cell fates. EMBO J. 31, 1879−1892.

Varjosalo, M., Taipale, J., 2008. Hedgehog: functions and mechanisms. Genes Dev. 22, 2454−2472.

Vastenhouw, N.L., Schier, A.F., 2012. Bivalent histone modifications in early embryogenesis. Curr. Opin. Cell Biol. 24, 374−386.

Vessey, J.P., Amadei, G., Burns, S.E., Kiebler, M.A., Kaplan, D.R., Miller, F.D., 2012. An asymmetrically localized Staufen2-dependent RNA complex regulates maintenance of mammalian neural stem cells. Cell Stem. Cell 11, 517−528.

Voigt, P., Tee, W.W., Reinberg, D., 2013. A double take on bivalent promoters. Genes Dev. 27, 1318−1338.

Wang, H., Ge, G., Uchida, Y., Luu, B., Ahn, S., 2011a. Gli3 is required for maintenance and fate specification of cortical progenitors. J. Neurosci. 31, 6440−6448.

Wang, X., Tsai, J.W., Imai, J.H., Lian, W.N., Vallee, R.B., Shi, S.H., 2009. Asymmetric centrosome inheritance maintains neural progenitors in the neocortex. Nature 461, 947−955.

Wang, X., Tsai, J.W., Lamonica, B., Kriegstein, A.R., 2011b. A new subtype of progenitor cell in the mouse embryonic neocortex. Nat. Neurosci. 14, 555−561.

Wilkinson, G., Dennis, D., Schuurmans, C., 2013. Proneural genes in neocortical development. Neuroscience 253, 256−273.

Wilsch-Bräuninger, M., Peters, J., Paridaen, J.T.M.L., Huttner, W.B., 2012. Basolateral rather than apical primary cilia on neuroepithelial cells committed to delamination. Development 139, 95−105.

Wong, F.K., Fei, J.F., Mora-Bermudez, F., Taverna, E., Haffner, C., Fu, J., et al., 2015. Sustained Pax6 expression generates primate-like basal radial glia in developing mouse neocortex. PLoS Biol. 13, e1002217.

Wrobel, C.N., Mutch, C.A., Swaminathan, S., Taketo, M.M., Chenn, A., 2007. Persistent expression of stabilized beta-catenin delays maturation of radial glial cells into intermediate progenitors. Dev. Biol. 309, 285−297.

Xie, Y., Juschke, C., Esk, C., Hirotsune, S., Knoblich, J.A., 2013. The phosphatase PP4c controls spindle orientation to maintain proliferative symmetric divisions in the developing neocortex. Neuron 79, 254−265.

Yamamoto, H., Mandai, K., Konno, D., Maruo, T., Matsuzaki, F., Takai, Y., 2015. Impairment of radial glial scaffold-dependent neuronal migration and formation of double cortex by genetic ablation of afadin. Brain Res. 1620, 139−152.

Yingling, J., Youn, Y.H., Darling, D., Toyo-Oka, K., Pramparo, T., Hirotsune, S., et al., 2008. Neuroepithelial stem cell proliferation requires LIS1 for precise spindle orientation and symmetric division. Cell 132, 474−486.

MiRNA-DEPENDENT AND INDEPENDENT FUNCTIONS OF THE MICROPROCESSOR IN THE REGULATION OF NEURAL STEM CELL BIOLOGY

Andrea Erni, Chiara Rolando and Verdon Taylor

University of Basel, Basel, Switzerland

INTRODUCTION

Stem cells exist in different organs of the body where they are committed to generate tissue specific somatic cells. In the developing and adult mammalian brain, neural stem cells (NSCs) generate neurons and glia. NSCs build the whole brain during embryonic development and enable the adult brain to adapt to environmental changes and to contribute to certain forms of memory by means of adult neurogenesis. During mammalian neural development, the complex structures of the brain are formed from a single layer of neuroepithelial cells that line the vesicular surface of the neural tube. In mice at around embryonic day 9, the neuroepithelial cells give rise to the NSCs, which in turn differentiate into the different neuronal subtypes and glia in a tightly controlled spatiotemporal manner (Molyneaux et al., 2007). NSCs persist in restricted regions of the postnatal brain where they support neurogenesis throughout life thus allowing brain plasticity and adaptation (Ming and Song, 2012). NSC regulation involves a precise coordination of several intrinsic and extrinsic factors (Song, 2008). A crucial intrinsic mechanism to finely regulate gene expression involves microRNAs (miRNAs) (Ha and Kim, 2014). miRNAs are short noncoding RNAs that posttranscriptionally regulate gene expression by targeting complementary mRNAs, thereby inducing cleavage and degradation or by inhibiting translation (Bartel, 2004). miRNAs are involved in all cellular processes and are expressed in all cell types including NSCs, where they affect maintenance and differentiation (Shi et al., 2010; Lang and Shi, 2012; Kawahara et al., 2012; Meza-Sosa et al., 2014). Interestingly, recent findings revealed that miRNA biogenesis components have more complex functions than expected and they are not restricted to classical miRNA-directed inhibition of mRNA expression. In this chapter, we will summarize the current knowledge about the canonical and noncanonical miRNA pathways and their influence on NSC maintenance and differentiation (Fig. 6.1).

Essentials of Noncoding RNA in Neuroscience. DOI: http://dx.doi.org/10.1016/B978-0-12-804402-5.00006-6

(A)

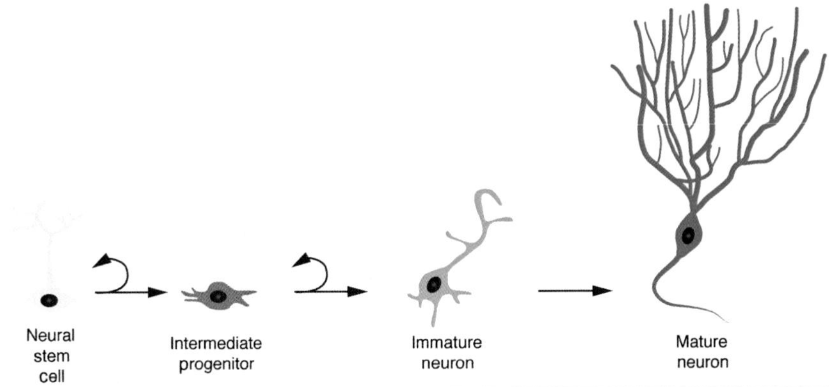

| | Neural stem cell | Intermediate progenitor | Immature neuron | Mature neuron |

(B)

	Proliferation		Differentiation		Maturation		
miRNA	miR-9 miR-92b miR-106b~25 miR-137 miR-184 miR-379-410	Laneve *et al.* 2010 Nowakowski *et al.* 2013 Brett *et al.* 2011 Szulwach *et al.* 2010 Liu *et al.* 2010 Winter 2015	let-7 b miR-9 miR-124 miR-137 miR-128 miR-184 miR-379-410	Nishino *et al.* 2008 Zhao *et al.* 2009 Lang and Shi 2012 Neo *et al.* 2014 Visvanathan 2007 Cheng *et al.* 2009 Makeyev *et al.* 2007 Sun *et al.* 2011 Silber *et al.* 2008 Zhang *et al.* 2016 Liu *et al.* 2010 Winter 2015	miR-9 miR-200 miR-132 miR-137 miR-379-410	Dealoly *et al.* 2010 Choi *et al.* 2008 Luikart *et al.* 2011 Smrt *et al.* 2010 Winter 2015	Microprocessor-dependent
mRNA	Neurogenin 2	Knuckles *et al.* 2012	NeuroD1 NeuroD6 NFIB	Knuckles *et al.* 2012 Knuckles *et al.* 2012 Rolando *et al.* 2016			
alternative splicing			REST	Raj *et al.* 2011	Disabled-1 Neurexin PSD-95	Yano *et al.* 2010 Iijima *et al.* 2011 Zheng *et al.* 2012	Suggested to be microprocessor-dependent
snoRNA			SNORD 44, 47, 74-81	Schouten *et al.* 2012			

FIGURE 6.1 Influence of the microprocessor on neurogenesis

(A) Schematic of neurogenesis. NSCs self-renew and give rise to intermediate progenitors, which proliferate, and give rise to immature neurons (INs), which differentiate into mature neurons. (B) Table of RNA species processed by the microprocessor. The upper two layers represent the microprocessor-dependent miRNAs and mRNAs that affect neurogenesis. Several miRNAs are known to influence NSC proliferation, neuronal differentiation, and maturation. Recent studies discovered microprocessor-dependent regulation of mRNAs that are involved in NSCs proliferation and differentiation. The lower two lines highlighted in grey indicate potential microprocessor targets involved in the regulation of neurogenesis through alternative splicing or snoRNAs.

EMBRYONIC AND ADULT NEUROGENESIS

The mammalian neocortex is a complex six-layered structure, responsible for processing sensory information, coordinating motor output, and mediating cognitive functions (Greig et al., 2013). The neocortex is populated by neurons, inhibitory interneurons, and excitatory projection neurons and glial cells, astrocytes, oligodendrocytes, and microglia. The interneurons are generated from ventral

NSCs and migrate tangentially to the neocortex where they connect as inhibitors in local circuits (Wichterle et al., 2001; Cobos et al., 2001; Wonders and Anderson, 2005). In contrast, the excitatory projection neurons are generated from progenitors in the dorsal telencephalon and connect to local as well as distant brain regions (Gorski et al., 2002; Molyneaux et al., 2007). The cortical projection neurons originate from the telencephalic wall or ventricular zone (VZ), which is populated by undifferentiated neuroepithelial cells and then radial glial cells. These are the NSCs that establish the VZ (Haubensak et al., 2004). NSCs have a radial morphology and span the cortex from the apical luminal to the basal pial surface. This scaffold is used by newborn neuronal progeny that migrate along the radial processes into the growing cortical plate (Rakic, 1971). Early during neurogenesis, NSCs predominantly proliferate symmetrically to expand the stem cell pool, at later stages, they switch and preferentially proliferate asymmetrically to self-renew and give-rise to an intermediate progenitor or a neuron (Noctor et al., 2001; Miyata et al., 2001). Intermediate basal progenitors are not attached to the VZ and function as transient amplifying cells, populating the subventricular zone (SVZ). They undergo a limited number of proliferative divisions and mostly divide symmetrically to produce two immature neurons (Haubensak et al., 2004; Noctor et al., 2004) for more detail, see Chapter 5, The Cell Biology of Neural Stem and Progenitor Cells and Neocortex Expansion in Development and Evolution of this book by Huttner and colleagues. These newborn neurons migrate radially to generate the cortex in an inside-out fashion. Early-born neurons populate the deeper layers (Layers VI and V), whereas later-born neurons migrate through these deeper layers to progressively populate more superficial layers, until the 6 layers of the isocortex are formed by birth (Greig et al., 2013). After neurogenesis is complete, NSCs switch fate and start to differentiate into glia for more detail, see Chapter 10, Transcriptional and Epigenetic Control of Astrogliogenesis of this book by Berninger and colleagues. Astrogliogenesis during late embryonic and early postnatal periods is followed by a wave of oligodendrogenesis (Kessaris et al., 2006; Rowitch and Kriegstein, 2010) for more detail, see Chapter 11, microRNAs in Oligodendrocyte Myelination and Repair in the Central Nervous System of this book by Lu and colleagues. On the other hand, microglia have a nonneuronal origin and develop from haemotopoietic cells.

All of these complex NSC behaviors, maintenance, differentiation, fate switching and migration during cortical development need to be tightly coordinated in order to achieve proper brain formation and function. Among the key factors regulating these processes, Notch signaling is crucial in regulating neurogenesis (Gaiano and Fishell, 2002). Notch pathway activates the expression of the basic helix−loop−helix (bHLH) transcription factors Hes1 and Hes5, which are required for NSC maintenance by inhibiting the expression of the proneural factors including Neurogenin2 (Ngn2). In NSCs, the expression of the *Hes* and *Ngn2* genes oscillate out of phase. A sustained expression of Ngn2 initiates NSC differentiation into intermediate progenitors (Shimojo et al., 2008; Imayoshi et al., 2013). Intermediate progenitors subsequently differentiate into neurons upon expression of neural bHLH determination factors including NeuroD1 or NeuroD6 (Bond et al., 2012). The expression of the transcription factors in this cascade during neocortical development needs to be strictly modulated, which includes posttranscriptional regulation. The oscillatory behavior of Ngn2 for example can be explained by direct transcript degradation. Indeed, it has been shown that Ngn2 mRNA degradation plays a pivotal role in preventing aberrant accumulation of neurogenic factors that would otherwise result in abnormal and precocious neurogenesis (Knuckles et al., 2012).

NSCs self-renew in the embryo and produce neurons and glia until they transform into parenchymal astrocytes, ependymal cells, or remain as adult stem cells in the two adult niches, the SVZ of the wall of the lateral ventricles (Furutachi et al., 2015) and the subgranular zone of the hippocampal dentate gyrus (DG) (Kriegstein and Alvarez-buylla, 2011). Under physiological conditions, adult NSCs exhibit structural and biological markers of astrocytes. Adult NSCs proliferate slowly, retain the ability to self-renew throughout life, and generate actively dividing intermediate cells that function as transit amplifying progenitors (TAPs). NSCs have distinct features in the lateral ventricular and hippocampal germinative areas (Kriegstein and Alvarez-buylla, 2011). NSCs in the lateral ventricle produce immature neuroblasts migrating in chains to the olfactory bulb where they differentiate into local interneurons (Lois and Alvarez-Buylla, 1994; Hack et al., 2005), while in the hippocampus, NSCs generate glutamatergic granule neurons (Seri et al., 2001). Moreover, SVZ but not DG NSCs also generates myelinating oligodendrocytes (Menn et al., 2006; Lugert et al., 2010; Bonaguidi et al., 2011).

In the adult neurogenic niches, fine regulation of the balance between stem cell preservation and production of differentiated progeny is achieved by interactions between extrinsic signals and intrinsic pathways based on the activity of intrinsic determinants including transcription factors (Ihrie and Álvarez-Buylla, 2011). In addition, recent work has highlighted the role of epigenetic regulators in the control of adult neurogenesis (Sun et al., 2011b). Ultimately, epigenetic regulation could represent the link between external environmental influence and internal transcriptional and posttranscriptional control of gene expression in neural progenitors of the adult brain for more detail, see Chapters 7 and 10 of this book.

ROLES OF CANONICAL miRNAs DURING NEUROGENESIS
MiRNA BIOGENESIS

Embryonic and adult neurogenesis requires fine regulation of signaling pathways and gene expression. miRNAs are abundantly expressed in the brains of embryos and adults where they influence NSC maintenance and differentiation as well as the integration of neurons into complex circuits (Bartel, 2004; Ji et al., 2013). miRNA biogenesis starts when a long primary transcript (pri-miRNA) containing the local stem-loop structure of the miRNA sequence is processed by the microprocessor, a large complex including the RNase III Drosha and the RNA binding protein (RBP) DGCR8 (Pasha in flies and worms). The microprocessor crops the pri-miRNA and produces a 60−70 nucleotide (nt) stem-loop pre-miRNA (Lee et al., 2003). This pre-miRNA is subsequently exported to the cytoplasm and further processed by the RNase III Dicer, generating a 22-nt double-stranded RNA duplex (Bohnsack et al., 2004; Lund et al., 2004; Ketting et al., 2001). The mature single-stranded lead-miRNA binds to the RNA-induced silencing complex (RISC) directing it to complementary mRNA targets and results in transcript repression either through mRNA cleavage and degradation or translational repression (Hammond et al., 2001; Ha and Kim, 2014). In mammals, it is estimated that more than 60% of all mRNAs are under miRNA control (Bartel, 2009). Single miRNAs can target several mRNAs, and one mRNA can be regulated by different miRNAs (Bartel, 2009). Therefore, it is believed that miRNAs function to fine-tune gene expression. miRNAs have been shown to influence neurogenesis by regulating the transcripts of key proteins involved in progenitor proliferation and differentiation (Lang and Shi, 2012; Kawahara et al., 2012).

MiRNAs IN NSCs

First evidences for miRNA influencing neurogenesis came from the genetic ablation of *Dicer* in the neurogenic regions of the mouse brain. Conditional *Dicer* deletion in Emx1 expressing dorsal telencephalic NSCs causes an impaired neuronal differentiation and cell death of progenitors and neurons thus resulting in a smaller cortex (De Pietri Tonelli et al., 2008). Similar results were obtained by depleting Dicer in Foxg1 expressing NSCs, which also induced loss of NSCs and a failure of neuronal differentiation (Davis et al., 2008). Furthermore, Dicer deficiency results in abnormal development of the CNS including failure of proper morphogenesis of the cerebellum, midbrain and the cortex (De Pietri Tonelli et al., 2008; Kawase-Koga et al., 2009; Huang et al., 2010; Choi et al., 2008). Recently, different miRNAs were identified to be involved in NSC maintenance and differentiation. In the following paragraphs, we summarize a few of the miRNAs discovered to be involved in neurogenesis (Fig. 6.1).

The let-7 miRNA was one of the first miRNA discovered in *Caenorhabditis elegans* and is highly conserved throughout evolution. The let-7 family of miRNAs varies only in a few nucleotides whereupon let-7a, b, c, and e are expressed in the brain and are upregulated upon neuronal differentiation (Lang and Shi, 2012). let-7 controls neurogenesis through different mechanisms (Fig. 6.1). For example, let-7b induces neurogenesis by repressing the transcripts of the orphan nuclear receptor TLX and the cell cycle regulator cyclin D1 (Zhao et al., 2010). In addition, let-7b overexpression reduces NSC proliferation and induces neuronal differentiation by directly repressing high mobility group AT-hook 2 (Hmga2) expression (Nishino et al., 2008). let-7 miRNAs can be repressed by the RBP Lin28. During embryonic stem cell (ESC) commitment to the neural lineage, Lin28 inhibits let-7a expression through specific binding to pri-let-7a, thereby inhibiting processing of the pre-miRNA to pri-miRNA by the microprocessor (Rybak et al., 2008). Early during neuronal differentiation, the RBP Musashi1 potentiates the inhibitory effect of Lin28 on let-7 miRNA by enhancing the localization of Lin28 to the nucleus (Kawahara et al., 2011).

Another well-known miRNA involved in neurogenesis and highly expressed in the embryonic and adult mouse brain is miR-9 (Fig. 6.1). First evidence for miR-9 being involved in neurogenesis came from overexpression experiments, which led to decreased proliferation and induced differentiation of NSCs (Zhao et al., 2009). miR-9 targets multiple transcripts including those encoding TLX, FoxG1, Sirtuin, RE1-Silencing Transcription Factor (REST), Meis2, and Gsh2, thereby regulating differentiation and maintenance of NSCs in a cellular and context-specific manner (Shibata et al., 2011; Delaloy et al., 2010). Several mechanisms have been proposed to explain this contradictory function of miR-9 during regulation of NSCs. For example, TLX repression by miR-9 induces neurogenesis, but TLX itself represses miR-9 expression. Therefore, miR-9 inhibits NSC proliferation and induces differentiation via a feedback loop with TLX (Zhao et al., 2009). Another feedback regulation of miR-9 has been shown via REST. REST suppresses miR-9 during NSC proliferation by occupying the miR-9-2 promoter, but REST can be removed by cAMP response element-binding (CREB) during differentiation (Laneve et al., 2010). Thus, these are examples of mechanisms that allow the same miRNA to have different functions on neurogenesis depending on the cellular context.

miRNA-124 is an abundant neural miRNA that induces embryonic and adult neural differentiation through several mechanisms (Fig. 6.1). miR-124 directly targets the transcriptional repressor Ezh2 thereby promoting neuronal differentiation and inhibiting astrocytic differentiation of embryonic mouse NSCs (Neo et al., 2014). Another target of miR-124 is the Small CTD Phosphatase 1 (SCP1). SCP1 is expressed in non neuronal tissue and at low levels by NSCs. SCP1 repression by miR-124 induces neuronal differentiation in chick embryos (Visvanathan et al., 2007). Moreover, during adult neurogenesis, miR-124 is upregulated when TAPs differentiate into neuroblasts, and its expression remains high in differentiated olfactory bulb neurons. Furthermore, miR-124 targets the SRY-box containing gene 9 (Sox9) mRNA in adult NSCs, thereby promoting neurogenesis (Cheng et al., 2009). miR-124 also influences neural-specific alternative splicing by repressing the polypyrimidine tract binding protein 1 (PTBP1), which is a repressor of neural-specific splicing. PTBP1 repression by miR-124 induces an upregulation of PTBP2, which in turn favors neural-specific splicing and induces neuronal differentiation in embryonic NSCs (Makeyev et al., 2007). Like miR-9, miR-124 can also be repressed by REST (Conaco et al., 2006).

Another miRNA involved in neurogenesis is miR-137 (Fig. 6.1). miR-137 promotes the differentiation of adult SVZ NSCs (Silber et al., 2008). It represses lysine-specific histone demethylase 1 (LSD1), thereby inhibiting NSC proliferation and promoting neuronal differentiation. miR-137 in turn is suppressed by TLX and LSD1. This regulatory loop provides a coordinated expression of LSD1 and miR-137 during the transition of NSC from proliferation to differentiation, providing a control mechanism during neurogenesis (Sun et al., 2011a). On the other hand, overexpression of miR-137 in adult DG NSCs favors their proliferation by repressing Ezh2 (Szulwach et al., 2010). Moreover, miR-137 has been shown to be involved in neuronal maturation by repressing Mib1 and inhibiting dendritic morphogenesis (Smrt et al., 2010). Thus, similar to miR-9, miR-137 has divergent functions at different stages of neurogenesis depending on its targets and regulation.

During brain development, miRNAs are expressed in a spatiotemporal manner suggesting a contribution to neurogenesis at different stages. Therefore, miRNA expression needs to be finely tuned by regulatory networks. Thus, when studying miRNAs, it is important to consider the miRNA targets, since feedback regulatory circuits are often found in miRNA function and regulation. Several miRNAs often function with a complex synergistic interplay. Hence, it will be important to have comprehensive analyses of miRNA biogenesis, targets, and regulation during neurogenesis. A recent study took advantage of the miRNA deep sequencing methods to examine the profile of NSCs (Zhao et al., 2014). Zhao et al. (2014) sequenced miRNAs from rosette NSCs (R-NSCs) derived from Rhesus monkey embryonic stem cells (rmESCs) and compared their miRNA expression profile with rmESC, early and late passage R-NSCs and neural progenitor cells. They discovered 451 of the 466 annotated rhesus miRNAs were expressed in R-NSCs, whereas the different cell types expressed specific sets of miRNAs. This approach revealed several miRNAs that are expressed by neural progenitors but not known to be involved in neurogenesis including miR-374, miR-758, and miR-889 (Zhao et al., 2014). By comparing mRNA with miRNA sequence data, Zhao et al. (2014) proposed that miRNAs negatively regulate the expression of specific signaling pathways. They correlated high expression of two Hedgehog regulatory genes Growth Arrest-Specific 1 and Patched 1 with low miRNA expression targeting their transcripts, suggesting that specific sets of miRNAs regulate Hedgehog signaling during neurogenesis (Zhao et al., 2014). It will be important to unravel the interplay between the expression of specific miRNA classes, their mRNA targets and their regulators to fully understand the impact of miRNA on neurogenesis.

MiRNA-INDEPENDENT FUNCTIONS OF THE MICROPROCESSOR

miRNAs play pivotal roles during neurogenesis. However, components of the miRNA biogenesis pathway have a direct influence on neurogenesis without acting through the 22 nt mature miRNAs

(Knuckles et al., 2012). Drosha and DGCR8 build the core components of the microprocessor, catalyzing the nuclear step of miRNA biogenesis. However, the microprocessor also regulates stability of other RNA classes including mRNAs and small nucleolar RNAs (snoRNA) (Chong et al., 2010; Knuckles et al., 2012; Macias et al., 2012; Heras et al., 2013). Evidence for noncanonical functions of the microprocessor came from comparisons of Drosha/DGCR8- and Dicer-deficient cells. Some cell-types from distinct tissues show overlapping phenotypes when Drosha/DGCR8 and Dicer knockouts are compared, thus suggesting a common pathway (Chong et al., 2008; Teta et al., 2012; Bezman et al., 2010; Berdnik et al., 2008). However, this is not always the case. Importantly, Drosha- but not Dicer-deficiency results in precocious differentiation of NSCs in vivo (Knuckles et al., 2012). On the other hand, Dicer but not Drosha depletion in the eye leads to macular degeneration (Kaneko et al., 2011; Tarallo et al., 2012). Together, these results suggest independent functions of both enzymes. The first evidence that the microprocessor can act independent of miRNAs came from genome-wide comparisons of Drosha- and Dicer-knockdown in *Drosophila* Schneider S2 cells (Kadener et al., 2009). As expected, Drosha-knockdown leads to accumulation of several miRNA precursors but surprisingly also to mRNAs that are under the control of Drosha but which were not altered in Dicer-knockdown cells. Interestingly, Evofold hairpin predictions (Pedersen et al., 2006) revealed that some of the Drosha mRNA targets have strongly conserved structural hairpins in their sequences. Therefore, it was proposed that Drosha processing could affect specific coding genes (Kadener et al., 2009).

The first microprocessor mRNA-target identified was that of DGCR8/Pasha (Han et al., 2009; Kadener et al., 2009). Drosha-depletion leads to DGCR8 mRNA accumulation indicating that Drosha inhibits DGCR8 expression in an autoregulatory mechanism to control microprocessor levels (Han et al., 2009; Kadener et al., 2009). It has been shown that DGCR8 mRNA contains hairpins in the coding sequence and the 5′-UTR, which are conserved amongst organisms and that are targeted and processed by the microprocessor (Han et al., 2009). Taken together, these data suggests that miRNA-biogenesis pathway is autoregulated by a negative feedback loop where DGCR8 levels are the limiting factor (Han et al., 2009).

Additional transcriptional analyses have underlined miRNA-independent functions of Drosha. In thymocyte progenitors many transcripts are upregulated in Drosha but not in Dicer-deficient cells (Chong et al., 2010). Furthermore, many of these regulated mRNAs contained pri-miRNA-like structures that are cleaved by the microprocessor in a miRNA-independent way (Chong et al., 2010). In line with this observation, comparison of Drosha and Dicer deletion in dendritic cell progenitors also revealed a miRNA-independent role of Drosha, where Drosha controls the development of dendritic cells by targeting the hairpin-containing mRNAs of Myl9 and Todr1 thereby repressing their expression (Johanson et al., 2015). Interestingly, transcriptome-wide mRNA cleavage patterns revealed additional Drosha-dependent mRNA substrates. Comparison of wild-type and Drosha-knockout ESCs identified a variety of mRNA targets including DGCR8 and Calcipressin-3 transcripts. These putative targets are upregulated in Drosha-knockout ESCs thus suggesting that Drosha-mediated cleavage directly affects their expression (Karginov et al., 2010).

In addition, the novel high-throughput sequencing of RNA isolated by cross-linking immunoprecipitation (HITS-CLIP) identified several novel putative RNA species that the microprocessor complex binds. HITS-CLIP of DGCR8 from HEK 293T cells identified miRNAs, several long noncoding RNAs, snoRNAs, and mRNAs. Interestingly, several of the mRNA targets of DGCR8 contain predicted RNA secondary structures that resemble pri-miRNA. Some of these mRNAs have been shown to be cleaved by the microprocessor, thereby repressing their expression (Macias et al., 2012). HITS-CLIP of DGCR8 and Drosha from human ESCs identified mostly miRNAs bound to the microprocessor and only a few mRNAs and snoRNAs. However, several of these mRNAs were shown to be directly processed by the microprocessor, destabilizing these transcripts (Jung et al., 2014; Seong et al., 2014). Thus, the microprocessor has a complex role in regulating several classes of RNAs and can have different functions depending on the cell type and context.

The noncanonical functions of the microprocessor represent a rapid and efficient way to influence gene expression. During neurogenesis, a fast regulation of the transcriptome and proteome is essential for the maintenance and differentiation of NSCs. The noncanonical functions of the microprocessor are predominant in the early regulation of embryonic neurogenesis (Knuckles et al., 2012). Loss of Drosha or DGCR8 in NSCs of the forebrain results in a loss of NSCs and precocious neuronal differentiation, whereas Dicer-deficiency does not. Drosha binds to and negatively regulates the stability of the proneural gene Ngn2 and the neural determination factor NeuroD6, thereby maintaining NSCs in concert with Notch signaling. Ngn2 and NeuroD6 contain evolutionarily conserved hairpins resembling pri-miRNA structures, which can be bound by Drosha. 3'RACE revealed Drosha-dependent cleavage of Ngn2 mRNA (Knuckles et al., 2012). Interestingly, comparison of Drosha- and Dicer-deficient NSCs did not reveal significant changes in miRNA profile, suggesting that the miRNAs are relatively stable and that microprocessor-induced phenotypes are miRNA-independent. These data indicate that the microprocessor facilitates embryonic NSC maintenance by directly blocking the accumulation of mRNAs encoding for critical differentiation factors (Fig. 6.1). Interestingly, the microprocessor is not only crucial for embryonic neurogenesis but also affects maintenance and fate restriction of adult NSCs (Rolando et al., 2016). Here, Drosha acts independently of miRNAs to regulate neuronal versus glial cell fate acquisition from adult NSCs by directly targeting mRNAs essential for gliogenic differentiation (Fig. 6.1).

Pri-miRNA transcripts are often located within the introns of genes. Moreover, the microprocessor and the spliceosome, which are responsible for pre-mRNA splicing, could interact with the same sequences as the microprocessor physically associates with the spliceosome (Gregory et al., 2004). Drosha has been shown to enhance exon splicing in vitro and in vivo. For example, the alternatively spliced eIF4H exon 5 is predicted to form a hairpin loop that resembles a Drosha substrate. The microprocessor can indeed bind and cleave exon 5 of eIF4H thus precluding its inclusion in the mRNA (Kataoka et al., 2009; Havens et al., 2014). This indicates that the microprocessor has a role in splicing that is distinct from its role in miRNA biogenesis. In addition, microprocessor-dependent alternative splicing can produce mirtrons from the spliced-out introns that mature into functional miRNAs (Okamura et al., 2007; Ladewig et al., 2012; Wen et al., 2015). Moreover, the microprocessor regulates retrotransposable elements, which are mobile DNA elements. The microprocessor can bind and cleave the retrotransposable elements LINE-1 and Alu-containing pre-miRNA like stem-loop structures, thereby acting as a defense against human genome integrity (Heras et al., 2013).

In addition to the miRNA-independent effects of the microprocessor on transcript regulation, the microprocessor could directly influence transcription by either promoting or delaying the transcriptional initiation (Gromak et al., 2013; Wagschal et al., 2012). It has been shown that binding of the microprocessor to promoter-proximal regions of human genes leads to an upregulation of transcription through Drosha binding to the RNA Polymerase II (Gromak et al., 2013). Furthermore, the microprocessor is involved in RNA Polymerase II pausing and premature transcriptional termination by opening the transcript for exonucleolytic degradation by Xrn2 and Rrp6 (Wagschal et al., 2012). However, the mechanisms underlying microprocessor-dependent activation or inhibition of mRNA transcription are still relatively unclear.

Taken together, findings over the last few years uncovered different and unexpected noncanonical roles of the microprocessor. These versatile functions are involved in a broad range of biological processes including direct transcriptional regulation and splicing. However, our understanding of the mechanisms underlying these alternate functions of the microprocessor is limited and it needs further investigation. It would be of major interest to understand how the multifaced microprocessor orchestrates brain development and homeostasis.

ALTERNATIVE DROSHA AND DGCR8 COMPLEXES

DGCR8 and Drosha-knockouts exhibit different phenotypes indicating that they may also function separately and possibly interact within other complexes (Macias et al., 2012; Luhur et al., 2014). HITS-CLIP experiments for DGCR8 revealed mRNAs, lncRNAs, snoRNAs, and retrotransposable elements as putative targets (Macias et al., 2012). The discovery that DGCR8 controls snoRNA stability in a Drosha-independent manner confirmed the existence of an alternative DGCR8 complex in association with other nucleases than Drosha. Recently, novel proteins have been found associated with DGCR8 and Drosha using mass spectrometry analysis of DGCR8 or Drosha coimmunoprecipitation assays. This study revealed that DGCR8 forms a complex with the nuclear exosome that targets and degrades mature snoRNAs (Macias et al., 2015). DGCR8 only interacts with the exonuclease when it is localized within the nucleolus, suggesting that, in the nucleoplasm, DGCR8 processes pri-miRNA in complex with Drosha whereas in the nucleolus DGCR8 induces degradation of snoRNAs by interacting with the exonuclease. Moreover, some snoRNAs can be further processed into functional miRNAs (Ender et al., 2008; Scott and Ono, 2011).

The binding between Drosha and DGCR8 can be modulated by other proteins, which can be expressed in a cell and time-specific manner. The transcriptional repressor MeCP2 is implicated in Rett syndrome and autism spectral disorders and MeCP2 binds methylated DNA and recruits histone deacetylase complex (HDAC) (Chahrour and Zoghbi, 2007; Ramocki et al., 2009; Guy et al., 2011). However, MeCP2 can regulate gene expression posttranscriptionally by suppressing miRNA processing (Cheng et al., 2014). MeCP2 competes with Drosha to bind DGCR8 resulting in a reduction of miRNA biogenesis. Deep sequencing of MeCP2-knockout hippocampal tissue revealed an upregulation of mature miRNAs (Cheng et al., 2014). In line with this, overexpression of MeCP2 in mouse cortical neurons represses miRNA maturation and inhibits dendritic and spine growth by suppressing miR-134 which targets CREB, LIMK1, and Pumilio2 that play critical roles during neurodevelopment (Cheng et al., 2014).

Drosha also interacts with different binding partners that could potentially orchestrate its cleavage substrates. One example for an alternative RBP partner for Drosha is the TAR DNA-binding protein 43 (TDP-43). It has been shown that TDP-43 can directly interact with Drosha and a TDP-43 loss of function reduces Drosha in human neuroblastoma cells in vitro. Interestingly, TDP-43 is also involved in Drosha substrate recognition (Di Carlo et al., 2013). Interaction between Drosha and TDP-43 is required for Drosha-dependent cleavage of Ngn2 mRNA but not DCGR8 mRNA. However, this mechanism still needs to be evaluated in vivo (Di Carlo et al., 2013).

These data support the hypothesis that the microprocessor exists in different complexes, thereby operating on different RNA targets. It will be of further interest to elucidate the alternate Drosha and DGCR8 complexes in NSCs and their functions on neurogenesis.

CONCLUSIONS

Neurogenesis is controlled by a hypostable transcriptome (Hsieh, 2012). miRNAs represent an efficient way to induce translational repression by blocking translation or inducing cleavage of specific transcripts. miRNA-independent functions of the microprocessor through direct binding and cleavage of specific mRNAs add an additional layer of regulation to neurogenesis (Fig. 6.1). The microprocessor affects embryonic NSCs maintenance and prevents differentiation, thus allowing normal brain development (Knuckles et al., 2012). Moreover, our data indicate that Drosha targets several mRNAs and modulates NSC differentiation in the adult brain in regions with active neurogenesis (Rolando et al., 2016). Therefore, the microprocessor is crucial in controlling mRNA levels of key genes involved in NSC maintenance and differentiation by cleaving mRNAs harboring stem loops with characteristics of pri-miRNA. Whether the Drosha-processed hairpin can be further processed by Dicer and RISC to produce a silencing miRNA-like molecule as has been shown for some mirtrons (spliced introns) and snoRNAs remains unknown. Thus, it will be of interest to elucidate whether noncanonical microprocessor cleavage leads to the production of functional miRNAs from processed mRNAs.

Recent studies have identified different and unexpected functions of the microprocessor. Our knowledge is based on experiments performed on immortalized cell lines, and it is a priority to understand how the diverse microprocessor functions are tissue and cell-type specific. Interestingly, some of the alternative microprocessor target RNA species are known to be involved during neurogenesis including alternative splicing variants and snoRNAs (Fig. 6.1). However, it is unclear if the microprocessor's regulation of neurogenesis involves these mechanisms. Alternative splice variants are particularly prominent in the nervous systems and play important roles during neurogenesis (Norris and Calarco, 2012) (Fig. 6.1). Alternative splicing allows multiple mRNA isoforms to be generated through the use of different and alternate splice sites, which is an important mechanism of gene regulation that contributes to transcriptome and proteome diversity (Nilsen and Graveley, 2010; Wang et al., 2008). It will be of interest to address whether Drosha is involved in processing neural-specific splice variants.

One of the alternatively spliced transcripts crucial for neuronal differentiation and maturation is the transcriptional repressor REST. In NSCs, REST occurs in the active isoform and it promotes proliferation, whereas in neurons it exists in an inactive isoform (Raj et al., 2011). PSD-95, an important scaffolding protein essential for synaptic maturation and plasticity of excitatory neurons, is another example of neural-specific splicing. PTBP1/2 induce increased exon skipping of PSD-95 leading to the degradation of the protein. Interestingly, PTBP1 is highly expressed in NSCs where it inhibits PSD-95 accumulation, whereas it is absent in neurons to allow functional PSD-95 expression for synapse formation. Interestingly, PTBP1 binds to Drosha in human cell lines and, therefore, Drosha could be involved in PSD-95 splice regulation (Macias et al., 2015).

Another alternatively processed RNA family regulated by DGCR8 is that of the snoRNAs. snoRNAs posttranscriptionally process RNA by methylation and are involved in various biological processes including rRNA modifications, alternative splicing, transcriptional regulation, genomic imprinting, and cell-cycle regulation (Mehler and Mattick, 2007). snoRNAs are expressed tissue and context specific and are especially abundant in the brain where they are believed to be involved in neurogenesis (Schouten et al., 2012). Growth arrest-specific 5, for example, is expressed in adult DG NSCs and harbors several snoRNAs, SNORD 44,47,74−81, which are assumed to be involved in NSC maintenance and differentiation (Smith and Steitz, 2010; Schouten et al., 2012) (Fig. 6.1). These data suggest that DGCR8 could have influence on neurogenesis via snoRNAs.

The composition of different microprocessor complexes, target recognition during the noncanonical functions, and how RNAs containing pre-miRNA-like loop structures escape microprocessor-mediated cleavage, is still open questions. The microprocessor is almost ubiquitously expressed but it is able to process distinct RNAs in specific cell-types and compartments. For example, the noncanonical function of DGCR8-dependent snoRNA production is restricted to the nucleolus, whereas pre-miRNA processing is carried out in the nucleoplasm (Macias et al., 2015). Therefore, it is possible that Drosha and DGCR8 interact with specific partners that trigger compartmental and cell-type specific functions in RNA processing. Moreover, specific proteins can compete for hairpin containing microprocessor RNA recognition sites and protect the transcripts from processing. One example is Lin28, which inhibits let-7 miRNA maturation by protecting the pre- and pri-miRNA structure from RNAse III cleavage (Thornton and Gregory, 2012; Heo et al., 2009). Moreover, Lin28 can bind to mRNAs with a GGAGA sequence within loop structures that are enriched within exons and untranslated regions of mRNAs including its own and that of other RBPs (Wilbert et al., 2012). Interestingly, a recent study supports the hypothesis that Lin28 inhibits microprocessor targets by demonstrating that Drosha directly mediates the destabilization of Lin28 mRNA targets via their Lin28-responsive elements (Qiao et al., 2012). Therefore, it is possible that a similar mechanism involving Lin28 or some of the other >1500 annotated RBPs in the genome could protect microprocessor RNA targets from cleavage, thereby stabilizing their transcripts when needed.

The identification of noncanonical functions for the microprocessor complex opens new perspectives in the field of NSC biology. Further analysis will aim to provide new insights into the complex role of the microprocessor in controlling gene expression during neurogenesis.

REFERENCES

Bartel, D.P., 2004. MicroRNAs: genomics, biogenesis, mechanism, and function. Cell 116, 281−297.

Bartel, D.P., 2009. MicroRNAs: target recognition and regulatory functions. Cell 136, 215−233.

Berdnik, D., Fan, A.P., Potter, C.J., Luo, L., 2008. MicroRNA processing pathway regulates olfactory neuron morphogenesis. Curr. Biol. 18, 1754−1759.

Bezman, N.A., Cedars, E., Steiner, D.F., Blelloch, R., Hesslein, D.G.T., Lanier, L.L., 2010. Distinct requirements of microRNAs in NK cell activation, survival, and function. J. Immunol. 185, 3835−3846.

Bohnsack, M.T., Czaplinski, K., Görlich, D., 2004. Exportin 5 is a RanGTP-dependent dsRNA-binding protein that mediates nuclear export of pre-miRNAs. RNA 10, 185−191.

Bonaguidi, M.A., Wheeler, M.A., Shapiro, J.S., Stadel, R.P., Sun, G.J., Ming, G.L., et al., 2011. In vivo clonal analysis reveals self-renewing and multipotent adult neural stem cell characteristics. Cell 145, 1142−1155.

Bond, A.M., Bhalala, O.G., Kessler, J.A., 2012. The dynamic role of bone morphogenetic proteins in neural stem cell fate and maturation. Dev. Neurobiol. 72, 1068−1084.

Chahrour, M., Zoghbi, H.Y., 2007. The story of Rett syndrome: from clinic to neurobiology. Neuron 56, 422−437.

Cheng, L.-C., Pastrana, E., Tavazoie, M., Doetsch, F., 2009. miR-124 regulates adult neurogenesis in the subventricular zone stem cell niche. Nat. Neurosci. 12, 399−408.

Cheng, T.L., Wang, Z., Liao, Q., Zhu, Y., Zhou, W.H., Xu, W., et al., 2014. MeCP2 suppresses nuclear microRNA processing and dendritic growth by regulating the DGCR8/Drosha complex. Dev. Cell 28, 547−560.

Choi, P.S., Zakhary, L., Choi, W.Y., Caron, S., Alvarez-Saavedra, E., Miska, E.A., et al., 2008. Members of the miRNA-200 family regulate olfactory neurogenesis. Neuron 57, 41−55.

Chong, M.M.W., Rasmussen, J.P., Rudensky, A.Y., Rundensky, A.Y., Littman, D.R., 2008. The RNAseIII enzyme Drosha is critical in T cells for preventing lethal inflammatory disease. J. Exp. Med. 205, 2005−2017.

Chong, M.M.W., Zhang, G., Cheloufi, S., Neubert, T.A., Hannon, G.J., Littman, D.R., 2010. Canonical and alternate functions of the microRNA biogenesis machinery. Genes Dev. 24, 1951−1960.

Cobos, I., Puelles, L., Martínez, S., 2001. The avian telencephalic subpallium originates inhibitory neurons that invade tangentially the pallium (dorsal ventricular ridge and cortical areas). Dev. Biol. 239, 30−45.

Conaco, C., Otto, S., Han, J.-J., Mandel, G., 2006. Reciprocal actions of REST and a microRNA promote neuronal identity. Proc. Natl. Acad. Sci. U. S. A. 103, 2422−2427.

Davis, T.H., Cuellar, T.L., Koch, S.M., Barker, A.J., Harfe, B.D., McManus, M.T., et al., 2008. Conditional loss of Dicer disrupts cellular and tissue morphogenesis in the cortex and hippocampus. J. Neurosci. 28, 4322−4330.

De Pietri Tonelli, D., Pulvers, J.N., Haffner, C., Murchison, E.P., Hannon, G.J., Huttner, W.B., 2008. miRNAs are essential for survival and differentiation of newborn neurons but not for expansion of neural progenitors during early neurogenesis in the mouse embryonic neocortex. Development (Cambridge, England) 135, 3911−3921.

Delaloy, C., Liu, L., Lee, J.A., Su, H., Shen, F., Yang, G.Y., et al., 2010. MicroRNA-9 coordinates proliferation and migration of human embryonic stem cell-derived neural progenitors. Cell Stem Cell 6, 323−335.

Di Carlo, V., Grossi, E., Laneve, P., Morlando, M., Dini Modigliani, S., Ballarino, M., et al., 2013. TDP-43 regulates the microprocessor complex activity during in vitro neuronal differentiation. Mol. Neurobiol. 952−963.

Ender, C., Krek, A., Friedländer, M.R., Beitzinger, M., Weinmann, L., Chen, W., et al., 2008. A human snoRNA with microRNA-like functions. Mol. Cell 32, 519−528.

Furutachi, S., Miya, H., Watanabe, T., Kawai, H., Yamasaki, N., Harada, Y., et al., 2015. Slowly dividing neural progenitors are an embryonic origin of adult neural stem cells. Nat. Neurosci. 18, 657−665.

Gaiano, N., Fishell, G., 2002. The role of notch in promoting glial and neural stem cell fates. Annu. Rev. Neurosci. 25, 471−490.

Gorski, Ja, Talley, T., Qiu, M., Puelles, L., Rubenstein, J.L.R., Jones, K.R., 2002. Cortical excitatory neurons and glia, but not GABAergic neurons, are produced in the Emx1-expressing lineage. J. Neurosci. 22, 6309−6314.

Gregory, R.I., Yan, K.-P., Amuthan, G., Chendrimada, T., Doratotaj, B., Cooch, N., et al., 2004. The microprocessor complex mediates the genesis of microRNAs. Nature 432, 235−240.

Greig, L.C., Woodworth, M.B., Galazo, M.J., Padmanabhan, H., Macklis, J.D., 2013. Molecular logic of neocortical projection neuron specification, development and diversity. Nat. Rev. Neurosci. 14, 755−769.

Gromak, N., Dienstbier, M., Macias, S., Plass, M., Eyras, E., Cáceres, J.F., et al., 2013. Drosha regulates gene expression independently of RNA cleavage function. Cell Rep. 5, 1499−1510.

Guy, J., Cheval, H., Selfridge, J., Bird, A., 2011. The role of MeCP2 in the Brain. Annu. Rev. Cell Dev. Biol. 27, 631−652.

Ha, M., Kim, V.N., 2014. Regulation of microRNA biogenesis. Nat. Rev. Mol. Cell Biol. 15, 509−524.

Hack, Ma, Saghatelyan, A., de Chevigny, A., Pfeifer, A., Ashery-Padan, R., Lledo, P.-M., et al., 2005. Neuronal fate determinants of adult olfactory bulb neurogenesis. Nat. Neurosci. 8, 865−872.

Hammond, S.M., Boettcher, S., Caudy, A.A., Kobayashi, R., Hannon, G.J., 2001. Argonaute2, a link between genetic and biochemical analyses of RNAi. Science (New York, N.Y.) 293, 1146−1150.

Han, J., Pedersen, J.S., Kwon, S.C., Belair, C.D., Kim, Y.-K., Yeom, K.-H., et al., 2009. Posttranscriptional crossregulation between Drosha and DGCR8. Cell 136, 75−84.

Haubensak, W., Attardo, A., Denk, W., Huttner, W.B., 2004. Neurons arise in the basal neuroepithelium of the early mammalian telencephalon: a major site of neurogenesis. PNAS 101, 3196−3201.

Havens, M.A., Reich, A.A., Hastings, M.L., 2014. Drosha promotes splicing of a pre-microRNA-like alternative exon. PLoS Genet. 10, e1004312.

Heo, I., Joo, C., Kim, Y.K., Ha, M., Yoon, M.J., Cho, J., et al., 2009. TUT4 in concert with Lin28 suppresses microRNA biogenesis through pre-microRNA uridylation. Cell 138, 696−708.

Heras, S.R., Macias, S., Plass, M., Fernandez, N., Cano, D., Eyras, E., et al., 2013. The microprocessor controls the activity of mammalian retrotransposons. Nat. Struct. Mol. Biol. 20, 1173−1181.

Hsieh, J., 2012. Orchestrating transcriptional control of adult neurogenesis. Genes Dev. 26, 1010−1021.

Huang, T., Liu, Y., Huang, M., Zhao, X., Cheng, L., 2010. Wnt1-cre-mediated conditional loss of Dicer results in malformation of the midbrain and cerebellum and failure of neural crest and dopaminergic differentiation in mice. J. Mol. Cell. Biol. 2, 152−163.

Ihrie, R.A., Álvarez-Buylla, A., 2011. Lake-front property: a unique germinal niche by the lateral ventricles of the adult brain. Neuron 70, 674−686.

Imayoshi, I., Isomura, A., Harima, Y., Kawaguchi, K., Kori, H., Miyachi, H., et al., 2013. Oscillatory control of factors determining multipotency and fate in mouse neural progenitors. Science 342, 1203−1208.

Ji, F., Lv, X., Jiao, J., 2013. The role of microRNAs in neural stem cells and neurogenesis. J. Genet. Genomics 40, 61−66.

Johanson, T.M., Keown, Aa, Cmero, M., Yeo, J.H.C., Kumar, A., Lew, A.M., et al., 2015. Drosha controls dendritic cell development by cleaving messenger RNAs encoding inhibitors of myelopoiesis. Nat. Immunol. 16, 1134−1141.

Jung, E., Seong, Y., Seo, J.H., Kwon, Y.S., Song, H., 2014. Cell cycle-dependent regulation of Aurora kinase B mRNA by the microprocessor complex. Biochem. Biophys. Res. Commun. 446, 241−247.

Kadener, S., Rodriguez, J., Abruzzi, K.C., Khodor, Y.L., Sugino, K.E.N., Ii, M.T.M., et al., 2009. Genome-wide identification of targets of the drosha − pasha/DGCR8 complex 537−545, RNA 15.

Kaneko, H., Dridi, S., Tarallo, V., Gelfand, B.D., Fowler, B.J., Cho, W.G., et al., 2011. DICER1 deficit induces Alu RNA toxicity in age-related macular degeneration. Nature 471, 325−330.

Karginov, F.V., Cheloufi, S., MMW, Chong, Stark, A., Smith, A.D., Hannon, G.J., 2010. Diverse endonucleolytic cleavage sites in the mammalian transcriptome depend upon microRNAs, Drosha, and additional nucleases. Mol. Cell 38, 781−788.

Kataoka, N., Fujita, M., Ohno, M., 2009. Functional association of the microprocessor complex with the spliceosome. Mol. Cell. Biol. 29, 3243−3254.

Kawahara, H., Imai, T., Okano, H., 2012. MicroRNAs in neural stem cells and neurogenesis. Front. Neurosci. 6, 1−13.

Kawahara, H., Okada, Y., Imai, T., Iwanami, A., Mischel, P.S., Okano, H., 2011. Musashi1 cooperates in abnormal cell lineage protein 28 (Lin28)-mediated let-7 family microRNA biogenesis in early neural differentiation. J. Biol. Chem. 286, 16121–16130.

Kawase-Koga, Y., Otaegi, G., Sun, T., 2009. Different timings of dicer deletion affect neurogenesis and gliogenesis in the developing mouse central nervous system. Dev. Dyn. 238, 2800–2812.

Kessaris, N., Fogarty, M., Iannarelli, P., Grist, M., Wegner, M., Richardson, W.D., 2006. Competing waves of oligodendrocytes in the forebrain and postnatal elimination of an embryonic lineage. Nat. Neurosci. 9, 173–179.

Ketting, R.F., Fischer, S.E.J., Bernstein, E., Sijen, T., Hannon, G.J., Plasterk, R.H.A., 2001. *C. elegans*. Genes Dev. 2654–2659.

Knuckles, P., Vogt, Ma, Lugert, S., Milo, M., Chong, M.M.W., Hautbergue, G.M., et al., 2012. Drosha regulates neurogenesis by controlling neurogenin 2 expression independent of microRNAs. Nat. Neurosci. 15, 962–969.

Kriegstein, A., Alvarez-buylla, A., 2011. The glial nature of embryonic and adult neural stem cells. Annu. Rev. Great Lakes 149–184.

Ladewig, E., Okamura, K., Flynt, A.S., Westholm, J.O., Lai, E.C., 2012. Discovery of hundreds of mirtrons in mouse and human small RNA data. Genome Res. 22, 1634–1645.

Laneve, P., Gioia, U., Andriotto, A., Moretti, F., Bozzoni, I., Caffarelli, E., 2010. A minicircuitry involving REST and CREB controls miR-9-2 expression during human neuronal differentiation. Nucleic Acids Res. 38, 6895–6905.

Lang, M.F., Shi, Y., 2012. Dynamic roles of microRNAs in neurogenesis. Front. Neurosci. 1–10. Available from: http://dx.doi.org/10.3389/fnins.2012.00071, eCollection 2012.

Lee, Y., Ahn, C., Han, J., Choi, H., Kim, J., Yim, J., et al., 2003. The nuclear RNase III Drosha initiates microRNA processing. Nature 425, 415–419.

Lois, C., Alvarez-Buylla, A., 1994. Long-distance neuronal migration in the adult mammalian brain. Science 264, 1145–1148.

Lugert, S., Basak, O., Knuckles, P., Haussler, U., Fabel, K., Götz, M., et al., 2010. Quiescent and active hippocampal neural stem cells with distinct morphologies respond selectively to physiological and pathological stimuli and aging. Cell Stem Cell 6, 445–456.

Luhur, A., Chawla, G., Wu, Y.-C., Li, J., Sokol, N.S., 2014. Drosha-independent DGCR8/Pasha pathway regulates neuronal morphogenesis. Proc. Natl. Acad. Sci. U. S. A. 111, 1421–1426.

Lund, E., Güttinger, S., Calado, A., Dahlberg, J.E., Kutay, U., 2004. Nuclear export of microrna precursors. Science 303, 95–98.

Macias, S., Cordiner, R.A., Gautier, P., Plass, M., Cáceres, J.F., 2015. DGCR8 acts as an adaptor for the exosome complex to degrade double-stranded structured RNAs. Mol. Cell 60, 873–885.

Macias, S., Plass, M., Stajuda, A., Michlewski, G., Eyras, E., Cáceres, J.F., 2012. DGCR8 HITS-CLIP reveals novel functions for the Microprocessor. Nat. Struct. Mol. Biol. 19, 760–766.

Makeyev, E.V., Zhang, J., Carrasco, M.A., Maniatis, T., 2007. The microRNA miR-124 promotes neuronal differentiation by triggering brain-specific alternative pre-mRNA splicing. Mol. Cell 27, 435–448.

Mehler, M.F., Mattick, J.S., 2007. Noncoding RNAs and RNA editing in brain development, functional diversification, and neurological disease. Physiol. Rev. 87, 799–823.

Menn, B., Garcia-Verdugo, J.M., Yaschine, C., Gonzalez-Perez, O., Rowitch, D., Alvarez-Buylla, A., 2006. Origin of oligodendrocytes in the subventricular zone of the adult brain. J. Neurosci. 26, 7907–7918.

Meza-Sosa, K.F., Pedraza-Alva, G., Pérez-Martínez, L., 2014. microRNAs: key triggers of neuronal cell fate. Front. Cell. Neurosci. 8, 175–275.

Ming, G.-l, Song, H., 2012. Adult neurogenesis in the mammalian brain: significant answers and significant questions. Neuron 70, 687–702.

Miyata, T., Kawaguchi, A., Okano, H., Ogawa, M., 2001. Asymmetric inheritance of radial glial fibers by cortical neurons. Neuron 31, 727−741.

Molyneaux, B.J., Arlotta, P., Menezes, J.R.L., Macklis, J.D., 2007. Neuronal subtype specification in the cerebral cortex. Nat. Rev. Neurosci. 8, 427−437.

Neo, W.H., Yap, K., Lee, S.H., Looi, L.S., Khandelia, P., Neo, S.X., et al., 2014. MicroRNA miR-124 controls the choice between neuronal and astrocyte differentiation by fine-tuning Ezh2 expression. J. Biol. Chem. 289, 20788−20801.

Nilsen, T.W., Graveley, B.R., 2010. Expansion of the eukaryotic proteome by alternative splicing. Nature 463, 457−463.

Nishino, J., Kim, I., Chada, K., Morrison, S.J., 2008. Hmga2 promotes neural stem cell self-renewal in young but not old mice by reducing p16Ink4a and p19Arf expression. Cell 135, 227−239.

Noctor, S.C., Flint, A.C., Weissman, T.A., Dammerman, R.S., Kriegstein, A.R., 2001. Neurons derived from radial glial cells establish radial units in neocortex. Nature 409, 714−720.

Noctor, S.C., Martínez-Cerdeño, V., Ivic, L., Kriegstein, A.R., 2004. Cortical neurons arise in symmetric and asymmetric division zones and migrate through specific phases. Nat. Neurosci. 7, 136−144.

Norris, A.D., Calarco, J.A., 2012. Emerging roles of alternative pre-mRNA splicing regulation in neuronal development and function. Front. Neurosci. 1−11.

Okamura, K., Hagen, J.W., Duan, H., Tyler, D.M., Lai, E.C., 2007. The mirtron pathway generates microRNA-class regulatory RNAs in Drosophila. Cell 130, 89−100.

Pedersen, J.S., Bejerano, G., Siepel, A., Rosenbloom, K., Lindblad-Toh, K., Lander, E.S., et al., 2006. Identification and classification of conserved RNA secondary structures in the human genome. PLoS Comp. Biol. 2, 0251−0262.

Qiao, C., Ma, J., Xu, J., Xie, M., Ma, W., Huang, Y., 2012. Drosha mediates destabilization of Lin28 mRNA targets. Cell Cycle 11, 3590−3598.

Raj, B., O'Hanlon, D., Vessey, J.P., Pan, Q., Ray, D., Buckley, N.J., et al., 2011. Cross-regulation between an alternative splicing activator and a transcription repressor controls neurogenesis. Mol. Cell 43, 843−850.

Rakic, P., 1971. Guidance of neurons migrating to the fetal monkey neocortex. Brain Res. 33, 471−476.

Ramocki, M.B., Peters, S.U., Tavyev, Y.J., Zhang, F., Carvalho, C.M.B., Schaaf, C.P., et al., 2009. Autism and other neuropsychiatric symptoms are prevalent in individuals with MECP2 duplication syndrome. Ann. Neurol. 66, 771−782.

Rolando, C., Erni, A., Grison, A., Beattie, R., Engler, A., Gokhale, P.J., et al., 2016. Multipotency of adult hippocampal NSCs in vivo is restricted by Drosha/NFIB. Cell Stem Cell 19, 653−662.

Rowitch, D.H., Kriegstein, A.R., 2010. Developmental genetics of vertebrate glial-cell specification. Nature 468, 214−222.

Rybak, A., Fuchs, H., Smirnova, L., Brandt, C., Pohl, E.E., Nitsch, R., et al., 2008. A feedback loop comprising lin-28 and let-7 controls pre-let-7 maturation during neural stem−cell commitment. Nat. Cell Biol. 10, 987−993.

Schouten, M., Renate Buijink, M., Lucassen, P.J., Fitzsimons, C.P., 2012. New neurons in aging brains: Molecular control by small non-coding RNAs. Front. Neurosci. 1−13. Available from: http://dx.doi.org/10.3389/fnins.2012.00025, eCollection 2012.

Scott, M.S., Ono, M., 2011. From snoRNA to miRNA: Dual function regulatory non-coding RNAs. Biochimie 93, 1987−1992.

Seong, Y., Lim, D.H., Kim, A., Seo, J.H.O., Lee, Y.S.I., Song, H., et al., 2014. Global identification of target recognition and cleavage by the Microprocessor in human ES cells. Nucleic Acids Res. 42, 12806−12821.

Seri, B., García-Verdugo, J.M., McEwen, B.S., Alvarez-Buylla, a, 2001. Astrocytes give rise to new neurons in the adult mammalian hippocampus. J. Neurosci. 21, 7153−7160.

Shi, Y., Zhao, X., Hsieh, J., Wichterle, H., Impey, S., Banerjee, S., et al., 2010. MicroRNA regulation of neural stem cells and neurogenesis. J. Neurosci. 30, 14931−14936.

Shibata, M., Nakao, H., Kiyonari, H., Abe, T., Aizawa, S., 2011. MicroRNA-9 regulates neurogenesis in mouse telencephalon by targeting multiple transcription factors. J. Neurosci. 31, 3407−3422.

Shimojo, H., Ohtsuka, T., Kageyama, R., 2008. Oscillations in notch signaling regulate maintenance of neural progenitors. Neuron 58, 52−64.

Silber, J., Lim, Da, Petritsch, C., Persson, A.I., Maunakea, A.K., Yu, M., et al., 2008. miR-124 and miR-137 inhibit proliferation of glioblastoma multiforme cells and induce differentiation of brain tumor stem cells. BMC Med. 6, 1741−7015.

Smith, C.M., Steitz, J.A., 2010. Classification of GAS5 as a multi-small-nucleolar-RNA (snoRNA) host gene and a member of the 5 J-terminal oligopyrimidine gene family reveals common features of snoRNA host genes. Mol. Cell Biol. 18, 6897−6909.

Smrt, R.D., Szulwach, K.E., Pfeiffer, R.L., Li, X., Guo, W., Pathania, M., et al., 2010. MicroRNA miR-137 regulates neuronal maturation by targeting ubiquitin ligase mind bomb-1. Stem Cells (Dayton, Ohio) 28, 1060−1070.

Song, H., 2008. Extrinsic and intrinsic mechanisms regulating adult mammalian neural stem cells and neurogenesis. Cell Res. 18, 88.

Sun, G., Ye, P., Murai, K., Lang, M.-F., Li, S., Zhang, H., et al., 2011a. miR-137 forms a regulatory loop with nuclear receptor TLX and LSD1 in neural stem cells. Nat. Commun. 2, 529−629.

Sun, J., Sun, J., Ming G-l, Song, H., 2011b. Epigenetic regulation of neurogenesis in the adult mammalian brain. Eur. J. Neurosci. 33, 1087−1093.

Szulwach, K.E., Li, X., Smrt, R.D., Li, Y., Luo, Y., Lin, L., et al., 2010. Cross talk between microRNA and epigenetic regulation in adult neurogenesis. J. Cell Biol. 189, 127−141.

Tarallo, V., Hirano, Y., Gelfand, B.D., Dridi, S., Kerur, N., Kim, Y., et al., 2012. DICER1 loss and Alu RNA induce age-related macular degeneration via the NLRP3 inflammasome and MyD88. Cell 149, 847−859.

Teta, M., Choi, Y.S., Okegbe, T., Wong, G., Tam, O.H., Chong, M.M.W., et al., 2012. Inducible deletion of epidermal Dicer and Drosha reveals multiple functions for miRNAs in postnatal skin. J. Cell Sci. 125, 1405−1416.

Thornton, J.E., Gregory, R.I., 2012. How does Lin28 let-7 control development and disease?. Trends Cell Biol. 22, 474−482.

Visvanathan, J., Lee, S., Lee, B., Lee, J.W., Lee, S.-k, 2007. The microRNA miR-124 antagonizes the anti-neural REST/SCP1 pathway during embryonic CNS development The microRNA miR-124 antagonizes the anti-neural REST/SCP1 pathway during embryonic CNS development. Genes Dev. 21, 744−749.

Wagschal, A., Rousset, E., Basavarajaiah, P., Contreras, X., Harwig, A., Laurent-chabalier, S., et al., 2012. Co-operate to induce premature termination of transcription by RNAPII. Cell 150, 1147−1157.

Wang, Y., Baskerville, S., Shenoy, A., Babiarz, J.E., Baehner, L., Blelloch, R., 2008. Embryonic stem cell-specific microRNAs regulate the G1-S transition and promote rapid proliferation. Nat. Genet. 40, 1478−1483.

Wen, J., Ladewig, E., Shenker, S., Mohammed, J., Lai, E.C., 2015. Analysis of nearly one thousand mammalian mirtrons reveals novel features of dicer substrates. PLoS Comp. Biol. 11, 1−29.

Wichterle, H., Turnbull, D.H., Nery, S., Fishell, G., Alvarez-Buylla, A., 2001. In utero fate mapping reveals distinct migratory pathways and fates of neurons born in the mammalian basal forebrain. Development (Cambridge, England) 128, 3759−3771.

Wilbert, M.L., Huelga, S.C., Kapeli, K., Stark, T.J., Liang, T.Y., Chen, S.X., et al., 2012. LIN28 binds messenger RNAs at GGAGA motifs and regulates splicing factor abundance. Mol. Cell 48, 195−206.

Wonders, C., Anderson, Sa, 2005. Cortical interneurons and their origins. The neuroscientist 11, 199−205.

Zhao, C., Sun, G., Li, S., Shi, Y., 2009. A feedback regulatory loop involving microRNA-9 and nuclear receptor TLX in neural stem cell fate determination. Nat. Struct. Mol. Biol. 16, 365−371.

Zhao, C., Sun, G., Li, S., Lang, M.-F., Yang, S., Li, W., et al., 2010. MicroRNA let-7b regulates neural stem cell proliferation and differentiation by targeting nuclear receptor TLX signaling. Proc. Natl. Acad. Sci. U. S. A. 107, 1876−1881.

Zhao, Y., Ji, S., Wang, J., Huang, J., Zheng, P., 2014. mRNA-seq and microRNA-seq whole-transcriptome analyses of rhesus monkey embryonic stem cell neural differentiation revealed the potential regulators of rosette neural stem cells. DNA Res. 21, 1−14.

FURTHER READING

Brett, J.O., Renault, Vr.M., Rafalski, V.A., Webb, A.E., Brunet, A., 2011. The microRNA cluster miR-106b∼25 regulates adult neural stem/progenitor cell proliferation and neuronal differentiation. Aging 3, 108−124.

Iijima, T., Wu, K., Witte, H., Hanno-Iijima, Y., Glatter, T., Richard, S., et al., 2011. SAM68 regulates neuronal activity-dependent alternative splicing of neurexin-1. Cell 147, 1601−1614.

Liu, C., Teng, Z.Q., Santistevan, N.J., Szulwach, K.E., Guo, W., Jin, P., et al., 2010. Epigenetic regulation of miR-184 by MBD1 governs neural stem cell proliferation and differentiation. Cell Stem Cell 6, 433−444.

Luikart, B.W., Bensen, A.L., Washburn, E.K., Perederiy, J.V., Su, K.G., Li, Y., et al., 2011. MiR-132 mediates the integration of newborn neurons into the adult dentate gyrus. PLoS ONE 6, e19077.

Nowakowski, T.J., Fotaki, V., Pollock, a, 2013. MicroRNA-92b regulates the development of intermediate cortical progenitors in embryonic mouse brain. PNAS 110, 7056−7061.

Winter, J., 2015. MicroRNAs of the miR379−410 cluster: New players in embryonic neurogenesis and regulators of neuronal function. Neurogenesis 2, e1004970-10−e1004970-17.

Yano, M., Hayakawa-Yano, Y., Mele, A., Darnell, R.B., 2010. Nova2 regulates neuronal migration through an RNA switch in disabled-1 signaling. Neuron 66, 848−858.

Zhang, W., Kim, P.J., Chen, Z., Lokman, H., Qiu, L., Zhang, K., et al., 2016. MiRNA-128 regulates the proliferation and neurogenesis of neural precursors by targeting PCM1 in the developing cortex. eLife 5, 1−22.

Zheng, S., Gray, E.E., Chawla, G., Porse, B.T., O'Dell, T.J., Black, D.L., 2012. PSD-95 is post-transcriptionally repressed during early neural development by PTBP1 and PTBP2. Nat. Neurosci. 15, 381−388.

EPIGENETIC REGULATION OF NEUROGENESIS BY MicroRNAs

Yangjian Liu, Daniel G. Abernathy and Andrew S. Yoo

Washington University in St. Louis School of Medicine, St. Louis, MO, United States

INTRODUCTION

During embryogenesis, sequential developmental cues progressively specify cells from a single zygote into a full body plan consisting of hundreds of cell types. In 1957, Conrad Waddington proposed an "epigenetic landscape" as a metaphor for cell fate specification in developing embryonic tissues (Waddington, 1957). In this model, Waddington illustrated the differentiation of pluripotent stem cells as a number of marbles rolling down a mountain littered with small hills. Each marble (stem cell), progressively separated from other marbles as it rolled down the mountain (differentiation) until it settled at the lowest point of a valley (terminal differentiation). The hills between valleys represent epigenetic barriers that prevent the transition of a differentiated cell fate into another. This model illustrates the importance of epigenetics in shaping and maintaining a pluripotent cell as it attains its final differentiated state. Although the initial implication of "epigenetics" by Waddington differs from its current use (Rajagopal and Stanger, 2016), the model remains relevant today as it demonstrates the complexity and dynamics of gene regulation required to establish and maintain cell identity during development. In addition, Waddington's "epigenetic landscape" presumes that epigenetic barriers must be overcome to enable terminally differentiated cell fate switching. The importance of this hypothesis was acknowledged in 2012 when Sir John B. Gurdon and Shinya Yamanaka were awarded the Nobel Prize in Physiology or Medicine for their discoveries demonstrating that mature differentiated cells can be reprogramed to become pluripotent.

In modern biology, epigenetics is defined as heritable changes in gene expression without alterations in the DNA sequence (Berger et al., 2009). Epigenetic changes typically modify the chromatin architecture leading to selective activation or repression of genes to establish genetic networks inherent in specific cell types. The disruption of epigenetic homeostasis that results in abnormal activation or silencing of genes has been shown to be responsible for a variety of human diseases such as Rett syndrome and cancer (Brookes and Shi, 2014). The ability of epigenetic regulation to alter individual or broad changes in gene expression enables extensive complexity which is thought to concurrently enable cellular diversity.

Widespread cellular diversity is readily apparent in the intricate mammalian nervous system where numerous, highly specialized cells, develop, integrate and cooperate with one another to construct functional networks. Numerous studies have been conducted regarding the development, function and dysfunction of the mammalian nervous system yet, much remains to be known. Mammalian neurogenesis begins as multipotent neural progenitor cells (NPCs) exit cell cycle and acquire the postmitotic neuronal fate, followed by functional maturation over time (Edlund and Jessell, 1999; Moore et al., 2009; Temple, 2001). Increasing evidence suggests the importance of epigenetic controls as mechanisms that coordinate neurogenesis (Ma et al., 2010; Yao and Jin, 2014). This regulation, in part, depends on the activity of microRNAs (miRNAs) that control genetic pathways that influence chromatin regulation. Taken together, understanding the intersection of miRNA biology and epigenetic regulation may provide further insights into the development of functional neurons and may shed light on neurological diseases where neuronal function is impaired (Ronan et al., 2013; Tang et al., 2013). In this chapter, we will discuss the role of a number of key miRNAs that feed into epigenetic circuits during neurogenesis and draw attention to their importance in neuronal development.

ESSENTIAL ROLE OF miRNAs IN NEUROGENESIS

The biogenesis of miRNAs involves consecutive processing of primary miRNA transcripts by Drosha/DGCR8 and precursors by Dicer (for more detail, see Chapter 1: Making and Maintaining MicroRNAs in Animals of this book by Pasquinelli and colleagues and see Chapter 2: Essentials of miRNA-dependent Control of mRNA Translation and Decay, MiRNA Targeting Principles and Methods for Target Identification by Rajewsky and colleagues). Systemic ablation of miRNAs via genetic deletions of Dicer or DGCR8 in vertebrates impaired brain morphogenesis, suggesting the essential role of miRNAs during neurogenesis (Barca-Mayo and De Pietri Tonelli, 2014). The contribution of miRNAs during neural development to neuronal differentiation and subtype specification has been reviewed in depth elsewhere (for more detail, see Chapter 6: MiRNA-Dependent and Independent Functions of the Microprocessor in the Regulation of Neural Stem Cell Biology of this book by Taylor and colleagues) (Lopez-Ramirez and Nicoli, 2014; Sun et al., 2013). Below, we describe expression patterns of a subset of miRNAs, miR-9, miR-9* (the 3′ strand miRNA from the miR-9 precursor), miR-124, and let-7 family that have been shown to be important at the onset of neurogenesis. We then follow with in-depth discussions about how these and other miRNAs control epigenetic events during neural development.

miR-9/9* miR-9, a highly enriched miRNA in the brain (He et al., 2012), is first expressed during mid-embryogenesis when the major brain subdivisions are specified (Krichevsky et al., 2003; Lagos-Quintana et al., 2002; Miska et al., 2004). Strikingly, miR-9 expression predominantly appears in neural progenitors within ventricular areas of the developing brain (Bonev et al., 2011; Leucht et al., 2008; Shibata et al., 2008) and the neurogenic areas of the adult brain (Kapsimali et al., 2007), although miR-9 is also detected in differentiated neurons (He et al., 2012; Shibata et al., 2011). The expression pattern of miR-9*, the other strand of 22-base pair duplex after Dicer cropping, is very similar to that of miR-9 except that the miR-9 level slightly decreases during NPCs differentiating into neurons while miR-9* level remains stable (Krichevsky et al., 2006; Staahl et al., 2013). This expression pattern suggests a prominent role of miR-9 and miR-9* in the regulation of neural progenitor states. In fact, overexpression of miR-9 in the zebrafish embryo

(Leucht et al., 2008) and mouse embryonic cortex (Zhao et al., 2009) results in a reduction of proliferating progenitors accompanied by an increase of neuronal differentiation (Leucht et al., 2008; Zhao et al., 2009). By contrast, depletion of miR-9 increases the proliferation capability of neural progenitors or mouse adult neural stem cells (Bonev et al., 2011; Coolen et al., 2012; Shibata et al., 2011; Zhao et al., 2009). These results suggest that miR-9 negatively regulates the proliferation while promoting neuronal differentiation of neural progenitors.

miR-124 miR-124 is another miRNA highly expressed in brain (He et al., 2012). In contrast to miR-9's expression in early neurogenesis, the upregulation of miR-124 occurs during neuronal differentiation (Krichevsky et al., 2003; Sempere et al., 2004). While Drosophila miR-124 regulates the proliferation of neural stem cells, miR-124 in vertebrates promotes cell cycle exit in neural stem cells and favors neuronal differentiation (Sun et al., 2013). Overexpression of miR-124 enhances neuronal differentiation of mouse embryonic stem (ES) cells (Krichevsky et al., 2006), P19 embryonic carcinoma cells (Makeyev et al., 2007), primary NPCs (Arvanitis et al., 2010), primary cortical precursor cells (Maiorano and Mallamaci, 2009), progenitor cells in the developing chick spinal cord (Visvanathan et al., 2007), and neural stem cells within the subventricular zone of the adult brain (Akerblom et al., 2012; Cheng et al., 2009). Conversely, depletion of miR-124 inhibits neuronal differentiation (Arvanitis et al., 2010) and reduces the total number of postmitotic neurons (Akerblom et al., 2012; Cheng et al., 2009). Interestingly, ectopic overexpression of miR-124 in HeLa cells decreases transcription of genes active in nonneuronal cells, shifting the transcriptome toward that of neurons (Lim et al., 2005). Taken together, these studies suggest an instructive role of miR-124 in neuronal fate acquisition.

let-7 The first miRNA identified in *C. elegans* (Lee et al., 1993; Reinhart et al., 2000), let-7 has been shown to be widely expressed in multiple organisms and a range of tissues including the brain (He et al., 2012; Pasquinelli et al., 2000). Members of the let-7 family share a common seed sequence but display selectivity on target genes in development, presumably due to the sequences at the 3′ region (Bartel, 2009; Pasquinelli, 2012). Within the let-7 family, let-7b and let-7d have been explicitly studied in the context of neural development. Let-7b and let-7d reduced progenitor self-renewal and induce neuronal differentiation by targeting the tailless homologue nuclear receptor TLX and the cell cycle regulator Cyclin D1 (Zhao et al., 2010; Zhao et al., 2013). While miR-9 can also target TLX's 3′-UTR, TLX was previously shown to repress the transcription of miR-9 primary miRNA, forming a negative regulatory loop (Zhao et al., 2009). Therefore, the neurogenic activity of let-7 at least partially functions via the miR-9 downstream effectors demonstrating synergism between two neuronal miRNAs on the same genetic pathway. Whereas let-7i also inhibits progenitor cell proliferation, it represses neuronal differentiation by targeting two proneural genes MASH1 and NGN1. The progenitor proliferation deficit by let-7 can be reversed by the overexpression of LIN28, a suppressor blocking the biogenesis of let-7 miRNAs (Cimadamore et al., 2013). Interestingly, TRIM32, an E3 ubiquitin ligase, inhibits NPC proliferation, promotes neuronal differentiation and has also been shown to modulate the activity of let-7 by interacting with AGOs through the RNA helicase DDX6 (Nicklas et al., 2015; Schwamborn et al., 2009). The studies described above demonstrate the complexity afforded by miRNAs alone during neurogenesis. Let-7 has been shown to regulate a transcriptional repressor, a cell cycle regulator, and proneural transcription factors. Given the intricate networks modulated by let-7, further studies are required to dissect the specific role of let-7 in neurogenesis (Fig. 7.1).

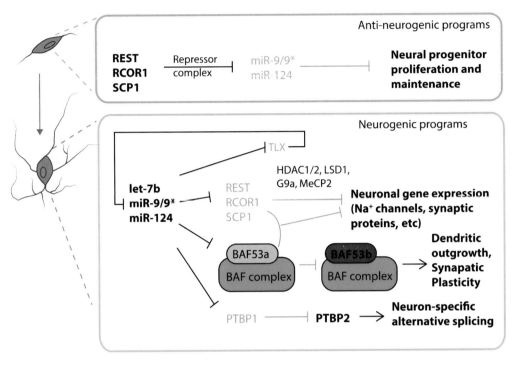

FIGURE 7.1

A schematic diagram of genetic pathways involving neuronal miRNAs, let-7b, miR-9/9*, and miR-124 at the onset of neurogenesis. In neural progenitors, the expression of key miRNAs is repressed by chromatin modifiers and transcription factors, for example, REST repressor complexes. During neurogenesis, miR-9/9* and miR-124 are activated as a result of the repression of REST complexes, and lead to the activation of neurogenic programs involving neuron-specific BAF complexes and derepression of the target genes of REST. The repression of PTBP1 by miR-124 also plays an important role to activate neuronal PTB, PTBP2 which promotes neuron-specific alternative splicing of numerous transcripts.

MiRNA-MEDIATED EPIGENETIC CONTROLS IN NEUROGENESIS

The highly compact nature of chromatin requires multiple levels of epigenetic regulation to interact with each other in order to regulate gene expression. There are broadly three modes of chromatin regulation: DNA methylation, histone modifications, and chromatin remodeling complexes (Brookes and Shi, 2014; Rajender et al., 2011). In addition, posttranscriptional regulation of gene expression via alternative splicing has also been implicated in modulating multiple developmental processes including neurogenesis (Grabowski, 2011; Raj and Blencowe, 2015; Vuong et al., 2016). In this chapter, these processes will be reviewed in relation to gene regulation, followed by discussions about miRNAs that control the components of chromatin regulation during neural development.

DNA METHYLATION

One of the prominent levels of chromatin regulation occurs at the level of DNA itself through the methylation of cytosine bases. In this chemical process, a methyl group is catalytically added to the cytosine of a CpG dinucleotide to form 5-methylcytosine (5mC) by enzymes belonging to the DNA methyltransferases (DNMTs) family (Robertson, 2002). In the genome of mammals, a high percentage of CpG dinucleotides (CpG islands) are found to localize next to the promoter regions of genes (Fatemi et al., 2005). These methyl groups within CpG islands project into the major groove of DNA altering the interaction with transcription machinery, normally leading to gene repression (Bird, 2002). DNA methylation occurs actively during neurogenesis, where existing hemimethylated and unmethylated DNA become methylated by DNMT1 and DNMT3a/b, respectively (Wang et al., 2016). Depletion of DNMT1 in neural precursor cells resulted in DNA hypomethylation and activated the JAK-STAT pathway skewing differentiation of neural progenitors towards astrocytes (Fan et al., 2005). Mice lacking functional DNMT3a in the nervous system were born normal, but displayed loss of motor neurons and impairment of neuromuscular endplate structure (Nguyen et al., 2007). These findings suggest the critical role of DNMTs in neurogenesis.

Methyl-CpG-binding proteins (MBDs) recognize methylated DNA and recruit chromatin remodeling complexes to repress gene expression (Defossez and Stancheva, 2011). Two MBD proteins, MBD1 and MeCP2, are highly expressed in the brain and have been shown to play important roles in neurogenesis (Fan and Hutnick, 2005). MBD1 repressed the expression of stem cell mitogen FGF-2 within NPCs. MBP1 loss-of-function increased FGF-2 expression, resulting in a neurogenic deficit (Li et al., 2008). MBD1 can also repress the expression of miR-195, a miRNA expressed in neurons. MiR-195, while activated, targets and represses the 3′-UTR of MBD1 forming a negative feedback loop (Liu et al., 2013). Overexpression of miR-195 in adult neural progenitors reduced neuronal differentiation, mimicking the differentiation arrest phenotype in MBD1 null mice (Li et al., 2008). Another MBD MeCP2 promotes neuronal development by inducing the expression of brain-derived neurotrophic factor (Zhou et al., 2006). The expression of MeCP2 has been shown to be repressed by miR-132. Overexpression of miR-132 decreased MeCP2 levels and impaired neuronal maturation (Klein et al., 2007; Smrt et al., 2007), whereas ablation of MeCP2 induced the expression of miR-137 (Szulwach et al., 2010) demonstrating an additional feedback loop between miRNAs and chromatin modifiers.

Methyl groups can also be removed from DNA. This process requires the 5mC residue to be converted to a nonmethylated form by a family of 10−11 translocation (TET) hydroxylases (including TET1, TET2 and TET3) (Pastor et al., 2013; Shen et al., 2014). Binding of these proteins to CpG rich regions lead to sequential oxidization of 5-mC to 5-hmC (5-hydroxymethylcytosines), 5-fC (5-formylcytosine), and 5-caC (5-carboxylcytosine) through their hydroxylase activity (Pastor et al., 2013; Shen et al., 2014). The 5-fC modification can also be converted back to unmodified cytosine through base excision repair mechanism mediated by thymine DNA glycosylase (Pastor et al., 2013; Shen et al., 2014). The TET proteins have been shown to be involved in many cellular processes including transcriptional activation and repression (TET1), tumor suppression (TET2), and cell differentiation (TET3) (Pastor et al., 2013; Shen et al., 2014). Whereas TET1 has been implicated to impair hippocampal neurogenesis in the adult mouse brain (Zhang et al., 2013), less is known about the role of TET2 or TET3 in NPCs. TET2/3 are upregulated during neuronal differentiation and knocking down TET2/3 by shRNAs resulted in abnormal progenitor cell development

(Hahn et al., 2013). Interestingly, miR-15b repressed TET3 by directly binding to its 3′-UTR, which is believed to mediate the role of miR-15b in cortical NPC cell-cycle exit and neuronal differentiation (Lv et al., 2014).

HISTONE MODIFICATIONS

Histones are repeating subunits of nucleosome, the basic unit of chromatin in which DNA wraps around the octamer of histone proteins. The N-terminal tail of histones extrudes freely outside the nucleosomal core and is subjected to extensive posttranslational modifications including acetylation, methylation, ubiquitination, phosphorylation, ribosylation, and SUMOylation (Kooistra and Helin, 2012). These modifications, especially acetylation and methylation, define the "histone code" (Jenuwein and Allis, 2001), which affects gene activation through altering the recognition and accessibility of chromatin.

Histone acetylation and deacetylation are catalyzed by histone acetyltransferases and histone deacetylases (HDACs), respectively. Acetylation neutralizes the positive charge of lysine on histone tails which loosens the chromatin by weakening the interaction of negatively charged DNA on neighboring nucleosomes. Histone acetylation is normally linked to the activation of proximal genes, while deacetylation leads to the inactivation of associated genes (Kooistra and Helin, 2012). The mammalian genome contains 18 HDACs, of which many (HDAC2, 4,5,7,9) are found in differentiating neural stem cells and mature neurons (Ajamian et al., 2003; Majdzadeh et al., 2008). Whereas HDAC1 and HDAC2 work redundantly in NPCs, deletion of both forms impaired neuronal differentiation and resulted in excessive cell death (Montgomery et al., 2009). The recruitment of HDACs to their target genes is mediated in part by TLX (Sun et al., 2007), a nuclear receptor transcription factor targeted by miR-9 and let-7 (Zhao et al., 2010; Zhao et al., 2009; Zhao et al., 2013). Collectively, these studies suggest miRNAs may regulate HDAC activity during neurogenesis via targeting an intermediate step, although this remains to be tested.

In contrast to histone acetylation, the effect of histone methylation on chromatin is much more complex and depends on the site and extent of modification. The common rule of thumb is that methylation of histone 3 at lysine 9 (H3K9) or lysine 27 (H3K27) marks silent DNA commonly found in heterochromatin and inactive genetic loci, whereas methylation of histone 3 at lysine 4 (H3K4) is a marker for active genes (Egger et al., 2004). The methylation of a histone on lysine or arginine residues is catalyzed by a variety of methyltransferases, which share a catalytic SET domain (Suppressor of variegation, Enhancer of zeste, Trithorax) (Greer and Shi, 2012). The methylation on H3 and H4 can also be removed by histone lysine demethylases (KDMs). The KDM family, including the first discovered histone demethylase lysine-specific demethylase 1 (LSD1), uses an amine oxidase mechanism to demethylate lysine with flavin adenine dinucleotide as a cofactor (Kooistra and Helin, 2012; Mosammaparast and Shi, 2010). LSD1 catalyzes the removal of methyl group(s) from mono- and dimethylated H3K4 and H3K9, whereas the other family member LSD2 acts only on mono- and dimethylated H3K4 (Kooistra and Helin, 2012; Mosammaparast and Shi, 2010).

The epigenetic chromatin modifier Polycomb Repressive Complex 2 (PRC2) has been shown to repress the expression of many genes important for mouse ES cells differentiation (Boyer et al., 2006). The expression of EZH2, a H3K27 methytranferase and the key subunit of PRC2, decreases during neurogenesis with high levels in cortical progenitor cells and low levels in mature neurons

(Pereira et al., 2010). Temporal deletion of EZH2 in the developing cortex preceding neurogenesis resulted in the acceleration of neuronal differentiation at the cost of neural progenitor depletion (Pereira et al., 2010). In a neuroblastoma cell culture model, EZH2 has been suggested to be the direct target of miR-124. Overexpression of EZH2 containing a miR-124 insensitive 3'-UTR promoted differentiation of embryonic mouse neural stem cells towards astrocytes (Neo et al., 2014). Another brain-enriched miRNA, miR-137, also has been suggested to regulate the expression of EZH2. In the adult NPCs, overexpression of miR-137 led to decrease of EZH2 level and a global reduction of H3K27 trimethylation (Szulwach et al., 2010). Brain-enriched miR-137 also negatively regulated the expression of histone LSD1 to promote neural stem cell differentiation. Interestingly, expression of miR-137 is repressed by transcription factor TLX which recruits LSD1 to the promoter of miR-137 (Sun et al., 2011). This feedback regulatory loop works in concert with the let-7 circuit to regulate NPC fate determination. These studies, again, exemplify multiple genetic circuit loops incorporating miRNAs during neural development.

CHROMATIN REMODELING COMPLEXES

Besides DNA methylation and histone modifications, the chromatin state can be regulated by the ATP- and DNA-dependent chromatin-remodeling complexes which facilitate the unpacking of DNA elements such as enhancers and promoters. Nucleosome remodeling complexes do not covalently modify DNA or histones. Instead, they slide along DNA to reposition nucleosomes and regulate the accessibility of gene loci for transcription machinery (Narlikar et al., 2002). There are mainly four families of chromatin remodeling complexes based on the homology of their ATPase domains: BRG1/BRM-associated factor (BAF), imitation switch, chromo helicase DNA binding, and inositol auxotroph 80. These chromatin remodeling complexes have specialized domains that recognize diverse histone modifications, recruit selective transcription factors, and exert their chromatin remodeling activities via an ATPase helicase domain (Clapier and Cairns, 2009). Here, we focus on BAF remodeling complexes, their functional interaction with RE1 Silencing Transcription Factor (REST also known as NRSF) repressive complex and the miRNAs known to influence their activities during neurogenesis.

BAF COMPLEX

The BAF complexes resemble the yeast SWI/SNF ATP-dependent chromatin remodelers with multimeric subunit compositions homologous to the yeast subunits (Ho and Crabtree, 2010; Ronan et al., 2013). A core ATPase subunit of either BRG1 (also called SMARCA4) or BRM (also called SMARCA2) associates with 14 other subunits encoded by 28 genes (Ronan et al., 2013; Staahl and Crabtree, 2013) to from a multisubunit complex. BRG1/BRM associates with AT rich DNA binding protein BAF200 or BAF250, zinc and PHD finger-containing protein BAF45, actin-related protein BAF53 and other subunits, which together facilitate the targeting and remodeling of chromatin (Lessard et al., 2007; Ronan et al., 2013). Interestingly, in mammals, many of these subunits are encoded by multiple homologs allowing combinatorial assemblies of the complexes in different cell types (Ronan et al., 2013). For example, BAF45a and BAF53a, highly expressed subunits in neural progenitors, are substituted by neuron-specific BAF45b/c and BAF53b subunits during neuronal differentiation to confer cell-type specific function (Lessard et al., 2007; Wu et al., 2007).

BAF complexes have essential roles in neural progenitor self-renewal and differentiation. Mice with conditional Brg1 deletion died at birth with microcephaly resulting from failure of neural progenitor self-renewal and excessive apoptosis (Lessard et al., 2007). Heterozygous mice with deletion in Baf155 fail to close the neural tube and overproduce neuronal epithelial cells (Kim et al., 2001). In neuronal progenitors, Brg1 physically interacted with the transcription factor Pax6 to promote neurogenesis. Genetic deletion of either Pax6 or Brg1 strikingly shifted the neural fate from neuronal progenitors to glia (Ninkovic et al., 2013).

Importantly, the transition from progenitors to neurons is accompanied by the switch of BAF subunits from neural-progenitor-specific BAF complex to a neuron-specific BAF complex. During this switch, the progenitor subunit BAF53a is replaced by the neuronal BAF53b subunit (Lessard et al., 2007). Interestingly, miRNAs promote subunit switching during neural development; miR-9/9* and miR-124 facilitate this process by decreasing the expression of BAF53a via targeting its 3′-UTR (Yoo et al., 2009). The upregulation of miR-9/9* and miR-124 during neurogenesis and the synergy of these three miRNAs on targeting BAF53a to allow BAF53b expression suggest the importance of subunit switching within the BAF complex during neurogenesis. However, whether the replacement of BAF45a by BAF45b and BAF45c (Lessard et al., 2007) and SS18 by CREST (Staahl et al., 2013) is mediated by miRNAs during neurogenesis remains to be tested.

REST COMPLEX

During embryonic development, reduction of REST activity is necessary for the expression of neuronal genes (Ballas et al., 2005; Chen et al., 1998; Chong et al., 1995). REST is a transcription factor exclusively expressed in nonneuronal cells. After binding to a conserved 23-base pair sequence repressor element 1 (RE1), REST recruits several corepressors including RCOR1 and Sin3A to form a large repressor complex that silences neuronal gene expression (Chen et al., 1998; Chong et al., 1995; Lunyak et al., 2002; You et al., 2001). In addition, the RCOR1 complex also contains LSD1 and G9a (Ooi and Wood, 2007). The deacetylation of H3K9 by HDAC1/2 provides a substrate for G9a-mediated methylation (Roopra et al., 2004) and stimulates LSD1 to remove di- and monomethylation marks from H3K4 (Lee et al., 2005). The recruitment of REST and its associated corepressors to RE1 sites results in the removal of active chromatin modifications and addition of repression marks on neuronal genes in nonneuronal cell types (Ooi and Wood, 2007). Further, REST complex is found to colocalize with PRC2 on the promoter regions of a subset of neuronal genes (Mozzetta et al., 2014). The recruitment of PRC2 complex requires the binding and enzymatic activity of G9a but not vice versa (McGann et al., 2014; Mozzetta et al., 2014). These studies provide compelling evidence for the crosstalk between REST complex and Polycomb Repressive Complex to regulate genes important for development and differentiation (Dietrich et al., 2012).

The reduced REST level observed during neurogenesis leads to derepression of neuronal genes (Ballas et al., 2005). Functional studies revealed that many components of REST complex are targets of miR-9. Many seed matching sites for miR-9 and miR-9* are found in the 3′-UTR of REST and its corepressor RCOR1 respectively (Packer et al., 2008). Repression by REST/RCOR1 complex inhibits the cell fate switch from neural stem cells to neurons by recruiting repressive histone modifiers to the loci of neuronal genes (Ballas and Mandel, 2005). During neuronal differentiation, increased expression of miR-9/9* decreases the level of REST and RCOR1 therefore dismantling the repressor complex, allowing the expression of neuronal genes during neuronal cell-fate specification. The induction of miR-124 has also been suggested to contribute to the dismantling of the REST complex by targeting its associated protein SCP1 which stabilizes REST by dephosphorylation on residues S861/864 (Nesti et al., 2014; Visvanathan et al., 2007). Similar to REST, SCP1 is exclusively expressed in nonneuronal cell types. After localizing to the RE1 elements on neuronal genes, REST recruited SCP1 to silence neuronal gene expression (Yeo et al., 2005). Strikingly, the expression pattern of SCP1 is mutually exclusive with miR-124 expression in the developing neural tube. Concordantly, antagonizing SCP1 expression by targeting its 3′-UTR with miR-124 induced neurogenic gene expression in P19 cells (Visvanathan et al., 2007).

Interestingly, REST interacts with BAF subunits, BAF57, BAF170 and BRG1, and this interaction appears to be required for its repressor function (Battaglioli et al., 2002). BRG1, a core subunit of BAF complex, was shown to increase the interaction of REST with local chromatin loci, presumably by repositioning nucleosomes to allow REST to form a more stable interactions with the DNA (Ooi et al., 2006). The occupancy of BRG1 at RE1 sites enhances the REST-mediated repression of the target genes, whereas inhibiting BRG1 activity reduces the binding of REST to RE1 sites, demonstrating a synergistic relationship between BAF complex and REST in controlling neuronal gene expression. The enhancement of binding REST to chromatin is dependent on the bromodomain of BRG1 and is facilitated by histone acetylation (Ooi et al., 2006). These findings demonstrate the integration of multiple levels of neuronal gene regulation downstream of miRNAs.

MESSENGER RNA SPLICING

About 90% of human genes are subjected to alternative splicing (Wang et al., 2008). The nervous system heavily relies on RNA splicing to produce protein complexity for its cellular and functional activity. The regulation of RNA splicing plays an important role in neuronal development and differentiation (Vuong et al., 2016). The splicing repressor PTBP1 has been shown to inhibit neuronal exon skipping in nonneuronal cells. PTBP1 known to regulate alternative splicing by interacting with U2AF and binding to CU-rich sequence element in the 3′ splice site of transcripts (Singh et al., 1995; Wagner and Garcia-Blanco, 2001). It is suggested that about a quarter of human genes are substrates of PTBP1 (Xue et al., 2009). These substrates include components of REST complex, EZH2, LSD1, and PHF21A, suggesting the orchestration of alternative splicing and the miRNA-REST axis on suppressing neuronal gene programs (Xue et al., 2009, 2013). Interestingly, PTBP1 also targets itself and a neuronal homologue PTBP2 (Wollerton et al., 2004; Xue et al., 2009).

During neurogenesis, the expression of nonneuronal PTBP1 is downregulated and accompanied by the expression of its neuronal homolog PTBP2 (also called nPTB), allowing the production of proteins with neuronal specific exons in the mRNA (Boutz et al., 2007; Makeyev et al., 2007; Vuong et al., 2016; Zheng et al., 2012). The upregulation of miR-124 during neurogenesis targets the 3′-UTR of PTBP1 and is thought to be responsible for its downregulation, leading to the programed switch of PTBP1 to PTBP2 (Makeyev et al., 2007). Whereas repression of PTBP1 in mouse embryonic fibroblasts induced neuron-like cells (Xue et al., 2013), overexpression of PTBP1 blocked miR-124-mediated neuronal differentiation (Makeyev et al., 2007). These results suggest the important role of miR-124-PTBP1 pathway in neurogenesis.

APPLICATION OF miRNAs TO NEURONAL REPROGRAMING

Multiple epigenetic processes involving BAF complexes, REST complexes, Polycomb Repressive Complexes, and PTBP1 play critical roles in regulating neuronal development. Among numerous miRNAs expressed in the brain, miR-9/9* and miR-124 are of particular interest as they target the components of all the above processes, thereby regulating the expression of neuronal genes, perhaps through mediating changes in the epigenetic landscape. Consistent with this notion, ectopically expressing miR-124 in HeLa cells was shown to shift the transcriptome toward that of neuronal tissue (Lim et al., 2005). Moreover, overexpression of miR-124 promoted cell cycle exit in neural progenitors and their differentiation into neurons (Makeyev et al., 2007; Visvanathan et al., 2007), collectively suggesting the instructive role of miR-124 during neurogenesis.

miR-9/9* AND miR-124 IN NEURONAL REPROGRAMING

Recently, the neurogenic extent of miR-9/9* and miR-124 was tested in human somatic cells to explore the capability of these miRNAs to convert (reprogram) the nonneuronal cell fate of human fibroblasts into neurons, thereby allowing the generation of human neurons (Yoo et al., 2011). When the primary culture of human postnatal and adult fibroblasts ectopically expressed the precursors of miR-9/9* and miR-124 on a single lentiviral vector (hence referred to as miR-9/9*-124), these cells readily exited cell cycle and eventually transitioned into postmitotic neurons that displayed neuronal morphologies and markers. It was also shown that miR-9/9*-124 synergized with neural transcription factors to robustly generate functionally mature neurons (Yoo et al., 2011). Importantly, it was also shown that when the neural transcription factors alone were expressed without miR-9/9*-124, these factors alone did not display robust reprograming activities (Yoo et al., 2011); NEUROD2, ASCL1 and MYT1L (DAM) appeared to facilitate the maturation of converted neurons mediated by miR-9/9*-124 and generate a mixed population of cortical neurons, whereas DAM factors without miR-9/9*-124 failed to produce MAP2 positive cells (Yoo et al., 2011). The converted human neurons were positive for neuronal makers such as SCN1a, Synapsin 1 and MAP2, and displayed membrane properties characteristic of mature functional neurons (generation of multiple spike action potentials and stimulation-dependent calcium influxes) (Yoo et al., 2011). This conversion is referred to as "direct" neuronal conversion (reprograming) as the cells undergoing reprograming bypass pluripotent and multipotent stem cell stages (Fig. 7.2).

The same study also dissected the synergistic relationship between miR-9/9* and miR-124. Interestingly, expressing miR-9/9* or miR-124 individually was not sufficient to generate neurons in contrast to simultaneously expressing miR-9/9* and miR-124 together (Yoo et al., 2011). These results imply an important aspect of miRNA function; (1) multiple miRNAs either target multiple components of genetic pathways converging onto the shared biological output (in this case, neurogenesis), or (2) multiple miRNAs converge on to a critical target to optimally elicit a biological process. It is highly plausible that these two scenarios coexit downstream of miRNAs to regulate genetic networks towards a biological output. Also importantly, the neuronal identity conferred by miR-9/9*-124 is stable and not reversible upon the withdrawal of exogenous miRNA expression (Yoo et al., 2011).

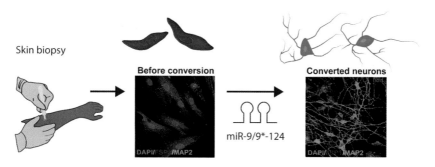

FIGURE 7.2

An example of human fibroblasts converted by miR-9/9*-124. Human postnatal fibroblasts were transduced with lentivirus expressing miR-9/9*-124 under the control of a doxycycline-inducible promoter. Photographs show starting adult human fibroblasts (left) and cells reprogramed by miR-9/9*-124 (right). Converted cells were fixed at four weeks posttransduction and immunostained with a fibroblast marker, FSP1 (red) and a neuronal marker, MAP2 (green).

Figures were adapted from Yoo A.S., Sun A.X., Li L., et al.: MicroRNA-mediated conversion of human fibroblasts to neurons, Nature 476: 228–231, 2011.

SUBTYPE CONTROL DURING miR-9/9*-124-MEDIATED NEURONAL REPROGRAMING

The above studies highlighted the potential of utilizing miRNAs to generate human neurons by directly converting skin fibroblasts for regenerative medicine. In particular, the ability to generate neurons from the individuals with neurological disorders would provide novel opportunities to model the disease in tissue culture using patient-specific cells. Towards this goal, it is imperative to control the subtype specificity of generated neurons as neurological diseases often differently affect distinctive neuronal types, a complex problem due to the highly divergent nature of neuronal subtypes in the brain (Molyneaux et al., 2007). During development, the spatiotemporal expression of transcription factors have been shown to play a critical role in the neuronal subtype specification (Molyneaux et al., 2007). Furthermore, combinatorial interactions among transcription factors have been suggested to constitute important final cell-fate determinants (Bertrand et al., 2002; Ravasi et al., 2010). Based on this notion, the neurogenic state mediated by miR-9/9*-124 was shown to be "guided" into distinct neuronal subtypes by miR-9/9*-124 combined with transcription factors. For instance, DAM factors appeared to generate human neurons that express multiple markers of cortical neurons (Hobert et al., 2010; Yoo et al., 2011), whereas the miR-9/9*-124 could be combined with transcription factors enriched in the striatum, CTIP2 (also known as BCL11B), DLX1/2 and MYT1L (CDM) to generate striatal medium spiny neurons (MSN) (Richner et al., 2015; Victor et al., 2014). These converted MSNs are highly homogeneous, with more than 70% of cells are positive for markers of MSN including DARPP-32, GABA, DLX5, and FOXP1. Single cell gene profiling of these converted MSNs resembled that of human striatal cells isolated by laser microdissection. Importantly, these analyses, showed the enrichment of MSNs markers but devoid of markers for dopaminergic, cholinergic, glutamatergic, and serotoniergic neurons (Victor et al., 2014). After transplantation into mouse brain, these converted neurons

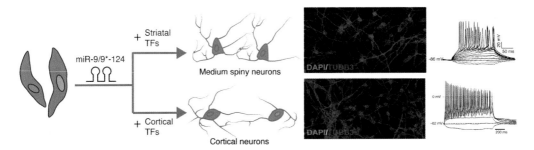

FIGURE 7.3

A schematic diagram illustrating the subtype-specific neuronal conversion of human adult fibroblasts. The miR-9/9*-124-induced neuronal state can be combined with neural transcription factors to generate specific types of human neurons. The photographs depict converted cells marked by the expression of a neuronal marker TUBB3. The traces on the right demonstrate the action potential generation of the converted cells.

Figures were adapted from Yoo A.S., Sun A.X., Li L., et al: MicroRNA-mediated conversion of human fibroblasts to neurons, Nature 476: 228–231, 2011; Victor M.B., Richner M., Hermanstyne T.O., et al.: Generation of human striatal neurons by microRNA-dependent direct conversion of fibroblasts, Neuron 84: 311–323, 2014 and Richner M., Victor M.B., Liu Y., et al.: MicroRNA-based conversion of human fibroblasts into striatal medium spiny neurons, Nat. Protoc. 10: 1543–1555, 2015.

survived and formed synapses with mouse neurons, displaying the electrophysiological characteristics of MSNs. Again, the generation of MSNs from human fibroblasts relied on the expression of miR-9/9*-124 as expressing CDM factors alone failed to produce cells with proper neuronal morphologies. These finding reinforces the neuronal fate priming activity of miR-9/9*-124, which can be guided to specific neuronal subtypes by coexpressing according subtype specific factors (Fig. 7.3).

CONCLUSION

Just as transcription factors work to establish genetic networks to specify cell types, miRNAs can target multiple genes to exert substantial activities to control cell fates. Increasing evidence from gain-of-function studies indicates that miRNAs exert cell-fate reprograming activities when introduced into an unrelated cell type, and this miRNA-mediated cell-fate transition likely occurs via the function of miRNAs in switching chromatin remodeling activities and repression of transcription factor activities that maintain the preexisting genetic programs. In this chapter, three neuronal miRNAs, miR-9/9* and miR-124, were mainly discussed in context of their activities to alter the activity of chromatin remodeling complexes and to downregulate the expression of anti-neurogenic transcription factors. Given the divergent modes of how miRNAs recognize their target genes, it is highly likely that additional miRNAs might exert similar activities to control neurogenesis.

ACKNOWLEDGMENTS

This work was supported by grants from NIH (NIH Director's New Innovator Award and Presidential Early Career Award for Scientists and Engineers), Phillip and Sima Needleman Fellowship and UM-Spinal Cord Injury and Disease Research Program.

REFERENCES

Ajamian, F., Suuronen, T., Salminen, A., et al., 2003. Upregulation of class II histone deacetylases mRNA during neural differentiation of cultured rat hippocampal progenitor cells. Neurosci. Lett. 346, 57−60.

Akerblom, M., Sachdeva, R., Barde, I., et al., 2012. MicroRNA-124 is a subventricular zone neuronal fate determinant. J. Neurosci. 32, 8879−8889.

Arvanitis, D.N., Jungas, T., Behar, A., et al., 2010. Ephrin-B1 reverse signaling controls a posttranscriptional feedback mechanism via miR-124. Mol. Cell. Biol. 30, 2508−2517.

Ballas, N., Mandel, G., 2005. The many faces of REST oversee epigenetic programming of neuronal genes. Curr. Opin. Neurobiol. 15, 500−506.

Ballas, N., Grunseich, C., Lu, D.D., et al., 2005. REST and its corepressors mediate plasticity of neuronal gene chromatin throughout neurogenesis. Cell 121, 645−657.

Barca-Mayo, O., De Pietri Tonelli, D., 2014. Convergent microRNA actions coordinate neocortical development. Cell. Mol. Life Sci. 71, 2975−2995.

Bartel, D.P., 2009. MicroRNAs: target recognition and regulatory functions. Cell 136, 215−233.

Battaglioli, E., Andres, M.E., Rose, D.W., et al., 2002. REST repression of neuronal genes requires components of the hSWI. SNF complex. J. Biol. Chem. 277, 41038−41045.

Berger, S.L., Kouzarides, T., Shiekhattar, R., et al., 2009. An operational definition of epigenetics. Genes Dev. 23, 781−783.

Bertrand, N., Castro, D.S., Guillemot, F., 2002. Proneural genes and the specification of neural cell types. Nat. Rev. Neurosci. 3, 517−530.

Bird, A., 2002. DNA methylation patterns and epigenetic memory. Genes Dev. 16, 6−21.

Bonev, B., Pisco, A., Papalopulu, N., 2011. MicroRNA-9 reveals regional diversity of neural progenitors along the anterior−posterior axis. Dev. Cell 20, 19−32.

Boutz, P.L., Stoilov, P., Li, Q., et al., 2007. A post-transcriptional regulatory switch in polypyrimidine tract-binding proteins reprograms alternative splicing in developing neurons. Genes Dev. 21, 1636−1652.

Boyer, L.A., Plath, K., Zeitlinger, J., et al., 2006. Polycomb complexes repress developmental regulators in murine embryonic stem cells. Nature 441, 349−353.

Brookes, E., Shi, Y., 2014. Diverse epigenetic mechanisms of human disease. Annu. Rev. Genet. 48, 237−268.

Chen, Z.F., Paquette, A.J., Anderson, D.J., 1998. NRSF/REST is required in vivo for repression of multiple neuronal target genes during embryogenesis. Nat. Genet. 20, 136−142.

Cheng, L.C., Pastrana, E., Tavazoie, M., et al., 2009. miR-124 regulates adult neurogenesis in the subventricular zone stem cell niche. Nat. Neurosci. 12, 399−408.

Chong, J.A., Tapia-Ramirez, J., Kim, S., et al., 1995. REST: a mammalian silencer protein that restricts sodium channel gene expression to neurons. Cell 80, 949−957.

Cimadamore, F., Amador-Arjona, A., Chen, C., et al., 2013. SOX2-LIN28/let-7 pathway regulates proliferation and neurogenesis in neural precursors. Proc. Natl. Acad. Sci. U. S. A. 110, E3017−E3026.

Clapier, C.R., Cairns, B.R., 2009. The biology of chromatin remodeling complexes. Annu. Rev. Biochem. 78, 273−304.

Coolen, M., Thieffry, D., Drivenes, O., et al., 2012. miR-9 controls the timing of neurogenesis through the direct inhibition of antagonistic factors. Dev. Cell 22, 1052−1064.

Defossez, P.A., Stancheva, I., 2011. Biological functions of methyl-CpG-binding proteins. Prog. Mol. Biol. Transl. Sci. 101, 377−398.

Dietrich, N., Lerdrup, M., Landt, E., et al., 2012. REST-mediated recruitment of polycomb repressor complexes in mammalian cells. PLoS Genet. 8, e1002494.

Edlund, T., Jessell, T.M., 1999. Progression from extrinsic to intrinsic signaling in cell fate specification: a view from the nervous system. Cell 96, 211−224.

Egger, G., Liang, G., Aparicio, A., et al., 2004. Epigenetics in human disease and prospects for epigenetic therapy. Nature 429, 457−463.

Fan, G., Hutnick, L., 2005. Methyl-CpG binding proteins in the nervous system. Cell Res. 15, 255−261.

Fan, G., Martinowich, K., Chin, M.H., et al., 2005. DNA methylation controls the timing of astrogliogenesis through regulation of JAK-STAT signaling. Development 132, 3345−3356.

Fatemi, M., Pao, M.M., Jeong, S., et al., 2005. Footprinting of mammalian promoters: use of a CpG DNA methyltransferase revealing nucleosome positions at a single molecule level. Nucleic Acids Res. 33, e176.

Grabowski, P., 2011. Alternative splicing takes shape during neuronal development. Curr. Opin. Genet. Dev. 21, 388−394.

Greer, E.L., Shi, Y., 2012. Histone methylation: a dynamic mark in health, disease and inheritance. Nat. Rev. Genet. 13, 343−357.

Hahn, M.A., Qiu, R., Wu, X., et al., 2013. Dynamics of 5-hydroxymethylcytosine and chromatin marks in Mammalian neurogenesis. Cell Rep. 3, 291−300.

He, M., Liu, Y., Wang, X., et al., 2012. Cell-type-based analysis of microRNA profiles in the mouse brain. Neuron 73, 35−48.

Ho, L., Crabtree, G.R., 2010. Chromatin remodelling during development. Nature 463, 474−484.

Hobert, O., Carrera, I., Stefanakis, N., 2010. The molecular and gene regulatory signature of a neuron. Trends Neurosci. 33, 435−445.

Jenuwein, T., Allis, C.D., 2001. Translating the histone code. Science 293, 1074−1080.

Kapsimali, M., Kloosterman, W.P., de Bruijn, E., et al., 2007. MicroRNAs show a wide diversity of expression profiles in the developing and mature central nervous system. Genome Biol. 8, R173.

Kim, J.K., Huh, S.O., Choi, H., et al., 2001. Srg3, a mouse homolog of yeast SWI3, is essential for early embryogenesis and involved in brain development. Mol. Cell. Biol. 21, 7787−7795.

Klein, M.E., Lioy, D.T., Ma, L., et al., 2007. Homeostatic regulation of MeCP2 expression by a CREB-induced microRNA. Nat. Neurosci. 10, 1513−1514.

Kooistra, S.M., Helin, K., 2012. Molecular mechanisms and potential functions of histone demethylases. Nat. Rev. Mol. Cell Biol. 13, 297−311.

Krichevsky, A.M., King, K.S., Donahue, C.P., et al., 2003. A microRNA array reveals extensive regulation of microRNAs during brain development. RNA 9, 1274−1281.

Krichevsky, A.M., Sonntag, K.C., Isacson, O., et al., 2006. Specific microRNAs modulate embryonic stem cell-derived neurogenesis. Stem Cells 24, 857−864.

Lagos-Quintana, M., Rauhut, R., Yalcin, A., et al., 2002. Identification of tissue-specific microRNAs from mouse. Curr. Biol. 12, 735−739.

Lee, M.G., Wynder, C., Cooch, N., et al., 2005. An essential role for CoREST in nucleosomal histone 3 lysine 4 demethylation. Nature 437, 432−435.

Lee, R.C., Feinbaum, R.L., Ambros, V., 1993. The *C. elegans* heterochronic gene lin-4 encodes small RNAs with antisense complementarity to lin-14. Cell 75, 843−854.

Lessard, J., Wu, J.I., Ranish, J.A., et al., 2007. An essential switch in subunit composition of a chromatin remodeling complex during neural development. Neuron 55, 201−215.

Leucht, C., Stigloher, C., Wizenmann, A., et al., 2008. MicroRNA-9 directs late organizer activity of the midbrain−hindbrain boundary. Nat. Neurosci. 11, 641−648.

Li, X., Barkho, B.Z., Luo, Y., et al., 2008. Epigenetic regulation of the stem cell mitogen Fgf-2 by Mbd1 in adult neural stem/progenitor cells. J. Biol. Chem. 283, 27644−27652.

Lim, L.P., Lau, N.C., Garrett-Engele, P., et al., 2005. Microarray analysis shows that some microRNAs downregulate large numbers of target mRNAs. Nature 433, 769−773.

Liu, C., Teng, Z.Q., McQuate, A.L., et al., 2013. An epigenetic feedback regulatory loop involving microRNA-195 and MBD1 governs neural stem cell differentiation. PLoS ONE 8, e51436.

Lopez-Ramirez, M.A., Nicoli, S., 2014. Role of miRNAs and epigenetics in neural stem cell fate determination. Epigenetics 9, 90−100.

Lunyak, V.V., Burgess, R., Prefontaine, G.G., et al., 2002. Corepressor-dependent silencing of chromosomal regions encoding neuronal genes. Science 298, 1747−1752.

Lv, X., Jiang, H., Liu, Y., et al., 2014. MicroRNA-15b promotes neurogenesis and inhibits neural progenitor proliferation by directly repressing TET3 during early neocortical development. EMBO Rep. 15, 1305−1314.

Ma, D.K., Marchetto, M.C., Guo, J.U., et al., 2010. Epigenetic choreographers of neurogenesis in the adult mammalian brain. Nat. Neurosci. 13, 1338−1344.

Maiorano, N.A., Mallamaci, A., 2009. Promotion of embryonic cortico-cerebral neuronogenesis by miR-124. Neural Dev. 4, 40.

Majdzadeh, N., Morrison, B.E., D'Mello, S.R., 2008. Class IIA HDACs in the regulation of neurodegeneration. Front. Biosci. 13, 1072−1082.

Makeyev, E.V., Zhang, J., Carrasco, M.A., et al., 2007. The MicroRNA miR-124 promotes neuronal differentiation by triggering brain-specific alternative pre-mRNA splicing. Mol. Cell 27, 435−448.

McGann, J.C., Oyer, J.A., Garg, S., et al., 2014. Polycomb- and REST-associated histone deacetylases are independent pathways toward a mature neuronal phenotype. Elife 3, e04235.

Miska, E.A., Alvarez-Saavedra, E., Townsend, M., et al., 2004. Microarray analysis of microRNA expression in the developing mammalian brain. Genome Biol. 5, R68.

Molyneaux, B.J., Arlotta, P., Menezes, J.R., et al., 2007. Neuronal subtype specification in the cerebral cortex. Nat. Rev. Neurosci. 8, 427−437.

Montgomery, R.L., Hsieh, J., Barbosa, A.C., et al., 2009. Histone deacetylases 1 and 2 control the progression of neural precursors to neurons during brain development. Proc. Natl. Acad. Sci. U. S. A. 106, 7876−7881.

Moore, A.R., Filipovic, R., Mo, Z., et al., 2009. Electrical excitability of early neurons in the human cerebral cortex during the second trimester of gestation. Cereb. Cortex 19, 1795−1805.

Mosammaparast, N., Shi, Y., 2010. Reversal of histone methylation: biochemical and molecular mechanisms of histone demethylases. Annu. Rev. Biochem. 79, 155−179.

Mozzetta, C., Pontis, J., Fritsch, L., et al., 2014. The histone H3 lysine 9 methyltransferases G9a and GLP regulate polycomb repressive complex 2-mediated gene silencing. Mol. Cell 53, 277−289.

Narlikar, G.J., Fan, H.Y., Kingston, R.E., 2002. Cooperation between complexes that regulate chromatin structure and transcription. Cell 108, 475−487.

Neo, W.H., Yap, K., Lee, S.H., et al., 2014. MicroRNA miR-124 controls the choice between neuronal and astrocyte differentiation by fine-tuning Ezh2 expression. J. Biol. Chem. 289, 20788−20801.

Nesti, E., Corson, G.M., McCleskey, M., et al., 2014. C-terminal domain small phosphatase 1 and MAP kinase reciprocally control REST stability and neuronal differentiation. Proc. Natl. Acad. Sci. U. S. A. 111, E3929−E3936.

Nguyen, S., Meletis, K., Fu, D., et al., 2007. Ablation of de novo DNA methyltransferase Dnmt3a in the nervous system leads to neuromuscular defects and shortened lifespan. Dev. Dyn. 236, 1663−1676.

Nicklas, S., Okawa, S., Hillje, A.L., et al., 2015. The RNA helicase DDX6 regulates cell-fate specification in neural stem cells via miRNAs. Nucleic Acids Res. 43, 2638−2654.

Ninkovic, J., Steiner-Mezzadri, A., Jawerka, M., et al., 2013. The BAF complex interacts with Pax6 in adult neural progenitors to establish a neurogenic cross-regulatory transcriptional network. Cell Stem Cell 13, 403−418.

Ooi, L., Wood, I.C., 2007. Chromatin crosstalk in development and disease: lessons from REST. Nat. Rev. Genet. 8, 544−554.

Ooi, L., Belyaev, N.D., Miyake, K., et al., 2006. BRG1 chromatin remodeling activity is required for efficient chromatin binding by repressor element 1-silencing transcription factor (REST) and facilitates REST-mediated repression. J. Biol. Chem. 281, 38974−38980.

Packer, A.N., Xing, Y., Harper, S.Q., et al., 2008. The bifunctional microRNA miR-9/miR-9* regulates REST and CoREST and is downregulated in Huntington's disease. J. Neurosci. 28, 14341–14346.

Pasquinelli, A.E., Reinhart, B.J., Slack, F., et al., 2000. Conservation of the sequence and temporal expression of let-7 heterochronic regulatory RNA. Nature 408, 86–89.

Pasquinelli, A.E., 2012. MicroRNAs and their targets: recognition, regulation and an emerging reciprocal relationship. Nat. Rev. Genet. 13, 271–282.

Pastor, W.A., Aravind, L., Rao, A., 2013. TETonic shift: biological roles of TET proteins in DNA demethylation and transcription. Nat. Rev. Mol. Cell. Biol. 14, 341–356.

Pereira, J.D., Sansom, S.N., Smith, J., et al., 2010. Ezh2, the histone methyltransferase of PRC2, regulates the balance between self-renewal and differentiation in the cerebral cortex. Proc. Natl. Acad. Sci. U. S. A. 107, 15957–15962.

Raj, B., Blencowe, B.J., 2015. Alternative splicing in the mammalian nervous system: recent insights into mechanisms and functional roles. Neuron 87, 14–27.

Rajagopal, J., Stanger, B.Z., 2016. Plasticity in the adult: how should the waddington diagram be applied to regenerating tissues? Dev. Cell 36, 133–137.

Rajender, S., Avery, K., Agarwal, A., 2011. Epigenetics, spermatogenesis and male infertility. Mutat. Res. 727, 62–71.

Ravasi, T., Suzuki, H., Cannistraci, C.V., et al., 2010. An atlas of combinatorial transcriptional regulation in mouse and man. Cell 140, 744–752.

Reinhart, B.J., Slack, F.J., Basson, M., et al., 2000. The 21-nucleotide let-7 RNA regulates developmental timing in *Caenorhabditis elegans*. Nature 403, 901–906.

Richner, M., Victor, M.B., Liu, Y., et al., 2015. MicroRNA-based conversion of human fibroblasts into striatal medium spiny neurons. Nat. Protoc. 10, 1543–1555.

Robertson, K.D., 2002. DNA methylation and chromatin—unraveling the tangled web. Oncogene 21, 5361–5379.

Ronan, J.L., Wu, W., Crabtree, G.R., 2013. From neural development to cognition: unexpected roles for chromatin. Nat. Rev. Genet. 14, 347–359.

Roopra, A., Qazi, R., Schoenike, B., et al., 2004. Localized domains of G9a-mediated histone methylation are required for silencing of neuronal genes. Mol. Cell 14, 727–738.

Schwamborn, J.C., Berezikov, E., Knoblich, J.A., 2009. The TRIM-NHL protein TRIM32 activates microRNAs and prevents self-renewal in mouse neural progenitors. Cell 136, 913–925.

Sempere, L.F., Freemantle, S., Pitha-Rowe, I., et al., 2004. Expression profiling of mammalian microRNAs uncovers a subset of brain-expressed microRNAs with possible roles in murine and human neuronal differentiation. Genome Biol. 5, R13.

Shen, L., Song, C.X., He, C., et al., 2014. Mechanism and function of oxidative reversal of DNA and RNA methylation. Annu. Rev. Biochem. 83, 585–614.

Shibata, M., Kurokawa, D., Nakao, H., et al., 2008. MicroRNA-9 modulates Cajal-Retzius cell differentiation by suppressing Foxg1 expression in mouse medial pallium. J. Neurosci. 28, 10415–10421.

Shibata, M., Nakao, H., Kiyonari, H., et al., 2011. MicroRNA-9 regulates neurogenesis in mouse telencephalon by targeting multiple transcription factors. J. Neurosci. 31, 3407–3422.

Singh, R., Valcarcel, J., Green, M.R., 1995. Distinct binding specificities and functions of higher eukaryotic polypyrimidine tract-binding proteins. Science 268, 1173–1176.

Smrt, R.D., Eaves-Egenes, J., Barkho, B.Z., et al., 2007. Mecp2 deficiency leads to delayed maturation and altered gene expression in hippocampal neurons. Neurobiol. Dis. 27, 77–89.

Staahl, B.T., Crabtree, G.R., 2013. Creating a neural specific chromatin landscape by npBAF and nBAF complexes. Curr. Opin. Neurobiol. 23, 903–913.

Staahl, B.T., Tang, J., Wu, W., et al., 2013. Kinetic analysis of npBAF to nBAF switching reveals exchange of SS18 with CREST and integration with neural developmental pathways. J. Neurosci. 33, 10348–10361.

Sun, A.X., Crabtree, G.R., Yoo, A.S., 2013. MicroRNAs: regulators of neuronal fate. Curr. Opin. Cell Biol. 25, 215−221.

Sun, G., Ye, P., Murai, K., et al., 2011. miR-137 forms a regulatory loop with nuclear receptor TLX and LSD1 in neural stem cells. Nat. Commun. 2, 529.

Sun, G., Yu, R.T., Evans, R.M., et al., 2007. Orphan nuclear receptor TLX recruits histone deacetylases to repress transcription and regulate neural stem cell proliferation. Proc. Natl. Acad. Sci. U. S. A. 104, 15282−15287.

Szulwach, K.E., Li, X., Smrt, R.D., et al., 2010. Cross talk between microRNA and epigenetic regulation in adult neurogenesis. J. Cell Biol. 189, 127−141.

Tang, J., Yoo, A.S., Crabtree, G.R., 2013. Reprogramming human fibroblasts to neurons by recapitulating an essential microRNA-chromatin switch. Curr. Opin. Genet. Dev. 23, 591−598.

Temple, S., 2001. The development of neural stem cells. Nature 414, 112−117.

Victor, M.B., Richner, M., Hermanstyne, T.O., et al., 2014. Generation of human striatal neurons by microRNA-dependent direct conversion of fibroblasts. Neuron 84, 311−323.

Visvanathan, J., Lee, S., Lee, B., et al., 2007. The microRNA miR-124 antagonizes the anti-neural REST/SCP1 pathway during embryonic CNS development. Genes Dev. 21, 744−749.

Vuong, C.K., Black, D.L., Zheng, S., 2016. The neurogenetics of alternative splicing. Nat. Rev. Neurosci. 17, 265−281.

Waddington, C.H., 1957. The Strategy of the Genes, London. George Alien & Unwin Ltd.

Wagner, E.J., Garcia-Blanco, M.A., 2001. Polypyrimidine tract binding protein antagonizes exon definition. Mol. Cell. Biol. 21, 3281−3288.

Wang, E.T., Sandberg, R., Luo, S., et al., 2008. Alternative isoform regulation in human tissue transcriptomes. Nature 456, 470−476.

Wang, Z., Tang, B., He, Y., et al., 2016. DNA methylation dynamics in neurogenesis. Epigenomics 8, 401−414.

Wollerton, M.C., Gooding, C., Wagner, E.J., et al., 2004. Autoregulation of polypyrimidine tract binding protein by alternative splicing leading to nonsense-mediated decay. Mol. Cell 13, 91−100.

Wu, J.I., Lessard, J., Olave, I.A., et al., 2007. Regulation of dendritic development by neuron-specific chromatin remodeling complexes. Neuron 56, 94−108.

Xue, Y., Ouyang, K., Huang, J., et al., 2013. Direct conversion of fibroblasts to neurons by reprogramming PTB-regulated microRNA circuits. Cell 152, 82−96.

Xue, Y., Zhou, Y., Wu, T., et al., 2009. Genome-wide analysis of PTB-RNA interactions reveals a strategy used by the general splicing repressor to modulate exon inclusion or skipping. Mol. Cell 36, 996−1006.

Yao, B., Jin, P., 2014. Unlocking epigenetic codes in neurogenesis. Genes Dev. 28, 1253−1271.

Yeo, M., Lee, S.K., Lee, B., et al., 2005. Small CTD phosphatases function in silencing neuronal gene expression. Science 307, 596−600.

Yoo, A.S., Staahl, B.T., Chen, L., et al., 2009. MicroRNA-mediated switching of chromatin-remodelling complexes in neural development. Nature 460, 642−646.

Yoo, A.S., Sun, A.X., Li, L., et al., 2011. MicroRNA-mediated conversion of human fibroblasts to neurons. Nature 476, 228−231.

You, A., Tong, J.K., Grozinger, C.M., et al., 2001. CoREST is an integral component of the CoREST- human histone deacetylase complex. Proc. Natl. Acad. Sci. U. S. A. 98, 1454−1458.

Zhang, R.R., Cui, Q.Y., Murai, K., et al., 2013. Tet1 regulates adult hippocampal neurogenesis and cognition. Cell Stem Cell 13, 237−245.

Zhao, C., Sun, G., Li, S., et al., 2009. A feedback regulatory loop involving microRNA-9 and nuclear receptor TLX in neural stem cell fate determination. Nat. Struct. Mol. Biol. 16, 365−371.

Zhao, C., Sun, G., Li, S., et al., 2010. MicroRNA let-7b regulates neural stem cell proliferation and differentiation by targeting nuclear receptor TLX signaling. Proc. Natl. Acad. Sci. U. S. A. 107, 1876−1881.

Zhao, C., Sun, G., Ye, P., et al., 2013. MicroRNA let-7d regulates the TLX/microRNA-9 cascade to control neural cell fate and neurogenesis. Sci. Rep. 3, 1329.

Zheng, S., Gray, E.E., Chawla, G., et al., 2012. PSD-95 is post-transcriptionally repressed during early neural development by PTBP1 and PTBP2. Nat. Neurosci. 15 (381−388), S381.

Zhou, Z., Hong, E.J., Cohen, S., et al., 2006. Brain-specific phosphorylation of MeCP2 regulates activity-dependent BDNF transcription, dendritic growth, and spine maturation. Neuron 52, 255−269.

miRNAs IN MAMMALIAN ADULT OLFACTORY NEUROGENESIS

Antoine de Chevigny, Harold Cremer and Nathalie Coré

Aix-Marseille Université, CNRS, Marseille, France

ADULT OLFACTORY NEUROGENESIS IN MAMMALS

DESCRIPTION OF ADULT OB NEUROGENESIS

The subventricular zone (SVZ) is the largest germinal region in the adult mammalian brain and harbors stem cells that generate olfactory bulb (OB) interneurons (Fig. 8.1). The neural stem cells (NSCs) in this neurogenic niche are specialized astrocytes (also called type B1 cells) that give rise to rapidly dividing transit amplifying cells (type C cells) (Alvarez-Buylla and Garcia-Verdugo, 2002). These cells then generate neuroblasts (type A cells) that migrate along the rostral migratory stream (RMS) and differentiate into granule and periglomerular interneurons in the OB (Alvarez-Buylla and Garcia-Verdugo, 2002). These adult-generated neurons present a diversity of phenotypes. Although most of them are GABAergic inhibitory interneurons, an increasing heterogeneity in their subtype identity is arising (some corelease dopamine, others express Calretinin or Calbindin, Merkle et al., 2007). Glutamatergic excitatory neurons represent a small population of adult–born neurons (Brill et al., 2009). Interestingly, it was shown that NSCs in the different SVZ walls of the lateral ventricles (LV) are regionalized, indicating that they are already programed to generate a precise neuron type depending on their location (Merkle et al., 2007). Specifically, dorsal NSCs generate preferentially periglomerular and dopaminergic neurons in the OB while lateral NSCs produce mostly GABAergic granule neurons. Although much has been elucidated about the identity and lineage of SVZ stem cells, the regulatory mechanisms underlying in vivo stem cell self-renewal and differentiation are still largely unknown.

Maintenance and differentiation of SVZ NSCs is controlled by the balance between the relative amounts of key proteins that promote or inhibit the neurogenic program. MicroRNAs have the ability to regulate this balance since they can control the relative amounts of large numbers of such key proteins simultaneously. As such, they are attractive candidates for regulating stem cell lineage progression during OB neurogenesis (Fig. 8.1).

FACTORS INVOLVED IN ADULT OB NEUROGENESIS

A growing number of studies have identified genes involved in the control of the different steps of adult OB neurogenesis, from stem cell self-renewal and proliferation to fate specification and

Essentials of Noncoding RNA in Neuroscience. DOI: http://dx.doi.org/10.1016/B978-0-12-804402-5.00008-X

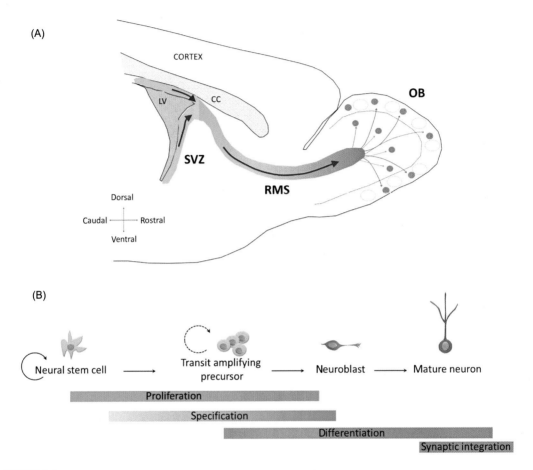

FIGURE 8.1

Olfactory bulb neurogenesis in rodents. (A) Schematic view of a sagittal section of the mouse forebrain. NSCs reside in the SVZ lining the walls of the LV. Neuroblasts generated in the lateral, dorsal, and septal aspects of the SVZ converge in the dorso-rostral part of SVZ and migrate tangentially in chains toward the OB through the RMS. In the OB, they detach from the chain and migrate radially to reach their final destination where they differentiate into granule neurons (orange circles) in the deep layer of the OB or into periglomerular neurons (brown hatched circles) at the periphery. Dotted line circles symbolize glomeruli. *CC*, corpus callosum. (B) Representation of the neurogenic sequence. NSCs divide asymmetrically to self-renew and generate rapidly dividing transit-amplifying precursors (or type C cells). After several rounds of division, transit precursor cells generate neuronal precursors (neuroblasts or type A cells) that migrate and differentiate into mature OB neurons. Finally, these newborn neurons integrate into the preexisting neuronal network by making synaptic contacts with others neurons.

functional integration of new neurons. The majority of the discovered determinants are transcription factors and signaling pathway molecules.

Because the expression levels of such factors need to be tightly controlled to ensure appropriate quantitative and qualitative OB neurogenesis, microRNAs are good candidates to finely control the system by modulating protein levels. Before describing the current knowledge about miRNA function during adult OB neurogenesis, we will focus in the next section on the role of their potential target genes in the system.

Factors regulating self-renewal and proliferation of SVZ NSCs

In the SVZ, the tailless orphan nuclear receptor (Tlx) is required for the conversion of neonatal radial glial NSCs into adult astrocyte-like NSCs. Tlx is crucial for the maintenance of adult NSC self-renewal since the specific invalidation of Tlx in adult NSCs leads to a complete loss of SVZ neurogenesis (Liu et al., 2008). Mechanistically, Tlx stimulates proliferation by activating the Wnt/β-catenin pathway through induction of Wnt7a expression in adult SVZ NSCs (Qu et al., 2010).

The involvement of Wnt-/β-catenin in adult SVZ proliferation is further demonstrated by the fact that stabilization of β-catenin promotes the proliferation of Mash1$^+$ transit-amplifying precursors (Adachi et al., 2007). However, depending on the context, β-catenin pathway can have opposite effects on proliferation and differentiation of SVZ progenitors. For instance, β-catenin signaling regulates the differentiation of progenitor cells in the presence of Hipk1, a protein which interacts with this pathway (Marinaro et al., 2012).

Notch signaling positively controls proliferation and long-term maintenance of SVZ NSCs. Inactivation of Rbpj, a downstream effector of Notch signaling, leads to loss of olfactory neurogenesis by exhausting the pool of quiescent NCS (Imayoshi et al., 2010).

Shh signaling also plays a role in SVZ NSC self-renewal and maintenance through its canonical pathway (Balordi and Fishell, 2007; Petrova et al., 2013).

Transcription factors such as Sox2 (Ferri et al., 2004), Sox9 (Scott et al., 2010), and Mash1 (Parras et al., 2004) play a dual role by both controlling proliferation in the adult SVZ and regulating neuronal differentiation.

Some transcription factors can control NSC proliferation in restricted regions of the SVZ. For example, Gsx2 promotes the activation of NSCs specifically in the dorso-lateral aspect of SVZ. By controlling proliferation of regionalized progenitors, Gsx2 specifically controls the production of olfactory interneurons subtypes (Lopez-Juarez et al., 2013).

Factors regulating neuronal commitment and subtype fate specification in olfactory neurogenesis

During olfactory neurogenesis, the induction of neuronal commitment and differentiation has been particularly associated with two transcription factors, Pax6 and Dlx2. Pax6 favors the commitment of progenitors toward the neuronal lineage in a dose-dependent manner (Dellovade et al., 1998). The Dlx1/2/5/6 family, and in particular Dlx2, is essential for the production of GABAergic interneurons in the OB (Qiu et al., 1995; Bulfone et al., 1998). Furthermore, Dlx2 expression is controlled by the epigenetic factor Mll1 in olfactory neurogenesis (Lim et al., 2009). Besides their generic neurogenic roles, Pax6 and Dlx2 are both necessary and sufficient to specify dopaminergic periglomerular neurons when their expression is maintained after neuroblasts leave the SVZ (Kohwi et al., 2005; Brill et al., 2008; De Chevigny et al., 2012a,b).

The glutamatergic olfactory lineage derives from Neurog2$^+$/Tbr1&2$^+$ progenitors but the dependency of glutamatergic fate on these transcription factors remains to be confirmed (Brill et al., 2009). In contrast, the involvement of Wnt signaling for glutamatergic fate has been demonstrated (Azim et al., 2014).

Factors regulating neuronal maturation and synaptic integration during olfactory neurogenesis

Once in the bulb, the newly born neurons have to differentiate, survive, and synaptically integrate into the neuronal network. The maturation and survival of new olfactory interneurons requires the transcription factors NeuroD1 (Boutin et al., 2010), Pax6 (Ninkovic et al., 2010), and the cAMP response element-binding (CREB) protein (Giachino et al., 2005; Herold et al., 2011). The transcription factors Sall3 (Harrison et al., 2008), SP8 (Li et al., 2011), and CoupTFI/II (Bovetti et al., 2013) are specifically required for the differentiation of subtypes of OB interneurons. Finally, the maturation of dendrites and dendritic spines necessary for functional synaptic integration requires the mRNA-binding molecule FMRP (Scotto-Lomassese et al., 2011), CREB (Herold et al., 2011), and the Agrin-signaling pathway (Burk et al., 2012) as key regulators.

In sum, OB neurogenesis is a tightly regulated process that ensures the production of the right amount of neuronal subtypes in order to allow proper functioning of the olfactory system.

This implies that the spatiotemporal expression pattern of the genes mentioned above must be strongly regulated. In this context, we will review the current knowledge about the role of microRNAs during postnatal and adult olfactory neurogenesis in mammals.

MiRNAs IN ADULT OLFACTORY NEUROGENESIS
GENERAL FUNCTION: THE DICER PHENOTYPE

Dicer is a key enzyme for the production of all mature miRNAs and therefore its knockout has been used in several genetic contexts to study the global role of miRNAs in different aspects of brain development (for detailed information on the Dicer phenotype in other brain regions see Chapters 6 and 7).

Relatively little is known concerning the impact of Dicer ablation on OB neurogenesis. First, analyzing Dicer-depleted postnatal NSCs using GFAP-Cre mice showed that miRNA deficiency results in reduced NSC proliferation in postnatal life (P15 and P40), suggesting a role of Dicer in the maintenance of long-term proliferating NSCs in adult SVZ (Nigro et al., 2012). Second, an in vitro study of adult SVZ-derived NSCs from Dicer conditional knock-out (cKO) mice revealed an increased expression of doublecortin (DCX) transcript and protein after addition of recombinant Cre protein (Cernilogar et al., 2015). Abnormally high DCX levels could be rescued in differentiating adult SVZ-derived NSCs by overexpression of miR-128, a miRNA normally expressed in adult NSCs.

In sum, the precise function of Dicer, and thus of the miRNA pathway in general, during adult OB neurogenesis has not yet been fully addressed. Nevertheless, current studies on Dicer mutants indicate a role of the miRNA pathway in the control of proliferation and differentiation of adult SVZ NSCs.

In the next section, we will review the works on specific miRNAs that regulate olfactory neurogenesis.

ROLE OF SPECIFIC miRNAs IN OB NEUROGENESIS

Figure 8.2 summarizes the developmental expression pattern of the five best studied miRNAs that are known to be involved in OB neurogenesis. In this section, we will review their specific roles as well as the roles of other miRNAs expressed in OB neurogenesis but assessed for function in other neurogenic systems.

NSC self-renewal and fate commitment in adult SVZ

Role of miR-124 in neuronal commitment during adult olfactoryneurogenesis

The first in vivo work on microRNAs function during adult olfactory neurogenesis was published in 2009 by the group of F. Doetsch and explored the role of the brain-enriched miRNA miR-124 (Cheng et al., 2009). The expression of miR-124 was characterized in the different cell populations of the adult SVZ niche. MiR-124 was undetectable in NSCs but highly enriched in neuroblasts, preferentially postmitotic, suggesting a role of miR-124 in controlling the transition between precursor amplification and neuronal commitment. A first approach using neurosphere differentiation assays showed that inhibiting miR-124 led to an increased number of dividing precursors and neuroblasts at the expense of postmitotic neurons. Conversely, miR-124 overexpression decreased cell proliferation and promoted neuronal differentiation by triggering cell-cycle exit. Similar results were observed in vivo when manipulating miR-124 expression in the SVZ of adult mice. Virus-mediated overexpression of miR-124 led to a significant reduction of astrocyte-like NSCs. In vivo knockdown was performed using chronic infusion of miR-124 antisense oligonucleotides into the LV. Experiments done in the SVZ in conditions of neuronal regeneration (using Ara-C ventricular infusions to transitorily block neurogenesis) confirmed the role of miR-124 in favoring the commitment of amplifying precursors into neurons.

To elucidate the mechanisms by which miR-124 could act in this process, in silico predicted target genes were validated for direct downregulation by miR-124 using in vitro assays. Two of these targets, the transcription factor Dlx2 and the Notch ligand Jagged-1, are known to be expressed in the SVZ and involved in the production of adult-generated GABAergic olfactory neurons and self-renewal of NSCs, respectively (Bulfone et al., 1998; Nyfeler et al., 2005). A third factor, the transcription factor Sox9, was also identified as a miR-124 target. Sox9 protein is expressed in all SVZ

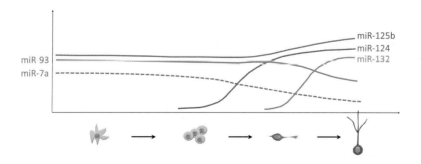

FIGURE 8.2

MicroRNAs expression during OB neurogenesis.

cell types except for neuroblasts in which it is only present at the mRNA level. Sox9 is necessary to maintain the NSC population and needs to be downregulated to allow the progression of neurogenesis. Finally, Cheng and colleagues demonstrate that the endogenous expression of the Sox9 protein is posttranscriptionally regulated in neuroblasts by miR-124.

A subsequent study investigated the role of miR-124 in postnatal olfactory neurogenesis by using a novel transgenic approach to analyze the precise expression and activity of miR-124 in the SVZ−RMS−OB system (Akerblom et al., 2012). The authors argued that studies of miRNA expression patterns in the brain had up to now mostly relied on in situ hybridization (ISH), which requires high stringency. In addition, discriminating between the pre-miRNA and mature miRNA is difficult through ISH, making it complicated to detect potential posttranscriptional regulation, which is thought to play a prominent role in controlling the activity of miRNAs (Obernosterer et al., 2006). To circumvent these issues, the authors decided to use a transgenic sensor system that can report the precise expression and activity of miR-124 in the mouse. This approach confirmed that miR-124 activity was coincident with the commitment of SVZ NSCs to neurogenesis. However, in contrast to the previous miR-124 study (Cheng et al., 2009), it indicated that miR-124 activity was already present in most transit-amplifying precursors and not restricted to neuroblasts (Fig. 8.2). This suggests that the sensor approach is more sensitive than ISH.

This scheme shows the expression levels of known miRNAs during OB neurogenesis. So far the expression of only 5 miRNAs has been described in this system. MiR-93 is expressed in all cell populations, with a slight decrease during neuronal maturation in the OB. MiR-125b is also expressed all along the differentiation sequence, but in contrast to miR-93 it is maintained in mature neurons. MiR-7a is expressed from SVZ to OB, and progressively decreases. MiR-124 expression is initiated in transit-amplifying precursors and its expression rises with neuronal differentiation. MiR-132 starts to be expressed in the rostral RMS at the OB level and is maintained in neurons of the granule cell layer.

Loss-of-function experiments were performed in the neonatal brain by injection of a lentivirus encoding a miR-124 sponge transcript into the LV. In this case, the authors introduced a GFP reporter upstream of the miR-124 sponge sequence to allow detection of transduced cells. Four weeks after miR-124 sponge lentivirus injection, very few GFP$^+$ cells were observed in the OB compared to controls, indicating that inhibition of miR-124 activity strongly impairs the production of olfactory neurons. In the SVZ, inhibition of miR-124 led to a global decrease in proliferation and to a reduced number of DCX$^+$ neuroblasts, indicating that miR-124 inhibition alters the transition from type B stem cells into rapid amplifying progenitors. Furthermore, some of the surviving transduced cells in the RMS and OB were GFAP$^+$/S100β$^+$ astrocytes, indicating that absence of miR-124 induces gliogenesis in the SVZ-OB system.

Conversely, gain-of-function experiments by overexpressing miR-124 via intraventricular lentiviral injection at P3 led to a gradual loss of newborn cells in the SVZ, RMS, and OB as a consequence of impaired NSC self-renewal in SVZ. Furthermore, the newborn neurons of the only cohort that was able to reach the OB were mispositioned with higher numbers of cells integrated in the deep part of the granule cell layer, suggesting a premature differentiation of the miR-124 overexpressing cells.

In agreement with the work of Cheng et al. (2009), this study confirmed the role of miR-124 in the control of neuronal fate commitment during olfactory neurogenesis (Table 8.1). However, Akerblom et al. (2012) proposed an earlier regulatory role of miR-124 in the transition from the

Table 8.1 List of microRNAs that Control Adult SVZ/OB Neurogenesis in Mice

miRNAs	Functions	Validated Targets	References
Let-7b[a]	Enhances differentiation of NSC	Tlx, CyclinD1	Zhao et al. (2010)
miR106−miR25[a]	NSC self-renewal, enhances differentiation	n.d.	Brett et al. (2011)
miR-124[a,b]	Favors neuronal commitment	Sox9	Cheng et al. (2009), Akerblom et al. (2012, 2014)
miR-125b[b]	Reduced neuronal integration	n.d.	Akerblom et al. (2014)
miR-132[b]	Enhances neuronal integration	n.d.	Pathania et al. (2012)
miR-137[a]	Promotes NSC proliferation	Ezh2	Szulwach et al. (2010)
miR-184[a]	Promotes NSC proliferation	Numbl	Liu et al. (2010)
miR-7a[b]	Specification of dopaminergic fate	Pax6	De Chevigny et al. (2012a,b)
miR-9[a]	Enhances differentiation of NSC	Tlx	Zhao et al. (2009)

[a]*From in vitro studies using forebrain or SVZ-derived NSCs.*
[b]*From OB neurogenesis studies. n.d., not determined.*

NSC to the transit amplifying precursor stage. Also, only the results from this second study suggest that miR-124 might regulate neuronal versus glial fate decisions in SVZ.

The function of miR-124 has also been addressed under pathological conditions, namely after stroke induction in adult brains. A miRNA expression profiling was performed in cultured rat NSCs subjected to focal cerebral ischemia (7 days after middle cerebral artery occlusion). MiR-124 expression was downregulated in ischemic SVZ NSCs compared to control NSCs (Liu et al., 2011). This was concomitant with increased progenitor cell proliferation. The over-proliferation phenotype could be rescued in vitro by reintroducing miR-124 into ischemia-derived progenitors using nanoparticle-mediated delivery of small RNA duplexes. Mechanistically, in vitro assays indicated that miR-124 targets the Jag1-Notch signaling pathway, known to be involved in stroke-induced neurogenesis (Wang et al., 2009a,b) and Dlx2, a key regulator of neuronal differentiation during SVZ neurogenesis (Brill et al., 2008).

In sum, a series of studies has demonstrated that miR-124 plays a pivotal role in inducing olfactory neurogenesis. Mechanistically, miR-124 acts, at least in part, by posttranscriptional inhibition of Sox9 in SVZ progenitors.

MiRNAs controlling adult NSC maintenance versus differentiation in the forebrain

Although very few studies have tested the implication of miRNAs during adult olfactory neurogenesis specifically, some miRNAs have been identified as key modulators of the balance between proliferation and differentiation of forebrain NSCs using in vitro or in vivo systems (see Table 8.1).

MiR-184 is a highly conserved single-copy miRNA expressed in the two neurogenic regions of the adult brain, the SVZ and the dentate gyrus of the hippocampus (Liu et al., 2010). Analyzing adult neural stem and progenitor cells isolated from the entire forebrain, which include SVZ NSCs, Liu and colleagues found that miR-184 expression in these cells is directly regulated by the Methyl-CpG binding protein 1 (MBD1). Overexpression of miR-184 promoted adult NSC proliferation and strongly reduced neuronal differentiation. Conversely, downregulation of endogenous miR-184 favored NSC differentiation (Liu et al., 2010). Numblike (Numbl) was identified as a miR-184 target in cultured adult forebrain NSCs. Numbl expression reversed the increased proliferation and neuronal differentiation deficits induced by miR-184 overexpression in NSCs. Thus, miR-184 downregulates Numbl expression in proliferating adult NSCs and inhibits cell differentiation.

MiR-137 is also a highly conserved single-copy microRNA that is expressed in adult forebrain NSCs and progenitors. MeCP2, a DNA methyl-CpG-binding protein just like MBD1, together with Sox2, maintains low level expression of miR-137 in proliferating progenitors through epigenetic regulation of the miR-137 locus (Szulwach et al., 2010). In vitro studies showed that the dosage of miR-137 is crucial for the proliferation/differentiation balance of NSCs isolated from the adult forebrain. Indeed, overexpression of miR-137 enhanced the proliferative capacity of NSC at the expense of differentiation. Inhibiting endogenous miR-137 activity had the opposite effect.

Interestingly, miR-184 and miR-137 are regulated at the transcriptional level by two members of the MBD protein family of epigenetic regulators: MBD1 and MeCP2, respectively. MiR-137, in turn, represses the expression of Ezh2, resulting in a global decrease in histone H3 trimethyl-lysine 27, uncovering an important cross talk between miRNAs and epigenetic regulation to modulate adult neurogenesis.

The miR-106b~25 cluster belongs to the miR-17 family and contains three microRNAs, miR-106b, miR-93, and miR-25, which are strongly expressed in the adult brain. Their expression and function have been addressed in primary cultures of NSCs isolated from the adult forebrain (Brett et al., 2011). All three miRNAs are expressed in both self-renewing NSCs and their differentiating progeny. As a whole, overexpression of the miR-106b~25 cluster promotes NSCs proliferation in self-renewal conditions but it also enhances neurogenesis in differentiation conditions. On the other hand, individual knockdowns indicated that only miR-25 had an effect on NSC proliferation. In contrast to miR-106b and miR-93 that share similar seed sequences, miR-25 has a unique seed sequence and therefore controls different mRNA targets.

MiR-9 is specifically expressed in neurogenic regions of the mammalian embryonic and adult brain, from two different loci, miR-9-1 and miR-9-2. In vivo studies have addressed the function of miR-9 during the development of the mouse cortex and showed that miR-9 prevents NSC proliferation and stimulates neural differentiation (Zhao et al., 2009). Similarly, miR-9 negatively regulated adult NSCs proliferation and accelerated neuronal differentiation in vitro, suggesting a role of miR-9 during adult neurogenesis. Mechanistically, mir-9 targeted the 3′-UTR of Tlx upon differentiation. Conversely, chromatin immunoprecipitation (ChIP) assays revealed that TLX protein binds to the 3′ genomic sequence of the miR-9-1 locus and is capable of repressing miR-9 expression. This implies that mir-9 and TLX participate in a negative feedback regulatory loop that controls the balance between NSC proliferation and differentiation during neurogenesis.

Using similar approaches, the same group showed that another microRNA, let-7b, behaves comparably to miR-9, preventing adult NSCs proliferation and enhancing neural differentiation by targeting the Tlx 3′-UTR (Zhao et al., 2010). In addition to TLX, Cyclin D1 was identified and validated as a let-7b target in the neurogenic process, at least in vitro.

Since TLX is required to maintain the renewal of astrocyte like-NSC in the adult SVZ and is critical for the production of neurons (Liu et al., 2008), it is likely that miR-9 and let-7b play a role in adult olfactory neurogenesis by modulating TLX expression.

MiRNAs involved in fate specification during adult OB neurogenesis

In the postnatal and adult SVZ of mammals, the NSCs surrounding the LV are regionalized and specified to generate different subtypes of olfactory interneurons depending on their localization (Merkle et al., 2007). The molecular mechanisms underlying this functional regionalization involve the restricted expression and activity of transcription factors and signaling pathways along the rostro-caudal and antero-posterior axes of the SVZ.

Although the mechanisms of this regulation remain poorly understood, one study based on Pax6 regulation suggests that miRNAs play an essential role in this general process.

Pax6 protein is restricted to progenitors located in the dorsal part of the SVZ, where it is required for the specification of dopaminergic periglomerular neurons (De Chevigny et al., 2012a; Kohwi et al., 2005). Although Pax6 protein is expressed specifically in dorsal progenitors, Pax6 mRNA is also found in the lateral part of SVZ, when it shows a shallow dorso-ventral expression gradient. In vivo studies demonstrated that the absence of Pax6 protein in lateral SVZ progenitors is caused by posttranscriptional silencing of Pax6 mRNA through binding of the miRNA miR-7a to its 3′-UTR (De Chevigny et al., 2012a). Expression analyses confirmed that miR-7a is expressed in lateral SVZ progenitors in a ventro-dorsal decreasing gradient opposing the dorso-ventral Pax6 gradient. Mir-7a loss and gain of function experiments were performed by in vivo postnatal brain electroporation (Boutin et al., 2008). Inhibition of endogenous miR-7a in lateral SVZ NSCs by electroporation with a sponge plasmid induced ectopic appearance of Pax6 protein from endogenously expressed mRNA and subsequently led to an increased generation of dopaminergic interneurons in the OB from the lateral progenitors. Conversely, overexpression of miR-7a in dorsally located NSCs reduced Pax6 expression. However, the reduction was not sufficient to impair dopaminergic fate. This observation, in agreement with the fact that Pax6$^{sey/+}$ heterozygous mutants lose olfactory dopaminergic neurons (Dellovade et al., 1998), indicates that the precise expression level of determinants rules cell fate and that miRNAs play an essential role in controlling this expression level.

In sum, this study demonstrated the role of miR-7a in controlling the specification of dopaminergic fate during olfactory neurogenesis, by sharply limiting the border of expression of its target, the transcription factor Pax6 (Table 8.1).

MiRNAs involved in neuronal maturation/synaptic integration during adult OB neurogenesis

After NSC division in the SVZ and migration in the RMS, newly generated olfactory neurons reach the OB to undergo final dendritic maturation and synaptic integration into the preexisting bulbar circuitry.

The two populations of embryonically versus adult−born neurons can be distinguished by the expression of miR-125b that is restricted to adult−born neurons (Akerblom et al., 2014). Expression of this microRNA in the SVZ−RMS−OB system (Fig. 8.2) was monitored using a GFP-sensor transgenic mouse line, where absence of GFP reflects miR-125 activity and GFP expression conversely reflects lack of miR-125.

Loss or reduction of miR-125 in the postnatal SVZ by lentiviral injection of a miR-125-sponge into the LV increased dendritic length and branching in newly born olfactory granule neurons without altering spine density, demonstrating that miR-125b plays a role in dendritic morphogenesis.

Furthermore, analysis of Fos expression in adult–born neurons upon odor exposure showed that miR-125 inhibition increased their activation. Thus, during adult olfactory neurogenesis mir-125b seems to finely control the integration of new neurons by slowing the process. To gain an insight into the underlying mechanisms, the authors bioinformatically analyzed argonaute HITS-CLIP data from Chi et al. (2009) and found that most miR-125 target genes in the P13 brain are related to synaptic function. This led to the conclusion that reduction of the integration process occurred at the level of the synapse. However, no miR-125 target has been validated to date in this system.

In agreement with the above findings, a study by Lattanzi et al. (2013) showed upregulation of miR-125b during neuronal commitment in cultures of adult SVZ NSCs, again by using GFP-sensor viruses. In addition, they examined miR-125b activity in vivo by injecting the GFP-sensor into the LV of postnatal mice. This showed that the microRNA appears precociously in SVZ cells and its expression is maintained during differentiation in all OB layers.

Another miRNA, miR-132, is implicated in the activity-dependent integration of newborn neurons during adult hippocampal neurogenesis by controlling the development of dendrites and spines (Vo, et al., 2005; Magill et al., 2010). One of the targets of miR-132 is the CREB (see above), which controls the maturation and survival of newborn OB neurons. Interestingly, miR-132, like CREB (Giachino et al., 2005), was found to be progressively expressed during OB neurogenesis at the onset of synaptic integration (Fig. 8.2) (Pathania et al., 2012), suggesting that it might control the maturation of adult–born olfactory interneurons. In vitro, inhibition of miR-132 in cultured OB neurons led to impaired dendritic morphogenesis. Similarly, in vivo, dendritic length and spine density were decreased in OB neurons after sequestration of miR-132 in progenitors and new neurons by electroporation of NSCs in the lateral SVZ with a sponge construct. This was accompanied by a significant decrease in the amplitude and frequency of excitatory postsynaptic currents (EPSC) and a tendency to a decreased frequency of spontaneous GABAergic inhibitory postsynaptic currents (IPSCs). Conversely, overexpressing miR-132 in SVZ NSCs and derived neurons increased dendritic complexity and length, spine density and the frequency of spontaneous IPSCs. In addition to enhancing synaptic integration, miR-132 also promoted long-term survival of OB newborn neurons when induced later, at 7 days post electroporation using a CreER^{T2} inducible approach. Mechanistically, miR-132 mRNA targets remain to be identified during OB neurogenesis. However, the gene encoding the Rho family GTPase-activating protein, p250GAP emerges as a potential direct target (Vo et al., 2005; Wayman et al., 2008).

In conclusion, miR-125b and miR-132 are two miRNAs involved in synaptic plasticity in olfactory as well as hippocampal neurons, with opposite effects on dendritic and spine morphogenesis (Table 8.1). Remarkably, both miRNAs interact with FMRP, an RNA-binding protein that interacts with the miRNA pathway to repress translation of specific proteins (Edbauer et al., 2010). Interestingly, FRMP targets p250GAP mRNA and is required to control the functional synaptic integration of OB interneurons (Scotto-Lomassese et al., 2011). Altogether, these studies suggest a functional interaction between miR-125b, miR-132, and FRMP in the regulation of the differentiation of newly born neurons in the adult OB.

In sum, the scarce functional studies that have been performed on the role of specific miRNAs during adult OB neurogenesis to date have revealed important roles in neuronal fate choice (miR-124), subtype specification (miR-7) and dendritic/synaptic morphogenesis (miR-125 and miR-132) (Fig. 8.3). Only some of the significant mRNA targets have been identified (Sox9 for miR-124, Pax6 for miR-7).

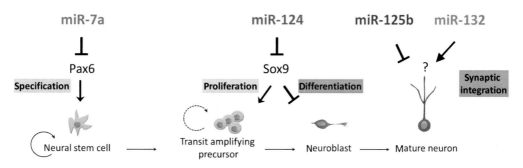

FIGURE 8.3

Summary of miRNA functional analyses during OB neurogenesis.

Gain and loss of function analysis revealed the role of miRNAs in controlling different steps of the OB neurogenic process. Mir-7a negatively controls the specification of NSCs toward the dopaminergic fate by targeting the transcription factor PAX6 in the lateral wall of the SVZ. MiR-124 favors the transition from proliferation to neuronal differentiation at least in part by inhibiting Sox9 in neuroblasts. In later stages, mir-125b and miR-132 control the maturation and synaptic integration of new neurons in the OB in an opposite manner, miR-125b reducing the integration while miR-132 stimulates it. For the moment, no mRNA targets have been identified for these two miRNAs in this system.

TECHNICAL APPROACHES FOR ADDRESSING miRNA FUNCTION IN OLFACTORY NEUROGENESIS

Because miRNAs are small and often expressed from different genomic loci, functional studies require specific experimental approaches. In the following section, we will focus on some of the recent developments that have been used in OB neurogenesis.

LOSS OF FUNCTION

Generation of knockout mice for individual miRNAs is often complicated since many miRNAs are encoded by more than one locus. This is the case for miR-7, miR-124, and miR125. Also, specific targeting of a single miRNA located within a cluster, such as miR-132 for example, makes the use of classical genetic tools difficult. Finally, investigating the function of miRNAs in a specific area of the brain or specific cell types would require generating conditional mutant alleles. To circumvent this issue several knockdown or inhibition approaches have been developed over the past few years (see also Chapter 4: Methodological Challenges in Functional Investigation and Therapeutic Use of microRNAs of this book).

A first alternative approach that has emerged is using antisense oligonucleotides (ASO). For example, to inhibit miR-124 during olfactory neurogenesis in vivo, penetratin-conjugated

$2'O$-methyl-RNAs was infused into the LV of adult brains for several days using osmotic mini-pumps to maintain downregulation of miR-124 during lineage progression (Cheng et al., 2009). Because the ASO-RNA is delivered locally into the ventricle, it can inhibit miRNA only in cells that are in contact with the ventricle, making it inadequate to sequester miRNAs expressed in more mature populations, in the RMS and the OB. These limitations make the technique complicated to use when targeting NSCs and progeny in vivo.

Another approach was to use microRNA sponges. Their main advantage is the possibility to achieve stable expression from integrated transgenes in vivo (Gentner et al., 2009). For this reason, the "sponge" strategy has often been used in the different OB neurogenesis studies to promote long-term miRNA loss-of-function (Akerblom et al., 2012, 2014; De Chevigny et al., 2012a; Pathania et al., 2012). Sponge transgenes have been delivered into the LV of postnatal brains by lentiviral vector injections (miR-124, miR-125) or plasmid electroporation (miR-7, miR-132), leading to maintained expression of the sponge inhibitor. Interestingly, miR-124 has been inactivated in two independent studies by ASO and sponge strategies (Cheng et al., 2009; Akerblom et al., 2012, 2014) and only the sponge condition led to a long-term phenotype in the OB (Akerblom et al., 2012), confirming the advantage of this technique.

GAIN OF FUNCTION

Electroporation or viral delivery of expression vectors encoding pri-miRNA transcripts under the control of strong promoters constituted the favored method to overexpress miRNAs in the SVZ-OB system. In contrast to RNA duplexes, this approach leads to stable expression of miRNAs, allowing long-term in vivo analysis of function and was fruitfully applied for studies of miR-124 (Cheng et al., 2009; Akerblom et al., 2012), miR-125b (Akerblom et al., 2014), and miR-7a (De Chevigny et al., 2012a). Besides manipulations of individual miRNAs, such vectors are also useful to overexpress multiple miRNAs at the same time, for example miRNAs belonging to the same cluster or to the same family.

For adult studies, infection by replication incompetent retroviruses is so far the unique way to drive in vivo expression of miRNA expression vectors. In this case, viruses are stereotactically injected into the SVZ (Cheng et al., 2009; Akerblom et al., 2012, 2014). Young postnatal animals offer an alternative possibility, which is to introduce expression plasmids specifically into SVZ NSCs by in vivo intraventricular electroporation. This approach has several advantages over viral injection. First, it is fast, easy, cost-effective, and does not require special safety precautions. Second, postnatal electroporation allows analyzing the behavior of a single neuronal cohort during OB neurogenesis. Third, a specific wall, and therefore a specific stem-cell pool, of the SVZ can be specifically targeted with relative ease (lateral, dorsal, or medial wall) (Fernandez et al., 2011; De Chevigny et al., 2012a). Therefore the role of miRNAs in the specification of regionalized NSCs (Merkle et al., 2007) can be investigated (De Chevigny et al., 2012a).

Overexpression at a specific developmental stage during OB neurogenesis can be performed by using inducible Cre-Lox based plasmid vector electroporation. This method allows the induction of miRNA overexpression in a time-dependent manner during the neurogenic process in WT mice. For example, Pathania et al. (2012) coelectroporated a ^{ERT2}CreER^{T2} vector with a stop-floxed-miR132 overexpression vector. Tamoxifen injection 7 days later induced miR-132 overexpression in maturing neurons, allowing the observation of an increased survival rate of adult–born olfactory neurons.

FUTURE PERSPECTIVES TO STUDY miRNAs DURING ADULT OB NEUROGENESIS

EXPRESSION STUDIES DURING OB NEUROGENESIS

Studies in fields other than adult olfactory neurogenesis have already used miRNA microarrays or deep sequencing to determine global profiles of miRNA expression from dissected tissues or purified cell populations in the brain (Bak et al., 2008; Chiang et al., 2010; Juhila et al., 2011). Very recently, a comprehensive spatiotemporal repertoire of miRNAs expressed during OB neurogenesis has been described, based on microdissection of different neurogenesis compartments and RNA-seq. This will be an important resource for future studies in this system (Beclin et al., 2016).

FUNCTIONAL STUDIES DURING OB NEUROGENESIS

Determining the gene targets of a given miRNA

In the future, systematic approaches will be required to determine all gene targets of specific miRNAs during OB neurogenesis. One such approach is depicted in an elegant study from Tan et al. (2013), which we consider to be a gold standard for future works. These authors generated floxed KO mice for their miRNA of interest, miR-128, and found that miR-128 ablation in striatal medium spiny neurons using the Drd1-Cre mouse line causes a severe epilepsy and motor phenotype. To determine the effective gene targets of miR-128 in these neurons, the authors combined two approaches. First, they use Translating Ribosome Affinity Purification (TRAP) to identify the ribosome-bound mRNAs enriched in Drd1-Cre; miR128 cKO neurons compared to control neurons. Secondly, they used high-throughput sequencing of RNA isolated by crosslinking immunoprecipitation (HITS-CLIP) for tagged-Ago2 to determine all mRNAs associated with the RISC complex and therefore subject to miRNA regulation (Licatalosi et al., 2008). The genes that share both upregulation in miR-128 deficient neurons and RISC-association are the bona fide in vivo targets of this miRNA. This approach is extremely powerful because it allows the determination of all gene targets of a given miRNA in a neuronal population of interest in vivo. Obviously, this method is very demanding since it requires miRNA deficient mice as well as the use of TRAP and HITS-CLIP approaches.

More recently, another highly efficient method based on a tagged Argonaute-interacting peptide, Ago-APP (referred to as "Ago protein Affinity Purification by Peptides"), has been described. This approach allows the efficient isolation of all Ago proteins and thus the identification of Ago-bound miRNAs and mRNAs in a given organism. On top of that, expression of this Ago-interacting peptide in living cells or mice leads to global miRNA inactivation, thus providing a powerful tool to study miRNA function on various levels (Hauptmann et al., 2015).

Loss-of-function with spatiotemporal control

cKO mice are classically engineered to invalidate gene products in a spatiotemporal manner. In the neurogenesis field, miR132/212^{flox} mice have been generated to ablate these miRNAs in newborn hippocampal neurons after injection of a Cre-expressing retrovirus into the dentate gyrus (Magill et al., 2010). However, this approach is not conceivable when a mature miRNA is expressed from numerous loci. Thus, alternative methods are needed that can block miRNA action rather than expression.

In the future, methods that allow the spatiotemporal control of sponge expression will be useful to dissect precise miRNA function during adult OB neurogenesis. Such methods have been successfully used in other neurogenic systems to conditionally downregulate miR-9 (Giusti et al., 2014) or miR-7 (Pollock et al., 2014).

Utilizing such Sponge$^{fl-Stop}$ mice and breeding them with Nestin-CreER^{T2} mice to control the timing of inhibition or with transgenic animals expressing Cre under cell type-specific promoters (BLBP, DCX, Camk2a, and DAT for NSCs, neuroblasts, granule neurons, and dopaminergic neurons, respectively) will be useful to study the precise contribution of miR-7, miR-9, and other miRNAs during adult OB neurogenesis.

Gain-of-function with spatiotemporal control

Overexpression in a specific cell population or developmental stage can be performed by using promoter-driven expression in plasmids, viruses or transgenic mice. An example from another field is the study from Hansen et al. (2010) who generated a tetracycline-inducible transgenic mouse overexpressing their miRNA of interest, miR-132, in a specific cell type, cortical glutamatergic neurons. To this aim, they crossed mice expressing tTA under the control of the cortical postmitotic neuron Camk2a promoter, with transgenic mice expressing miR-132 under the control of the TET-response element. Again, in analyzing OB neurogenesis, promoters like BLBP, DCX, or Camk2a can be used to overexpress miRNAs in NSCs, neuroblasts, or mature olfactory neurons, respectively.

CONCLUSION AND PERSPECTIVES

In conclusion, as yet, only very small numbers of microRNAs have been demonstrated to play a regulatory role during olfactory neurogenesis in mammals. Their main function appears to be in the control of stem cell maintenance, exit from this stage, neuronal subtype fate specification, and synaptic integration/function. For now, examples of microRNAs implicated in the control of the neuronal tangential or radial migration during OB neurogenesis are lacking, but it is likely that miRNAs are involved in this highly regulated process.

In the future, in addition to the discovery of new miRNAs involved in OB neurogenesis, the challenge will be to identify their functional endogenous mRNA targets and the nature of each miRNA/specific target function. The recent cutting-edge technologies described above, such as HITS-CLIP, should be helpful to this end (for further detail on these topic, see Chapters 2—4 of this book).

The recent discovery of natural miRNA sponges in the form of noncoding circular RNAs (circRNAs) further increases the complexity of posttranscriptional regulation by the microRNA pathway (Ebert and Sharp, 2010; Jens and Rajewsky, 2015) (see also Chapter 14: Circular RNAs Expression, Function and Regulation in Neural Systems of this book by Bozzoni and colleagues). Remarkably, one of these newly discovered circRNAs, CDR1as/ciRS-7, represents a potential endogenous sponge for miR-7 (Hansen et al., 2013; Memczak et al., 2013). It exhibits overlapping coexpression with miR-7 in neuronal tissues and strongly suppresses the miRNA activity, resulting in increased levels of miR-7 targets. Given that miR-7 is a key regulator during OB neurogenesis, open questions emerge regarding the putative function of natural sponges in this system.

REFERENCES

Adachi, K., Mirzadeh, Z., Sakaguchi, M., Yamashita, T., Nikolcheva, T., Gotoh, Y., et al., 2007. Beta-catenin signaling promotes proliferation of progenitor cells in the adult mouse subventricular zone. Stem Cells 25, 2827–2836.

Akerblom, M., Petri, R., Sachdeva, R., Klussendorf, T., Mattsson, B., Gentner, B., et al., 2014. microRNA-125 distinguishes developmentally generated and adult–born olfactory bulb interneurons. Development 141, 1580–1588.

Akerblom, M., Sachdeva, R., Barde, I., Verp, S., Gentner, B., Trono, D., et al., 2012. MicroRNA-124 is a subventricular zone neuronal fate determinant. J. Neurosci. 32, 8879–8889.

Alvarez-Buylla, A., Garcia-Verdugo, J.M., 2002. Neurogenesis in adult subventricular zone. J. Neurosci. 22, 629–634.

Azim, K., Fischer, B., Hurtado-Chong, A., Draganova, K., Cantu, C., Zemke, M., et al., 2014. Persistent Wnt/beta-catenin signaling determines dorsalization of the postnatal subventricular zone and neural stem cell specification into oligodendrocytes and glutamatergic neurons. Stem Cells 32, 1301–1312.

Bak, M., Silahtaroglu, A., Moller, M., Christensen, M., Rath, M.F., Skryabin, B., et al., 2008. MicroRNA expression in the adult mouse central nervous system. RNA 14, 432–444.

Balordi, F., Fishell, G., 2007. Hedgehog signaling in the subventricular zone is required for both the maintenance of stem cells and the migration of newborn neurons. J. Neurosci. 27, 5936–5947.

Beclin, C., Follert, P., Stappers, E., Barral, S., Nathalie, C., de Chevigny, A., et al., 2016. miR-200 family controls late steps of postnatal forebrain neurogenesis via Zeb2 inhibition. Sci. Rep. 6, 35729.

Boutin, C., Diestel, S., Desoeuvre, A., Tiveron, M.C., Cremer, H., 2008. Efficient in vivo electroporation of the postnatal rodent forebrain. PLoS ONE 3, e1883.

Boutin, C., Hardt, O., De Chevigny, A., Core, N., Goebbels, S., et al., 2010. NeuroD1 induces terminal neuronal differentiation in olfactory neurogenesis. Proc. Natl. Acad. Sci. U. S. A. 107, 1201–1206.

Bovetti, S., Bonzano, S., Garzotto, D., Giannelli, S.G., Iannielli, A., Armentano, M., et al., 2013. COUP-TFI controls activity-dependent tyrosine hydroxylase expression in adult dopaminergic olfactory bulb interneurons. Development 140, 4850–4859.

Brett, J.O., Renault, V.M., Rafalski, V.A., Webb, A.E., Brunet, A., 2011. The microRNA cluster miR-106b~25 regulates adult neural stem/progenitor cell proliferation and neuronal differentiation. Aging (Albany NY) 3, 108–124.

Brill, M.S., Ninkovic, J., Winpenny, E., Hodge, R.D., Ozen, I., Yang, R., et al., 2009. Adult generation of glutamatergic olfactory bulb interneurons. Nat. Neurosci. 12, 1524–1533.

Brill, M.S., Snapyan, M., Wohlfrom, H., Ninkovic, J., Jawerka, M., Mastick, G.S., et al., 2008. A dlx2- and pax6-dependent transcriptional code for periglomerular neuron specification in the adult olfactory bulb. J. Neurosci. 28, 6439–6452.

Bulfone, A., Wang, F., Hevner, R., Anderson, S., Cutforth, T., Chen, S., et al., 1998. An olfactory sensory map develops in the absence of normal projection neurons or GABAergic interneurons. Neuron 21, 1273–1282.

Burk, K., Desoeuvre, A., Boutin, C., Smith, M.A., Kroger, S., Bosio, A., et al., 2012. Agrin-signaling is necessary for the integration of newly generated neurons in the adult olfactory bulb. J. Neurosci. 32, 3759–3764.

Cernilogar, F.M., Di Giaimo, R., Rehfeld, F., Cappello, S., Lie, D.C., 2015. RNA interference machinery-mediated gene regulation in mouse adult neural stem cells. BMC Neurosci. 16, 60.

Cheng, L.C., Pastrana, E., Tavazoie, M., Doetsch, F., 2009. miR-124 regulates adult neurogenesis in the subventricular zone stem cell niche. Nat. Neurosci. 12, 399–408.

Chi, S.W., Zang, J.B., Mele, A., Darnell, R.B., 2009. Argonaute HITS-CLIP decodes microRNA–mRNA interaction maps. Nature 460, 479–486.

Chiang, H.R., Schoenfeld, L.W., Ruby, J.G., Auyeung, V.C., Spies, N., Baek, D., et al., 2010. Mammalian microRNAs: experimental evaluation of novel and previously annotated genes. Genes Dev. 24, 992–1009.

De Chevigny, A., Core, N., Follert, P., Gaudin, M., Barbry, P., Beclin, C., et al., 2012a. miR-7a regulation of Pax6 controls spatial origin of forebrain dopaminergic neurons. Nat. Neurosci. 15, 1120–1126.

De Chevigny, A., Core, N., Follert, P., Wild, S., Bosio, A., Yoshikawa, K., et al., 2012b. Dynamic expression of the pro-dopaminergic transcription factors Pax6 and Dlx2 during postnatal olfactory bulb neurogenesis. Front. Cell Neurosci. 6, 6.

Dellovade, T.L., Pfaff, D.W., Schwanzel-Fukuda, M., 1998. Olfactory bulb development is altered in small-eye (Sey) mice. J. Comp. Neurol. 402, 402–418.

Ebert, M.S., Sharp, P.A., 2010. Emerging roles for natural microRNA sponges. Curr. Biol. 20, R858–R861.

Edbauer, D., Neilson, J.R., Foster, K.A., Wang, C.F., Seeburg, D.P., Batterton, M.N., et al., 2010. Regulation of synaptic structure and function by FMRP-associated microRNAs miR-125b and miR-132. Neuron 65, 373–384.

Fernandez, M.E., Croce, S., Boutin, C., Cremer, H., Raineteau, O., 2011. Targeted electroporation of defined lateral ventricular walls: a novel and rapid method to study fate specification during postnatal forebrain neurogenesis. Neural Dev. 6, 13.

Ferri, A.L., Cavallaro, M., Braida, D., Di Cristofano, A., Canta, A., Vezzani, A., et al., 2004. Sox2 deficiency causes neurodegeneration and impaired neurogenesis in the adult mouse brain. Development 131, 3805–3819.

Gentner, B., Schira, G., Giustacchini, A., Amendola, M., Brown, B.D., Ponzoni, M., et al., 2009. Stable knockdown of microRNA in vivo by lentiviral vectors. Nat. Methods 6, 63–66.

Giachino, C., De Marchis, S., Giampietro, C., Parlato, R., Perroteau, I., Schutz, G., et al., 2005. cAMP response element-binding protein regulates differentiation and survival of newborn neurons in the olfactory bulb. J. Neurosci. 25, 10105–10118.

Giusti, S.A., Vogl, A.M., Brockmann, M.M., Vercelli, C.A., Rein, M.L., Trumbach, D., et al., 2014. MicroRNA-9 controls dendritic development by targeting REST. Elife 3.

Hansen, K.F., Sakamoto, K., Wayman, G.A., Impey, S., Obrietan, K., 2010. Transgenic miR132 alters neuronal spine density and impairs novel object recognition memory. PLoS ONE 5, e15497.

Hansen, T.B., Kjems, J., Damgaard, C.K., 2013. Circular RNA and miR-7 in cancer. Cancer Res. 73, 5609–5612.

Harrison, S.J., Parrish, M., Monaghan, A.P., 2008. Sall3 is required for the terminal maturation of olfactory glomerular interneurons. J. Comp. Neurol. 507, 1780–1794.

Hauptmann, J., Schraivogel, D., Bruckmann, A., Manickavel, S., Jakob, L., Eichner, N., et al., 2015. Biochemical isolation of Argonaute protein complexes by Ago-APP. Proc. Natl. Acad. Sci. U. S. A. 112, 11841–11845.

Herold, S., Jagasia, R., Merz, K., Wassmer, K., Lie, D.C., 2011. CREB signalling regulates early survival, neuronal gene expression and morphological development in adult subventricular zone neurogenesis. Mol. Cell Neurosci. 46, 79–88.

Imayoshi, I., Sakamoto, M., Yamaguchi, M., Mori, K., Kageyama, R., 2010. Essential roles of Notch signaling in maintenance of neural stem cells in developing and adult brains. J. Neurosci. 30, 3489–3498.

Jens, M., Rajewsky, N., 2015. Competition between target sites of regulators shapes post-transcriptional gene regulation. Nat. Rev. Genet 16, 113–126.

Juhila, J., Sipila, T., Icay, K., Nicorici, D., Ellonen, P., Kallio, A., et al., 2011. MicroRNA expression profiling reveals miRNA families regulating specific biological pathways in mouse frontal cortex and hippocampus. PLoS ONE 6, e21495.

Kohwi, M., Osumi, N., Rubenstein, J.L., Alvarez-Buylla, A., 2005. Pax6 is required for making specific subpopulations of granule and periglomerular neurons in the olfactory bulb. J. Neurosci. 25, 6997−7003.

Lattanzi, A., Gentner, B., Corno, D., Di Tomaso, T., Mestdagh, P., Speleman, F., et al., 2013. Dynamic activity of miR-125b and miR-93 during murine neural stem cell differentiation in vitro and in the subventricular zone neurogenic niche. PLoS ONE 8, e67411.

Li, X., Sun, C., Lin, C., Ma, T., Madhavan, M.C., Campbell, K., 2011. The transcription factor Sp8 is required for the production of parvalbumin-expressing interneurons in the olfactory bulb. J. Neurosci. 31, 8450−8455.

Licatalosi, D.D., Mele, A., Fak, J.J., Ule, J., Kayikci, M., Chi, S.W., et al., 2008. HITS-CLIP yields genome-wide insights into brain alternative RNA processing. Nature 456, 464−469.

Lim, D.A., Huang, Y.C., Swigut, T., Mirick, A.L., Garcia-Verdugo, J.M., Wysocka, J., 2009. Chromatin remodelling factor Mll1 is essential for neurogenesis from postnatal neural stem cells. Nature 458 (7237), 529−533.

Liu, C., Teng, Z.Q., Santistevan, N.J., Szulwach, K.E., Guo, W., Jin, P., et al., 2010. Epigenetic regulation of miR-184 by MBD1 governs neural stem cell proliferation and differentiation. Cell Stem Cell 6, 433−444.

Liu, H.K., Belz, T., Bock, D., Takacs, A., Wu, H., Lichter, P., et al., 2008. The nuclear receptor tailless is required for neurogenesis in the adult subventricular zone. Genes Dev. 22, 2473−2478.

Liu, K., Liu, Y., Mo, W., Qiu, R., Wang, X., Wu, J.Y., et al., 2011. MiR-124 regulates early neurogenesis in the optic vesicle and forebrain, targeting NeuroD1. Nucleic Acids Res. 39, 2869−2879.

Lopez-Juarez, A., Howard, J., Ullom, K., Howard, L., Grande, A., Pardo, A., et al., 2013. Gsx2 controls region-specific activation of neural stem cells and injury-induced neurogenesis in the adult subventricular zone. Genes Dev. 27, 1272−1287.

Magill, S.T., Cambronne, X.A., Luikart, B.W., Lioy, D.T., Leighton, B.H., Westbrook, G.L., et al., 2010. microRNA-132 regulates dendritic growth and arborization of newborn neurons in the adult hippocampus. Proc. Natl. Acad. Sci. U. S. A. 107, 20382−20387.

Marinaro, C., Pannese, M., Weinandy, F., Sessa, A., Bergamaschi, A., Taketo, M.M., et al., 2012. Wnt signaling has opposing roles in the developing and the adult brain that are modulated by Hipk1. Cereb. Cortex 22, 2415−2427.

Memczak, S., Jens, M., Elefsinioti, A., Torti, F., Krueger, J., Rybak, A., et al., 2013. Circular RNAs are a large class of animal RNAs with regulatory potency. Nature 495, 333−338.

Merkle, F.T., Mirzadeh, Z., Alvarez-Buylla, A., 2007. Mosaic organization of neural stem cells in the adult brain. Science 317, 381−384.

Nigro, A., Menon, R., Bergamaschi, A., Clovis, Y.M., Baldi, A., Ehrmann, M., et al., 2012. MiR-30e and miR-181d control radial glia cell proliferation via HtrA1 modulation. Cell Death Dis. 3, e360.

Ninkovic, J., Pinto, L., Petricca, S., Lepier, A., Sun, J., Rieger, M.A., et al., 2010. The transcription factor Pax6 regulates survival of dopaminergic olfactory bulb neurons via crystallin alphaA. Neuron 68, 682−694.

Nyfeler, Y., Kirch, R.D., Mantei, N., Leone, D.P., Radtke, F., Suter, U., et al., 2005. Jagged1 signals in the postnatal subventricular zone are required for neural stem cell self-renewal. EMBO J. 24, 3504−3515.

Obernosterer, G., Leuschner, P.J., Alenius, M., Martinez, J., 2006. Post-transcriptional regulation of microRNA expression. RNA 12, 1161−1167.

Parras, C.M., Galli, R., Britz, O., Soares, S., Galichet, C., Battiste, J., et al., 2004. Mash1 specifies neurons and oligodendrocytes in the postnatal brain. EMBO J. 23, 4495−4505.

Pathania, M., Torres-Reveron, J., Yan, L., Kimura, T., Lin, T.V., Gordon, V., et al., 2012. miR-132 enhances dendritic morphogenesis, spine density, synaptic integration, and survival of newborn olfactory bulb neurons. PLoS ONE 7, e38174.

Petrova, R., Garcia, A.D., Joyner, A.L., 2013. Titration of GLI3 repressor activity by sonic hedgehog signaling is critical for maintaining multiple adult neural stem cell and astrocyte functions. J. Neurosci. 33, 17490−17505.

Pollock, A., Bian, S., Zhang, C., Chen, Z., Sun, T., 2014. Growth of the developing cerebral cortex is controlled by microRNA-7 through the p53 pathway. Cell Rep. 7, 1184−1196.

Qiu, M., Bulfone, A., Martinez, S., Meneses, J.J., Shimamura, K., Pedersen, R.A., et al., 1995. Null mutation of Dlx-2 results in abnormal morphogenesis of proximal first and second branchial arch derivatives and abnormal differentiation in the forebrain. Genes Dev. 9, 2523−2538.

Qu, Q., Sun, G., Li, W., Yang, S., Ye, P., Zhao, C., et al., 2010. Orphan nuclear receptor TLX activates Wnt/ beta-catenin signalling to stimulate neural stem cell proliferation and self-renewal. Nat. Cell Biol. 12, 31−40, sup pp 1−9.

Scott, C.E., Wynn, S.L., Sesay, A., Cruz, C., Cheung, M., Gomez Gaviro, M.V., et al., 2010. SOX9 induces and maintains neural stem cells. Nat. Neurosci. 13, 1181−1189.

Scotto-Lomassese, S., Nissant, A., Mota, T., Neant-Fery, M., Oostra, B.A., Greer, C.A., et al., 2011. Fragile X mental retardation protein regulates new neuron differentiation in the adult olfactory bulb. J. Neurosci. 31, 2205−2215.

Szulwach, K.E., Li, X., Smrt, R.D., Li, Y., Luo, Y., Lin, L., et al., 2010. Cross talk between microRNA and epigenetic regulation in adult neurogenesis. J. Cell Biol. 189, 127−141.

Tan, C.L., Plotkin, J.L., Veno, M.T., Von Schimmelmann, M., Feinberg, P., Mann, S., et al., 2013. MicroRNA-128 governs neuronal excitability and motor behavior in mice. Science 342, 1254−1258.

Vo, N., Klein, M.E., Varlamova, O., Keller, D.M., Yamamoto, T., Goodman, R.H., et al., 2005. A cAMP-response element binding protein-induced microRNA regulates neuronal morphogenesis. Proc. Natl. Acad. Sci. U. S. A. 102, 16426−16431.

Wang, L., Chopp, M., Zhang, R.L., Zhang, L., Letourneau, Y., Feng, Y.F., et al., 2009a. The Notch pathway mediates expansion of a progenitor pool and neuronal differentiation in adult neural progenitor cells after stroke. Neuroscience 158, 1356−1363.

Wang, X., Mao, X., Xie, L., Greenberg, D.A., Jin, K., 2009b. Involvement of Notch1 signaling in neurogenesis in the subventricular zone of normal and ischemic rat brain in vivo. J. Cereb. Blood Flow Metab. 29, 1644−1654.

Wayman, G.A., Davare, M., Ando, H., Fortin, D., Varlamova, O., Cheng, H.Y., et al., 2008. An activity-regulated microRNA controls dendritic plasticity by down-regulating p250GAP. Proc. Natl. Acad. Sci. U. S. A. 105, 9093−9098.

Zhao, C., Sun, G., Li, S., Shi, Y., 2009. A feedback regulatory loop involving microRNA-9 and nuclear receptor TLX in neural stem cell fate determination. Nat. Struct. Mol. Biol. 16, 365−371.

Zhao, C., Sun, G., Li, S., Lang, M.F., Yang, S., Li, W., et al., 2010. MicroRNA let-7b regulates neural stem cell proliferation and differentiation by targeting nuclear receptor TLX signaling. Proc. Natl. Acad. Sci. U. S. A. 107, 1876−1881.

MicroRNA-MEDIATED REGULATION OF ADULT HIPPOCAMPAL NEUROGENESIS; IMPLICATIONS FOR HIPPOCAMPUS-DEPENDENT COGNITION AND RELATED DISORDERS?

Pascal Bielefeld, Ben Pustjens, Marijn Schouten and Carlos P. Fitzsimons

University of Amsterdam, Amsterdam, The Netherlands

INTRODUCTION INTO ADULT HIPPOCAMPAL NEUROGENESIS

Adult neurogenesis is the process in which new neurons are derived from neural stem cells (NSCs) throughout life, taking place in selecting few areas of the mammalian brain, such as the olfactory bulb and the subventricular zone (SVZ) (see also Chapters 6 and 8 of this book) and the subgranular zone (SGZ) of the dentate gyrus (DG) of the hippocampus, termed adult hippocampal neurogenesis (AHN), which will be the main topic of this chapter. Though most extensive knowledge about AHN comes from preclinical animal studies, it was recently shown that also adult human hippocampi possess extensive neurogenic capacity with one-third of all granule cells being replaced throughout life at a turnover rate of 1.75% per year, which remarkably resembles the neurogenic capacity in the middle-aged mouse (Knoth et al., 2010; Spalding et al., 2013). Though the exact role of AHN is still debated, it has become clear that newborn granule cells play a key role in hippocampus-dependent cognitive capacity (Aimone et al., 2014; Bizon et al., 2004; Deng et al., 2010; Ming & Song, 2011; Snyder et al., 2001) whereas it has also been hypothesized to have intrinsic therapeutic functions, responding to local loss of neurons by, e.g., disease or to pharmacological treatment (Arvidsson et al., 2002; Darsalia et al., 2005; Encinas et al., 2006; Jakubs et al., 2006; Santarelli, 2003; Scharfman and McCloskey, 2009; Ziv et al., 2007).

AHN can be divided into several distinct stages, each of which is under specific control by cell-intrinsic, cell-extrinsic, and environmental factors, making AHN a very tightly regulated process (Fig. 9.1). Quiescent Radial-Glia-like NSCs reside in the SGZ, which upon stimulation can reenter cell-cycle and become proliferative NSCs. SGZ NSCs have the potential to self-renew and give rise to early amplifying neuronal progenitors (NPCs) via asymmetric division several times before they terminally differentiate. Asymmetric division coincides with cell fate determination, driving the majority of newborn cells to the neuronal fate, whereas a small population of newborn cells

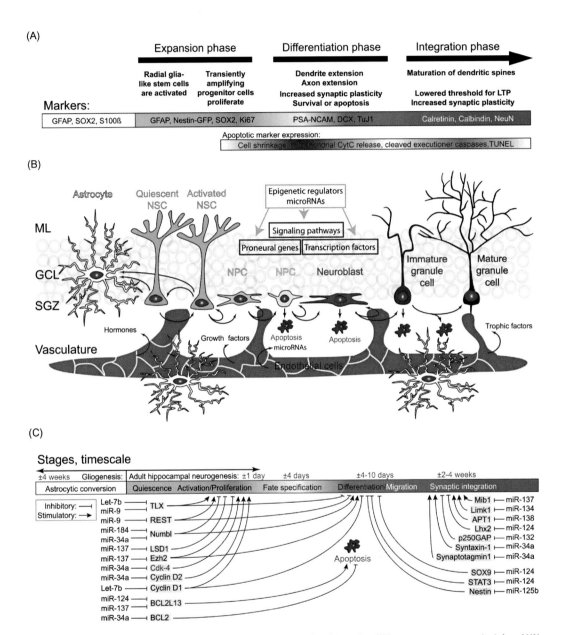

FIGURE 9.1 Schematic overview of the hippocampal neurogenic niche, the different processes underlying AHN, and its regulation by microRNAs.

(A) The different stages of AHN, in which specific (intermediate) cell types can be identified by a combination of presence and absence of different markers. (B) Overview of the neurogenic niche and the transition of an NSC

(Continued)

develops into mature astrocytes (Fig. 9.1). The NPCs then undergo symmetric division, triggering expansion of the neurogenic pool. Soon after birth, most NPCs are selected out by apoptosis, a process hypothesized to prevent excess production of new and/or unfit neurons (Biebl et al., 2000; Kuhn et al., 2005). NPCs that survive apoptotic selection give rise to neuroblasts, which generate immature neurons that after migration, differentiation, and maturation integrate as functional new granule cells in the DG (Kempermann et al., 2004; Ming and Song, 2005, 2011).

As mentioned before, AHN is controlled by a wide array of cell-intrinsic and cell-extrinsic factors. This complex interaction becomes clear when studying the fate of neural stem/progenitor cells (NSPCs) transplanted into ectopic locations. NSPCs derived from the SGZ will give rise to astrocytes when transplanted into nonneurogenic areas of the brain, whereas NSPCs derived from the SVZ will give rise to granule cells when transplanted into the SGZ (Seidenfaden et al., 2006; Shihabuddin et al., 2000). These observations indicate that the impact of the local environment, or neurogenic niche, on NSPC development is crucial for their neuronal development.

The hippocampal neurogenic niche comprises many different cell types, such as the NSPCs, neuroblasts and their progeny, and mature granule cells, but also astrocytes, GABAergic interneurons, microglia, macrophages, and endothelial cells connecting NSCs and their progeny to the vasculature (Fig. 9.1B) (Goldman and Chen, 2011; Palmer et al., 2000). The local vasculature and its associated extracellular matrix provide a means of both local cell−cell interaction, i.e., via β-catenin regulating mitotic spindle orientation (Chenn and Walsh, 2002; Perez-Moreno et al., 2003) and distant cell-extrinsic regulation of AHN, carrying and binding both proneurogenic growth factors and trophic factors released by both neuronal and nonneuronal cells, such as vascular endothelial growth factor (vEGF), basis fibroblast growth factor (bFGF), and brain-derived neurotrophic factor (BDNF) (Conover et al., 2000; Doetsch et al., 2002; Jin et al., 2002; Lu et al., 2002; Nakatomi et al., 2002). Together, all these elements provide the hippocampus with a finely tuned microenvironment permissive for adult neurogenesis.

Although the cellular and vascular composition of the niche provides the crucial structural organization, the local responsiveness of NSPCs and their progeny is under tight molecular control, i.e., by epigenetic regulation of chromatin states rendering cells permissive for regulation of gene transcription (Ballas et al., 2001; Lunyak et al., 2002; Takizawa et al., 2001).

Epigenetic control of AHN is a combination of NSPC intrinsic mechanisms and extrinsic regulation by non-NSPC cells within and even outside of the neurogenic niche (Ma et al., 2010). Conceptually, both temporal and spatial control of gene expression is crucial for progression of the different stages of AHN, and specifically transient gene repression is crucial for maintaining NSPC

◀ into a mature neuron. The hippocampal neurogenic niche comprises multiple cell types in close association with the vasculature, allowing for both local and distant cell−cell communication via factors released in the bloodstream, i.e., cell-extrinsic microRNAs, growth factors such as VEGF and bFGF, hormones, and trophic factors such as BDNF. Also depicted are cell-intrinsic factors, such as microRNAs, TLX signaling, WnT signaling, notch signaling, and REST (purple boxes), and cell-extrinsic factors, such as HDACs, DNA methylation, and microRNAs (pink box) that play a key role in AHN. (C) The transition between different stages of neurogenesis, as well as the maturation and integration of new neurons, are regulated by a.o. microRNAs. Depicted are microRNAs of which a clear link with neurogenesis has been identified, together with their targets through which the microRNAs might exert their effect.

pluripotency and proliferative capacity (Pichardo-Casas et al., 2012). Furthermore, cell fate specification and neuronal differentiation are partly determined by removing epigenetic repressive marks of NSPC differentiation-related genes, whereas noncell lineage specific genes are permanently silenced (Bertrand et al., 2002; Hu et al., 2012b; Lim et al., 2009). In short, epigenetic (transient) gene repression or silencing is key in controlling the switch from proliferation to neuronal differentiation in NSPCs (Ma et al., 2010). Epigenetic mechanisms known to play key roles in the regulation of AHN include chromatin modification. Chromatin can exist in an open transcribed state or a closed silent state, regulated by DNA methylation and histone modifications, such as acetylation and methylation. Chromatin can be further silenced by repressor proteins, including the Polycomb group proteins, resulting in complete silencing of a locus (Doetsch, 2003). More recently, noncoding RNAs, such as microRNAs, have been shown to play a central role in epigenetic regulation of NSPC, too (Jobe et al., 2012; Lang and Shi, 2012; Lopez-Ramirez and Nicoli, 2014; Schouten et al., 2012).

REGULATION OF ADULT HIPPOCAMPAL NEUROGENESIS BY MicroRNAs

microRNAs are small noncoding RNA molecules composed of approximately 22 nucleotides that play a role in RNA silencing and posttranscriptional regulation of gene expression (Kawahara et al., 2012). Gene silencing may take place via mRNA degradation or inhibition of mRNA translation; hence, microRNAs can only exert their function if the appropriate target mRNA is expressed as well. Given the biological characteristics of microRNAs, such as their potent ability to (transiently) repress mRNA translation, their rapid turnover, and their relatively high conservation between species, microRNAs have been hypothesized to play crucial roles in regulating the switch between different temporal stages of AHN (Hu et al., 2012b; Ma et al., 2010). As mentioned before, the transition of the proliferative state into the differentiating state is mainly regulated by repression of nonlineage-specific genes, and removing repression of lineage-specific genes. In recent years, multiple microRNAs have been identified that play crucial roles in these stage transitions (Fig. 9.1C, Table 9.1) by regulating key epigenetic regulators, such as Histone Deactylases (HDACs), Polycomb group proteins, and repressor element-1 silencing transcription factor (REST). Furthermore, it has become clear that microRNAs can be secreted into the bloodstream, allowing for cell-extrinsic regulation as well (Chen et al., 2012; Kosaka et al., 2010; Xin et al., 2012). Single microRNAs can regulate expression of multiple target genes and thus biological processes (Grimson et al., 2007). In order to discuss the complex role microRNAs play in regulating AHN, we here separate the biological functions of specific microRNAs into stage-specific effects based on the coexpression of microRNAs and their targets.

MicroRNA-Dependent Control of Stem-Cell Maintenance and Fate Specification

To maintain neurogenic capacity over time, it is crucial that NSPCs preserve their proliferative and self-renewal capacity. Several microRNAs have been identified that control the balance between self-renewal and generation of NSPCs.

Table 9.1 Overview of Identified microRNAs that Regulate (Hippocampal) Neurogenesis, Their Targets, and Their Role in Neurogenesis

		microRNAs Involved in Regulation of AHN	
	Target	**miR Function**	**References**
miR-184	Numbl	Inhibits differentiation, promotes proliferation, maintenance of the neurogenic stem-cell pool	Liu et al. (2010)
miR-34a	Numbl	Inhibits differentiation, promotes proliferation, maintenance of the neurogenic stem-cell pool	Fineberg et al. (2012)
	BCL-2	Promotes apoptosis	Wang et al. (2009)
	Cdk-4	Inhibits cell-cycle progression	Antonini et al. (2010)
	Cyclin D2	Inhibits cell-cycle progression	Antonini et al. (2010)
	Synaptotagmin1	Inhibits synaptic development	Agostini et al. (2011)
	Syntaxin-1A	Inhibits synaptic development	Agostini et al. (2011)
miR-124	Sox9	Promotes neuronal differentiation	Stolt et al. (2003)
	STAT3	Promotes neuronal differentiation	Krichevsky et al. (2006)
	BCL2L13	Inhibits apoptosis	Schouten et al. (2015)
	Lhx2	Promotes neurite outgrowth	Sanuki et al. (2011)
miR-137	BCL2L13	Inhibits apoptosis	Schouten et al. (2015)
	Mib-1	Inhibits dendritic growth and spine formation	Smrt et al. (2010)
	LSD1	Promotes proliferation	Sun et al. (2011)
	Ezh2	Inhibits differentiation	Szulwach et al. (2010)
miR-9	TLX	Promotes differentiation, inhibits differentiation	Zhao et al. (2009)
	REST	Promotes neuronal differentiation	Packer et al. (2008)
Let-7b	TLX	Inhibits proliferation, promotes differentiation	Zhao et al. (2010)
	Cyclin D1	Inhibits cell-cycle progression	Zhao et al. (2010)
miR-125b	Nestin	Promotes neuronal differentiation	Cui et al. (2012)
miR-138	APT1	Inhibits spinogenesis	Siegel et al. (2009)

(Continued)

Table 9.1 Overview of Identified microRNAs that Regulate (Hippocampal) Neurogenesis, Their Targets, and Their Role in Neurogenesis *Continued*

		microRNAs Involved in Regulation of AHN		
	Target	**miR Function**		**References**
miR-134	Limk1	Inhibits spinogenesis		Schratt et al. (2006)
miR-132	P250GAP	Promotes spinogenesis		Impey et al. (2010)

miR-184 plays a crucial role in maintaining the proliferative capacity of NSPCs over time, as high levels of miR-184 are associated with long-term impaired neurogenesis. miR-184 targets the 3′UTR of Numbl, a repressor of Notch signaling. Notch signaling is crucial for maintenance of the stem-cell pool, as it allows asymmetric division rendering both NSCs and NPCs. Absence of Numbl in NSCs results in increased proliferation, exhausting the stem-cell pool. miR-184 expression is tightly regulated by MBD1, a protein that mediates gene repression by binding to methylated sequences or through interaction with SETDB1, a histone H3k9 transferase (Liu et al., 2010). Like miR-184, miR-34a also regulates Numbl expression, as well as two other genes downstream in the Notch signaling, NeuroD1, and Mash1 (Fineberg et al., 2012), to maintain the proliferative capacity of the NSPCs, indicating that Numbl may work as a hub in microRNA-mediated regulation of NSPC proliferation.

The second crucial early aspect of AHN is cell fate specification during symmetrical division of adult NSPCs. Under normal physiological conditions, most adult NSPCs will give rise to cells of the neuronal lineage, though a small subset will give rise to mature astrocytes. Cell fate specification is largely dependent on repression of nonlineage-specific gene expression through DNA methylation, but some evidence also points towards a role for microRNAs. Most evidence for such fate specification regulation by microRNAs comes from studies done on miR-124. miR-124 is absent in NSPCs, but its expression increases when NSPCs become neuroblasts and stays high in differentiated neurons. One of the targets of miR-124 is SOX9, a transcription factor crucial for gliogenesis (Stolt et al., 2003), which is expressed in both NSPCs and astrocytes, but not in neuroblasts. Knockdown of miR-124 increased ectopic SOX9 expression in neuroblasts, whereas overexpression of a 3′UTR-deficient SOX9 construct inhibited neuronal differentiation. A second possible pathway via which miR-124 exerts its effects is the STAT3 pathway (Krichevsky et al., 2006), which is known to inhibit terminal neuronal differentiation and enhances the differentiation of NPCs towards the glial-lineage (Gu et al., 2005; Moon et al., 2002). Overexpression of miR-124 significantly reduced STAT3 phosphorylation, resulting in less glial and more neuronal differentiation (Krichevsky et al., 2006). Furthermore, another study showed that overexpression of miR-124 in HeLa cells (a nonneuronal cell-type) resulted in the expression of a wide array of neuronal genes, while silencing nonneuronal genes (Lim et al., 2005). Together, this points towards a role for miR-124 in stimulating neuronal differentiation by repressing glial-lineage specific genes.

Microrna-dependent control of apoptotic selection

NSPCs undergo an extensive numerical selection, which regulates adult neurogenesis levels and quality. A major population of all NSPCs' progeny will be selected out by apoptosis, though not much research has yet focused on the role of microRNAs in this process (Biebl et al., 2000; Kuhn et al., 2005). A recent study showed that miR-124 and miR-137 cooperatively regulate NSPC apoptosis, by fine-tuning the levels of the proapoptotic protein Bcl2L13. Together, miR-124 and miR-137 significantly decrease Bcl2L13 and concomitant cleaved Caspase-3 levels, indicating a reduction in apoptosis (Schouten et al., 2015), and indicating that BCL2L13 may work as a hub in microRNA-mediated regulation of NSPC apoptosis.

miR-34a is another microRNA commonly linked to regulation of apoptosis, as it targets among others the antiapoptotic Bcl-2, and several cell-cycle regulators involved in cell-cycle progression such as Cdk-4 and Cyclin D2 (Antonini et al., 2010; Aranha et al., 2010; Wang et al., 2009). Though these findings originate from nonneuronal cell-type studies, both miR-34a and some of its targets are coexpressed in neuronal cell lineages as well, providing a possible new regulation of apoptosis in AHN (Biebl et al., 2000; Fineberg et al., 2012).

Microrna-dependent control of the balance between proliferation and differentiation of nspcs

miR-9 is highly expressed in the adult brain and targets several genes that play key roles in neuronal differentiation, such as Foxg1, Gsh2, and REST (Conaco et al., 2006; Packer et al., 2008; Shibata et al., 2008), thereby promoting differentiation and inhibiting proliferation. miR-9 expression levels are tightly controlled by a negative-feedback regulatory loop with the nuclear receptor TLX (Zhao et al., 2009). TLX activates the Wnt/B-catenin pathway and is crucial for NSCs to maintain their self-renewal and proliferative capacity (Lie et al., 2005; Qu et al., 2010). NSPCs expressing TLX can proliferate, self-renew, and differentiate into all neuronal lineages, whereas NSPCs devoid of TLX fail to proliferate. This capacity can be reintroduced by reintroducing TLX into these NSPCs (Shi et al., 2004). The TLX mRNA contains a seed region for miR-9, and overexpression of miR-9 results in decreased expression levels of TLX and increased (premature) differentiation of NPCs, whereas miR-9 knockdown results in increased proliferation of NSPCs. Significantly, TLX itself represses the expression of miR-9 precursors, indicating a balanced negative-feedback regulatory loop (Zhao et al., 2009). Moreover, miR-9 also targets two members of the REST complex, which acts as a transcriptional repressor of neuronal genes, inhibiting neuronal differentiation (Packer et al., 2008).

Besides the regulatory loop between miR-9 and TLX, TLX is a part of a second regulatory loop involving miR-137. miR -137 stimulates proliferation by repressing LSD1 (Sun et al., 2011), a corepressor of TLX, while repressing differentiation by downregulating Ezh2 (Szulwach et al., 2010), a histone methyltransferase, and part of the Polycomb group proteins. Like miR-9, miR-137 expression levels are tightly regulated via closed regulatory loops involving TLX and its corepressor LSD1. Such closed regulatory loops provide the perfect mechanism for the fast and transient switch in gene expression needed to drive NPCs out of their proliferative state and into differentiation, without losing proliferative capacity on the long term.

A third microRNA important for the switch from proliferation to differentiation is Let-7b. Let-7b overexpression in NSCs derived from the adult mouse brain results in inhibition of proliferation and accelerated neural differentiation, whereas Let-7b knockdown increases proliferation (Schwamborn et al., 2009). Significantly, Let-7b also regulates TLX signaling by targeting its 3′UTR and its downstream effector Cyclin D1 (Zhao et al., 2010), thereby inhibiting progression into the s-phase of the cell-cycle and lengthening the G1-phase, stimulating the transition into differentiation (Lange et al., 2009). These observations indicate that TLX may work as a hub in microRNA-mediated regulation of NSPC proliferation and transition into differentiation.

Finally, miR-125b regulates the transition between proliferation and differentiation. Overexpression of miR-125b results in decreased proliferation and increased migration and differentiation of NSPCs, which can be rescued by mutations in the miR-125b binding site on the 3′UTR of the Nestin mRNA (an intermediate filament protein present in NSPCs but not in neurons (Cui et al., 2012). Along the same line of reasoning it was shown that among the numerous targets of miR-125b, there are several repressors of neuronal genes (Le et al., 2009), highlighting a role for miR-125b in inducing neuronal differentiation.

MicroRNA-dependent regulation of neuronal maturation and integration

Once newborn cells are committed to their neuronal fate and start differentiating into mature neurons they start to develop axonal and dendritic processes, as well as synaptic connections with afferent neurons. The development of strong synaptic connections is crucial, as absence of synaptic input will cause the immature neuron to go into apoptosis. Several microRNAs are known to regulate the processes involved in maturation and integration.

miR-34a negatively regulates neurite growth, dendritic branching, and overall dendritic complexity. Further electrophysiological characterization of neurons in which miR-34a was overexpressed revealed changes in synaptic development, indicating a reduction in spine functions. This decreased spine functioning can at least be partially explained by the fact that two key targets of miR-34a are synaptotagmin1 and syntaxin-1A, both essential for synaptic transmission (Agostini et al., 2011).

Similarly, miR-137, which is specifically enriched in neurons, also negatively regulates dendritic outgrowth, complexity, and spine formation by regulating Mib-1 expression, an ubiquitin ligase crucial for neuronal development (Smrt et al., 2010).

miR-125b also negatively regulates spine development, resulting in spines of increased length, but decreased width. Furthermore, miR-125b was shown to repress NMDA receptor subunit 2A, resulting in changed functionality of glutamatergic synapses (Edbauer et al., 2010).

Other microRNAs known to negatively regulate dendritic spine formation are miR-134 and miR-138. These are specifically localized in dendritic spines, where miR-134 was shown to downregulate Limk1, a transcription factor crucial for spinogenesis (Schratt et al., 2006), whereas miR-138 controls the expression of APT1, an enzyme that regulates the palmitoylation state of many synaptic proteins (Siegel et al., 2009).

microRNAs known to positively regulate neuronal maturation include miR-124 and miR-132. Removal of miR-124 by knocking out Rcnr3, the primary source of miR-124, results in severe neuronal malformation and aberrant axonal sprouting. These effects are mediated by Lhx2, a primary target of miR-124 (Sanuki et al., 2011). miR-132 is thought to shape dendritic spine formation in an activity-dependent manner. Expression of miR-132 is regulated by CREB expression and is necessary and sufficient for spine formation. Inhibition of miR-132 decreases mESPC frequency and GluR1-positive spines, pointing towards a key role for miR-132 in functional spinogenesis (Impey et al., 2010).

ALTERATIONS IN AHN UNDER PATHOLOGICAL CONDITIONS

Despite being tightly regulated, numerous internal and external factors can compromise AHN. This becomes strikingly clear when studying the neurogenic processes under pathological conditions in which it is affected, e.g., neurodegenerative diseases or epilepsy. AHN seems to be particularly vulnerable to such pathologies, resulting in numerous changes, which together are pooled under the umbrella term "aberrant AHN" (Jessberger et al., 2007b). Whether alterations in AHN underlie certain pathologies or are just one of the numerous consequences still deserves extensive research. However, there is already ample evidence connecting aberrant AHN and the concomitant cognitive deficits associated with epilepsy. In the next sections, we will discuss further the evidence supporting this statement.

EPILEPSY AND AHN

Epilepsy is a neurological disorder characterized by the occurrence of chronic spontaneous seizures and is often accompanied by cognitive deficits (Elger et al., 2004; Helmstaedter et al., 2003; Pitkänen et al., 2007, 2009). Though seizures can arise from numerous locations in the brain, the largest group of patients suffers from Temporal Lobe Epilepsy, which is characterized by seizures originating from the hippocampal region and is clinically diagnosed by visualization of severe hippocampal sclerosis visible on MRI (Elger et al., 2004; Engel, 1998; Helmstaedter et al., 2003).

Most nongenetic types of epilepsy seem to share a common developmental window characterized by a first insult, followed by a silent latent phase during which no seizures arise and eventually leading to a chronic epileptic phase, during which seizures spontaneously arise. It is commonly thought that the original insult triggers multiple reactive processes in the brain, which, in turn, contribute to the development of the chronic epileptic condition (Pitkänen et al., 2007, 2009).

One of the processes severely affected by seizures is AHN. It was already shown in 1997 that seizures strongly induce proliferation in the SGZ; however, this induction was not lasting, and even resulted in diminished proliferation in the long term (Parent et al., 1997). Besides changes in proliferation, newborn neurons born under seizure conditions show several morphological and functional differences from neurons born in a healthy hippocampus. Changes include somatic hypertrophy, presence of hilar basal dendrites (Murphy et al., 2012), ectopic location of newborn granule cells in the hilus or molecular layer (Parent et al., 1997), axonal sprouting towards the molecular layer forming recurrent circuits (Scharfman et al., 2000), and changes in spine density accompanied by abnormal excitability (Dashtipour et al., 2001; Murphy et al., 2011; Wood et al., 2011). All these changes result in a hyperexcitable recurrent network within the DG, raising questions regarding a possible causal role for aberrant AHN in the development of chronic epilepsy (Bielefeld et al., 2014).

Whether these alterations in AHN play a crucial role in the development of chronic epilepsy is not yet agreed upon, but several studies have already shown that AHN plays a significant role in epilepsy development in mouse models. Preventing aberrant AHN by completely removing AHN by pharmacological or genetic strategies proved to be sufficient to prevent or slow down the development of a chronic epileptic state, or decrease the severity of the chronic seizures compared to mice with functional neurogenesis, after induction of a severe seizure-inducing insult known as Status Epilepticus (SE) (Cho et al., 2015; Jung et al., 2004). Genetic induction of ectopic granule cells with altered dendritic trees, a common hallmark of the epileptic hippocampus, results in spontaneous interictal activity and significantly increases seizure-susceptibility in mice (Korn et al., 2016). Beyond a role in epilepsy development, it has been shown that SE-induced aberrant AHN underlies cognitive deficits at least in mouse models of epilepsy. In agreement with this conclusion, preventing SE-induced aberrant AHN results in improved hippocampus-dependent cognitive capacity (Cho et al., 2015; Jessberger et al., 2007a).

As described before, microRNAs play a crucial role in regulating AHN. Several studies have already been undertaken to profile microRNA expression, both in human patients as well as in animal models of epilepsy. Interpretation of human data on this topic needs caution, as most studies are done on tissue obtained from patients who are already chronically epileptic; it is impossible to extrapolate the microRNA profile obtained to the developmental stages of the disease.

More thorough data come from animal studies. The big advantage of these models is that we can induce the first seizure insult, known as SE, and follow the development of the disease over time. Multiple animal studies have provided us with differentially expressed microRNAs during different stages of the disease. Some interesting microRNAs related to seizure-induced aberrant AHN will be discussed in more detail below and are summarized in Table 9.2A.

Micrornas affecting the neurogenic pool capacity after SE

miR-184 has been identified both in human temporal lobe epilepsy (TLE) patients as well as in a mouse model as differentially expressed. In human focal TLE patients, miR-184 was significantly upregulated during the chronic stage of the disease (Haenisch et al., 2015). Caution should be taken here as it is unclear how different antiepileptic drugs could interfere with miR-expression profiles. One animal study that used seizure preconditioning also showed upregulation of miR-184 after seizures. As described before, miR-184 activates Notch signaling by targeting its repressor Numbl resulting in asymmetrical division and increased proliferation. Significantly, this study showed that miR-184 expression after seizures protects neurons from seizure-induced death, while not exploring Notch-signaling based proliferative effects (McKiernan et al., 2012). The second miR known to regulate Notch signaling, miR-34a, is also commonly found upregulated in animal models of epilepsy at different stages of the disease (Gorter et al., 2014; Jimenez-Mateos and Henshall, 2013). One study targeted miR-34a and showed a role for miR-34a in the induction of neuronal cell death after seizures. Again, Notch-dependent changes in proliferation were not assessed (Hu et al., 2012a).

MicroRNAs affecting stem-cell self-renewal, proliferation, apoptosis, and differentiation after SE

Expression of the Let-7 family is significantly affected by the initial seizure insult, though its direction changes in a temporal manner. Shortly after induction of SE, Let-7 expression decreases, followed by a peak expression 24 hours later. After this peak, expression was again downregulated lasting up till 50 days post-SE (Song et al., 2011). As described previously, Let-7 regulates expression of the TLX receptor, maintaining the balance between self-renewal and differentiation. Since

Table 9.2 Overview of Differentially Expressed microRNAs and Their Effect on Hippocampal Neurogenesis Under Pathological Condition, such as (A) Epilepsy and (B) Alzheimer's Disease

A. Differentially Expressed AHN-related microRNAs in Epilepsy

miR	Expression During Acute Stage	Expression During Chronic Stage	Biological Consequence	References
miR-184	Up	Up	Increased proliferation	Haenisch et al. (2015), McKiernan et al. (2012)
miR-34a	Up	Up	Increased proliferation Increased apoptosis	Gorter et al. (2014), Hu et al. (2012a), Jimenez-Mateos and Henshall (2013)
miR-124	Up	Up	Increased neuronal differentiation decreased apoptosis	Peng et al. (2013), Schouten et al. (2015)
miR-137	Up	Up	Increased proliferation Decreased apoptosis	Schouten et al. (2015), Song et al. (2011)
miR-9	Down	Up	Dynamic changes in proliferation and differentiation	Pichardo-Casas et al. (2012), Song et al. (2011)
Let-7b	Down	Down	Increased proliferation Decreased differentiation	Song et al. (2011)
miR-134	Up	Up	Decreased spinogenesis	Peng et al. (2013)

B. Differentially Expressed AHN-related microRNAs in Alzheimer's Disease

miR	Expression	Biological Consequence	References
miR-9	Upregulated Downregulated	Decreased proliferation, increased differentiation Increased proliferation, decreased differentiation	Sethi and Lukiw (2009) Hébert et al. (2008)
miR-34a	Upregulated	Neuronal cell death	Zovoilis et al. (2011), Wang et al. (2009)
miR-206	Upregulated	Lower BDNF levels, decreased spinogenesis	Lee et al. (2012)

TLX activation results in activation of the Wnt/B-catenin pathway, this results in increased proliferation and decreased differentiation. Hence, decreased Let-7 expression would stimulate proliferation shortly after the insult, whereas increased Let-7 expression after 24 hours could potentially function as an intrinsic control of proliferation.

Like the Let-7 family, miR-9 is also part of a regulatory loop with the TLX receptor, and it is also differentially expressed in a temporal manner. Shortly after SE, miR-9 is downregulated (Pichardo-Casas et al., 2012), whereas during the chronic epileptic stage miR-9 is upregulated

(Song et al., 2011). Again, this could explain the short-term increased proliferation, commonly found after SE-induction.

miR-137, which also comprises a regulatory loop with the TLX receptor to control proliferation, is found to be upregulated in both acute (Schouten et al., 2015) and chronic (Song et al., 2011) stages of animal SE models. Together, these data show the complexity of microRNA function. With multiple microRNAs controlling the same pathway, and expression profiles of these microRNAs being different after SE, this indicates complex control of the switch from proliferation to differentiation after seizures.

miR-124, known for its role in neuronal differentiation and apoptotic selection of newborn neuroblasts, is also upregulated during both the acute and chronic, but not the latent, stage of epilepsy development in mouse models (Peng et al., 2013). Schouten et al (2015) showed that both miR-124 and miR-137 work together to regulate NSPC apoptosis by inhibiting the proapoptotic protein Bcl2L13. Hence, upregulation of miR-124 and miR-137 shortly after SE could also provide an explanation of the increased survival of unfit newborn neurons commonly found in the epileptic hippocampus.

MicroRNAs affecting neuronal maturation after SE

From the microRNAs known to negatively regulate neuronal maturation, such as miR-34a, miR-137, miR-125b, miR-134, and miR-138 (Fig. 9.1C), several are differentially expressed after SE in animal models. miR-34a, miR-137, and miR-134 were found upregulated after SE (Gorter et al., 2014; Hu et al., 2012a; Jimenez-Mateos and Henshall, 2013; Peng et al., 2013; Schouten et al., 2015; Song et al., 2011). Together, upregulation of these microRNAs should result in decreased neurite outgrowth, dendritic complexity, and spinogenesis, hampering maturation of newborn neurons.

On the other hand, miR-124, which is known to promote neurite outgrowth, is also found upregulated after SE (Peng et al., 2013; Schouten et al., 2015), again showing the complex regulation of neuronal maturation by microRNAs.

ALZHEIMER'S DISEASE AND AHN

Alzheimer's disease (AD) is a neurodegenerative disorder characterized by the accumulation of B-amyloid resulting in senile plaques and the hyperphosphorylation of the microtubule-associated protein Tau resulting in intracellular neurofibrillary tangles (Ittner and Götz, 2011; Maccioni et al., 2001). AD patients concomitantly show severe and progressive memory deficits, atrophy of specific brain regions and gliosis. Several studies have shown that neuroinflammatory and neurodegenerative processes can severely affect adult neurogenesis (Fuster-Matanzo et al., 2013). As mentioned in the case of epilepsy, a possible relation with the cognitive deficits observed in AD patient could be partially driven by aberrant AHN.

Data from human AD patients are hard to interpret due to major challenges when it comes to tissue collection and huge differences due to severity of the individual pathology. However, most data seem to indicate a decrease in the number of NSPCs, compensated by increased proliferation of NPCs that however fail to become fully functional and integrated neurons (Li et al., 2008; Perry et al., 2012).

More data are available from preclinical animal studies, for which several transgenic mouse lines are available that mimic AD. In most of these models, familial mutations of either the APP or the PS1 gene are introduced into the mouse genome, resulting in the accumulation of B-amyloid and Tau and cognitive deficits as seen in human AD patients. Most of these studies have shown decreased proliferation, differentiation, and survival of newborn granule cells; however, this seems to be very dependent on the stage of the disease and the extent of the pathology (Brasnjevic et al., 2013; Chen et al., 2008; Lilja et al., 2013; Morgenstern et al., 2013).

MicroRNAs AND ABERRANT AHN IN AD

Numerous studies have focused on identifying roles for microRNAs in the processing of APP and Tau to further elucidate the developmental progress of the disease (reviewed in Fitzsimons et al., 2014; Van den Hove et al., 2014). However, contrary to the extensive literature available on seizure-induced changes in microRNA expression profiles resulting in changes in AHN, such profiling studies have not been widely undertaken yet in the AD field. Several implicated microRNAs will be discussed below and are summarized in Table 9.2B.

One study of major interest performed extensive microRNA profiling from blood samples obtained from AD patients, identifying numerous microRNAs differentially expressed in AD. Significantly, when performing Gene Ontology on the predicted targets of these microRNAs biological processes involving neural projection development, neural projection morphogenesis, neurogenesis, neuron development, and neuroblast division were identified. Though of course not all restricted to AHN, this study at least provides an indication that differential expression of microRNAs in AD patients could underlie several of the changes observed in AHN (Leidinger et al., 2013). Multiple other human studies have found deregulation of several AHN-related microRNAs including miR-9. Expression of miR-9 however differed between studies, implying caution when drawing conclusions (Hébert et al., 2008; Sethi and Lukiw, 2009). Preclinical studies showing direct effects of AD-related microRNAs on AHN are scarce; however, multiple studies have undertaken approaches to elucidate the role of miR-9 in AD-related AHN alterations. One study showed that administration of Aβ to primary hippocampal cultures significantly decreased miR-9 expression, indicating a possible role for Aβ-related changes in AHN (Schonrock et al., 2010).

Neuronal cell death, which is under certain circumstances regulated by microRNAs, such as miR-34a, is another process deregulated in AD. Significantly, upregulation of miR-34a is found both in human AD tissue as well as in brain samples from an AD mouse model, even at time points before Aβ deposition occurs (Wang et al., 2009; Zovoilis et al., 2011). Upregulation of miR-34a results in increased neuronal cell death, possibly providing a more causal role in AD pathology onset (Wang et al., 2009).

Furthermore, one study identified miR-206 to be upregulated both in whole-brain samples in a mouse model for AD as well as in temporal cortex tissue from human AD patients. Though miR-206 is not known to directly regulate gene-expression critical for AHN, it does regulate BDNF expression, one of the crucial trophic factors required for spinogenesis during AHN (Lee et al., 2012).

POSSIBILITY OF MicroRNA-BASED AHN-RELATED THERAPY FOR NEUROLOGICAL AND NEURODEGENERATIVE DISORDERS

Summarizing the scientific literature regarding microRNA control of neurogenesis in health and disease discussed before proves a daunting challenge given its complexity. How do different microRNAs achieve precise control of gene networks during AHN? In previous sections, we have mentioned some examples of targets that may function as hubs for microRNA regulation. This concept is in agreement with recent observations, which have indicated that convergent microRNA actions have the potential to be one of the keys to disclosing how microRNAs achieve the precise coordination of complex biological processes (Barca-Mayo and De Pietri Tonelli, 2014) (Fig. 9.2). These convergent actions remain a poorly understood layer of complexity in microRNA signaling, but some examples involving NSPC and AHN have been recently described (Schouten et al., 2015). With multiple microRNAs targeting the same biological pathways, complex regulatory loops within these pathways, and bidirectional differential expression patterns of involved microRNAs in pathological conditions, one might argue that microRNAs-based therapeutic approaches seem rather challenging. However, several studies have undertaken single miR-based therapeutic approaches in both preclinical epilepsy and AD models, though not primarily focusing on AHN.

MICRORNA-BASED THERAPY IN EPILEPSY

Several studies have undertaken miR-34a silencing approaches to rescue seizure-induced apoptosis. One study showed that antagomir-34a administration during SE successfully reduced neuronal apoptosis, but this was only assessed for the CA1 and CA3 region, not the DG, so caution should be taken when extrapolating these results to AHN (Hu et al., 2012a). Another study also applied antagomir-34a, but 24 hours post-SE, and failed to find any neuroprotective effects, indicating a potential time-dependent treatment window (Sano et al., 2012). Both studies also failed to show any protective effects of antagomir-34a administration on seizure severity or duration during SE.

More compelling evidence for possible miR-based therapy comes from several studies administering antagomir-134. Administration of antagomir-134 24 hours before SE-induction reduced kainic acid-induced SE by 50%−70%, whereas administration 1 hour after SE onset reduced the later occurrence of spontaneous chronic seizures by over 90%. Antagomir-134 furthermore reduced CA3 pyramidal spine density and mitigated other pathological features of TLE, such as neuronal cell loss and astrogliosis, but the effects on AHN were not addressed (Jimenez-Mateos et al., 2012; Jimenez-Mateos et al., 2015).

MicroRNA-based therapy in AD

The role of aberrant AHN in AD most likely is less severe compared to epilepsy, though cognitive deficits associated with AD might be caused by AHN dysregulation.

One study identified miR-34c to be upregulated in the hippocampus in both 24-month-old healthy mice and in an AD mouse model. Though not extensively studied, miR-34c seems to play a role in learning, targeting many learning-associated genes. Applying a miR-34 seed inhibitor in 12-month-old AD mice significantly rescued the cognitive deficits observed in nontreated mice

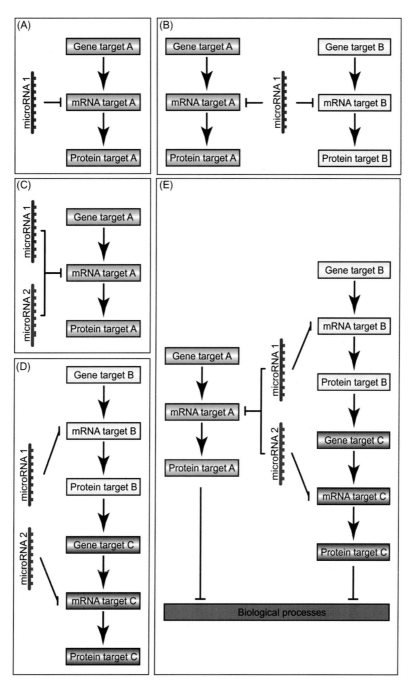

FIGURE 9.2 Schematic overview of the different mechanisms through which microRNAs can exert their effect on biological processes

(*Continued*)

(Zovoilis et al., 2011). Though this indicates a possible role for AHN, this was not further addressed in this study.

Another study aimed to target the upregulation of miR-206 observed both in human AD patients and AD mouse models. As described before, miR-206 does not directly regulate AHN, but targets BDNF, a trophic factor known to have beneficial effects on AHN. Administration of antagomir-206 either 1 or 3 weeks before behavioral testing in 12-month-old AD mice significantly increased hippocampus-dependent cognition compared to nontreated mice. In parallel, antagomir-206 administration resulted in increased synaptic density and increased neurogenesis as measured by the amount of immature neurons in the SGZ. This remarkable rescue of hippocampus-dependent cognition was not associated with any changes in amyloid pathology, indicating that the cognitive deficits in AD might be a comorbidity that could potentially be treated even when AD pathology is present (Lee et al., 2012).

FUTURE PERSPECTIVES

Given the relative small body of evidence currently available, it is still hard to predict the future of microRNA-based therapies targeting AHN for neurological and neurodegenerative disorders. However, preclinical epilepsy studies have shown a critical role for AHN in both disease development and cognitive comorbidity, whereas preclinical AD studies have only focused on AD-related cognitive deficits. If these cognitive deficits indeed depend on the development of aberrant AHN and microRNAs could revert or prevent this aberrant AHN, then microRNAs could emerge as feasible therapeutic targets. Several exciting approaches have already rendered significant success in preclinical epilepsy research, confirming the potential of microRNA-based therapy. However, extensive research is still needed to further explore and improve possible microRNA-based approached as antiepileptogenic drugs.

Most promising data come from studies applying microRNA-based approaches both in epilepsy and AD research to rescue disease progression and/or disease-associated cognitive deficits, a feature that significantly would improve life-quality of patients. AHN is now generally accepted to be crucial for hippocampal plasticity and new memory formation, making disease-associated aberrant AHN a prime target for future therapeutic approached.

Where most studies are still focusing on single microRNA-based approaches, fundamental research on miR-dependent regulation of AHN shows a more complex situation. The complex interplay of different microRNAs on the same biological process or signaling pathway, and the recent discovery of

◀ (A) Classic inhibition of a single-mRNA target by a single microRNA, (B) though one microRNA can potentially inhibit different mRNA targets. However, most physiological and pathological conditions studied point towards groups of microRNAs being differentially expressed, together with their targets. Furthermore, it seems that different microRNAs have a coordinated action on common targets. This coordinated action can be achieved via (C) cooperative action of a single-mRNA target, (D) convergence of different microRNAs on the same biological pathway, or (E) through convergence of different microRNAs on entire biological processes.

Adapted from (Barca-Mayo O., De Pietri Tonelli D.: Convergent microRNA actions coordinate neocortical development, Cell. Mol. Life Sci. 71(16):2975–2995, 2014).

microRNA-cooperativity or other forms of convergent action warrant caution when undertaking single microRNA-based therapeutic approaches (Barca-Mayo and De Pietri Tonelli, 2014). Thus, most likely, multi-microRNA-based approaches targeting several aspects of AHN will provide more promising avenues to successfully rescue disease-associated aberrant AHN; thereby, they could pave the way towards future clinical applications of microRNA-based therapies.

ACKNOWLEDGMENTS

This work was supported by an Innovational Research Incentives Scheme Vidi grant from the Netherlands Organization for Scientific Research (NWO) (grant number H64.09.016) and grants from the International Foundation for Alzheimer Research (ISAO) and Alzheimer Nederland to CPF.

REFERENCES

Agostini, M., Tucci, P., Steinert, J.R., et al., 2011. MicroRNA-34a regulates neurite outgrowth, spinal morphology, and function. Proc. Natl. Acad. Sci. U. S. A. 108 (52), 21099−21104.

Aimone, J.B., Li, Y., Lee, S.W., et al., 2014. Regulation and function of adult neurogenesis: from genes to cognition. Physiol. Rev. 94 (4), 991−1026.

Antonini, D., Russo, M.T., De Rosa, L., et al., 2010. Transcriptional repression of miR-34 family contributes to p63-mediated cell cycle progression in epidermal cells. J. Invest. Dermatol. 130 (5), 1249−1257.

Aranha, M.M., Santos, D.M., Xavier, J.M., et al., 2010. Apoptosis-associated microRNAs are modulated in mouse, rat and human neural differentiation. BMC Genomics 11, 514.

Arvidsson, A., Collin, T., Kirik, D., et al., 2002. Neuronal replacement from endogenous precursors in the adult brain after stroke. Nat. Med. 8 (9), 963−970.

Ballas, N., Battaglioli, E., Atouf, F., et al., 2001. Regulation of neuronal traits by a novel transcriptional complex. Neuron 31 (3), 353−365.

Barca-Mayo, O., De Pietri Tonelli, D., 2014. Convergent microRNA actions coordinate neocortical development. Cell. Mol. Life Sci. 71 (16), 2975−2995.

Bertrand, N., Castro, D.S., Guillemot, F., 2002. Proneural genes and the specification of neural cell types. Nat. Rev. Neurosci. 3 (7), 517−530.

Biebl, M., Cooper, C.M., Winkler, J., et al., 2000. Analysis of neurogenesis and programmed cell death reveals a self-renewing capacity in the adult rat brain. Neurosci. Lett. 291 (1), 17−20.

Bielefeld, P., van Vliet, E.A., Gorter, J.A., et al., 2014. Different subsets of newborn granule cells: a possible role in epileptogenesis? Eur. J. Neurosci. 39 (1), 1−11.

Bizon, J.L., Lee, H.J., Gallagher, M., 2004. Neurogenesis in a rat model of age-related cognitive decline. Aging Cell 3 (4), 227−234.

Brasnjevic, I., Lardenoije, R., Schmitz, C., et al., 2013. Region-specific neuron and synapse loss in the hippocampus of APP(SL)/PS1 knock-in mice. Transl. Neurosci. 4 (1), 8−19.

Chen, Q., Nakajima, A., Choi, S.H., et al., 2008. Adult neurogenesis is functionally associated with AD-like neurodegeneration. Neurobiol. Dis. 29 (2), 316−326.

Chen, X., Liang, H., Zhang, J., et al., 2012. Secreted microRNAs: a new form of intercellular communication. Trends Cell Biol. 22 (3), 125−132.

Chenn, A., Walsh, C.A., 2002. Regulation of cerebral cortical size by control of cell cycle exit in neural precursors. Science 297 (5580), 365−369.

Cho, K.-O., Lybrand, Z.R., Ito, N., et al., 2015. Aberrant hippocampal neurogenesis contributes to epilepsy and associated cognitive decline. Nat. Commun. 6, 6606.

Conaco, C., Otto, S., Han, J.-J., et al., 2006. Reciprocal actions of REST and a microRNA promote neuronal identity. Proc. Natl. Acad. Sci. U. S. A. 103 (7), 2422–2427.

Conover, J.C., Doetsch, F., Garcia-Verdugo, J.M., et al., 2000. Disruption of Eph/ephrin signaling affects migration and proliferation in the adult subventricular zone. Nat. Neurosci. 3 (11), 1091–1097.

Cui, Y., Xiao, Z., Han, J., et al., 2012. MiR-125b orchestrates cell proliferation, differentiation and migration in neural stem/progenitor cells by targeting Nestin. BMC Neurosci. 13, 116.

Darsalia, V., Heldmann, U., Lindvall, O., et al., 2005. Stroke-induced neurogenesis in aged brain. Stroke 36 (8), 1790–1795.

Dashtipour, K., Tran, P.H., Okazaki, M.M., et al., 2001. Ultrastructural features and synaptic connections of hilar ectopic granule cells in the rat dentate gyrus are different from those of granule cells in the granule cell layer. Brain Res. 890 (2), 261–271.

Deng, W., Aimone, J.B., Gage, F.H., 2010. New neurons and new memories: how does adult hippocampal neurogenesis affect learning and memory? Nat. Rev. Neurosci. 11 (5), 339–350.

Doetsch, F., Petreanu, L., Caille, I., et al., 2002. EGF converts transit-amplifying neurogenic precursors in the adult brain into multipotent stem cells. Neuron 36 (6), 1021–1034.

Doetsch, F., 2003. A niche for adult neural stem cells. Curr. Opin. Genet. Dev. 13 (5), 543–550.

Edbauer, D., Neilson, J.R., Foster, K.A., et al., 2010. Regulation of synaptic structure and function by FMRP-associated microRNAs miR-125b and miR-132. Neuron 65 (3), 373–384.

Elger, C.E., Helmstaedter, C., Kurthen, M., 2004. Chronic epilepsy and cognition. Lancet Neurol. 3 (11), 663–672.

Encinas, J.M., Vaahtokari, A., Enikolopov, G., 2006. Fluoxetine targets early progenitor cells in the adult brain. Proc. Natl. Acad. Sci. U. S. A. 103 (21), 8233–8238.

Engel, J., 1998. Etiology as a risk factor for medically refractory epilepsy: a case for early surgical intervention. Neurology 51 (5), 1243–1244.

Fineberg, S.K., Datta, P., Stein, C.S., et al., 2012. MiR-34a represses Numbl in murine neural progenitor cells and antagonizes neuronal differentiation. PLoS ONE 7 (6), e38562.

Fitzsimons, C.P., van Bodegraven, E., Schouten, M., et al., 2014. Epigenetic regulation of adult neural stem cells: implications for Alzheimer's disease. Mol. Neurodegener. 9, 25.

Fuster-Matanzo, A., Llorens-Martín, M., Hernández, F., et al., 2013. Role of neuroinflammation in adult neurogenesis and Alzheimer disease: therapeutic approaches. Mediators Inflamm. 260925, 2013.

Goldman, S.A., Chen, Z., 2011. Perivascular instruction of cell genesis and fate in the adult brain. Nat. Neurosci. 14 (11), 1382–1389.

Gorter, J.A., Iyer, A., White, I., et al., 2014. Hippocampal subregion-specific microRNA expression during epileptogenesis in experimental temporal lobe epilepsy. Neurobiol. Dis. 62, 508–520.

Grimson, A., Farh, K.K.H., Johnston, W.K., et al., 2007. MicroRNA targeting specificity in mammals: determinants beyond seed pairing. Mol. Cell 27 (1), 91–105.

Gu, F., Hata, R., Ma, Y.-J., et al., 2005. Suppression of Stat3 promotes neurogenesis in cultured neural stem cells. J. Neurosci. Res. 81 (2), 163–171.

Haenisch, S., Zhao, Y., Chhibber, A., et al., 2015. SOX11 identified by target gene evaluation of miRNAs differentially expressed in focal and non-focal brain tissue of therapy-resistant epilepsy patients. Neurobiol. Dis. 77, 127–140.

Hébert, S.S., Horré, K., Nicolaï, L., et al., 2008. Loss of microRNA cluster miR-29a/b-1 in sporadic Alzheimer's disease correlates with increased BACE1/beta-secretase expression. Proc. Natl. Acad. Sci. U. S. A. 105 (17), 6415–6420.

Helmstaedter, C., Kurthen, M., Lux, S., et al., 2003. Chronic epilepsy and cognition: a longitudinal study in temporal lobe epilepsy. Ann. Neurol. 54 (4), 425–432.

Hu, K., Xie, Y.-Y., Zhang, C., et al., 2012a. MicroRNA expression profile of the hippocampus in a rat model of temporal lobe epilepsy and miR-34a-targeted neuroprotection against hippocampal neurone cell apoptosis post-status epilepticus. BMC Neurosci. 13, 115.

Hu, X.-L., Wang, Y., Shen, Q., 2012b. Epigenetic control on cell fate choice in neural stem cells. Protein Cell 3 (4), 278−290.

Impey, S., Davare, M., Lesiak, A., et al., 2010. An activity-induced microRNA controls dendritic spine formation by regulating Rac1-PAK signaling. Mol. Cell. Neurosci. 43 (1), 146−156.

Ittner, L.M., Götz, J., 2011. Amyloid-β and tau − a toxic pas de deux in Alzheimer's disease. Nat. Rev. Neurosci. 12 (2), 65−72.

Jakubs, K., Nanobashvili, A., Bonde, S., et al., 2006. Environment matters: synaptic properties of neurons born in the epileptic adult brain develop to reduce excitability. Neuron 52 (6), 1047−1059.

Jessberger, S., Nakashima, K., Clemenson, G.D., et al., 2007a. Epigenetic modulation of seizure-induced neurogenesis and cognitive decline. J. Neurosci. 27 (22), 5967−5975.

Jessberger, S., Zhao, C., Toni, N., et al., 2007b. Seizure-associated, aberrant neurogenesis in adult rats characterized with retrovirus-mediated cell labeling. J. Neurosci. 27 (35), 9400−9407.

Jimenez-Mateos, E.M., Engel, T., Merino-Serrais, P., et al., 2015. Antagomirs targeting microRNA-134 increase hippocampal pyramidal neuron spine volume in vivo and protect against pilocarpine-induced status epilepticus. Brain Struct. Funct. 220 (4), 2387−2399.

Jimenez-Mateos, E.M., Engel, T., Merino-Serrais, P., et al., 2012. Silencing microRNA-134 produces neuroprotective and prolonged seizure-suppressive effects. Nat. Med. 18 (7), 1087−1094.

Jimenez-Mateos, E.M., Henshall, D.C., 2013. Epilepsy and microRNA. Neuroscience 238, 218−229.

Jin, K., Zhu, Y., Sun, Y., et al., 2002. Vascular endothelial growth factor (VEGF) stimulates neurogenesis in vitro and in vivo. Proc. Natl. Acad. Sci. U. S. A. 99 (18), 11946−11950.

Jobe, E.M., McQuate, A.L., Zhao, X., 2012. Crosstalk among epigenetic pathways regulates neurogenesis. Front. Neurosci. 6, 59.

Jung, K.-H., Chu, K., Kim, M., et al., 2004. Continuous cytosine-b-D-arabinofuranoside infusion reduces ectopic granule cells in adult rat hippocampus with attenuation of spontaneous recurrent seizures following pilocarpine-induced status epilepticus. Eur. J. Neurosci. 19 (12), 3219−3226.

Kawahara, H., Imai, T., Okano, H., 2012. MicroRNAs in neural stem cells and neurogenesis. Front. Neurosci. 6, 30.

Kempermann, G., Jessberger, S., Steiner, B., et al., 2004. Milestones of neuronal development in the adult hippocampus. Trends Neurosci. 27 (8), 447−452.

Knoth, R., Singec, I., Ditter, M., et al., 2010. Murine features of neurogenesis in the human hippocampus across the lifespan from 0 to 100 years. PLoS ONE 5 (1), e8809.

Korn, M.J., Mandle, Q.J., Parent, J.M., 2016. Conditional disabled-1 deletion in mice alters hippocampal neurogenesis and reduces seizure threshold. Front. Neurosci. 10, 63.

Kosaka, N., Iguchi, H., Yoshioka, Y., et al., 2010. Secretory mechanisms and intercellular transfer of microRNAs in living cells. J. Biol. Chem. 285 (23), 17442−17452.

Krichevsky, A.M., Sonntag, K.-C., Isacson, O., et al., 2006. Specific microRNAs modulate embryonic stem cell-derived neurogenesis. Stem Cells 24 (4), 857−864.

Kuhn, H.G., Biebl, M., Wilhelm, D., et al., 2005. Increased generation of granule cells in adult Bcl-2-overexpressing mice: a role for cell death during continued hippocampal neurogenesis. Eur. J. Neurosci. 22 (8), 1907−1915.

Lang, M.-F., Shi, Y., 2012. Dynamic Roles of microRNAs in Neurogenesis. Front. Neurosci. 6, 71.

Lange, C., Huttner, W.B., Calegari, F., 2009. Cdk4/cyclinD1 overexpression in neural stem cells shortens G1, delays neurogenesis, and promotes the generation and expansion of basal progenitors. Cell Stem Cell 5 (3), 320−331.

Le, M.T.N., Xie, H., Zhou, B., et al., 2009. MicroRNA-125b promotes neuronal differentiation in human cells by repressing multiple targets. Mol. Cell. Biol. 29 (19), 5290−5305.

Lee, S.-T., Chu, K., Jung, K.-H., et al., 2012. MiR-206 regulates brain-derived neurotrophic factor in Alzheimer disease model. Ann. Neurol. 72 (2), 269–277.

Leidinger, P., Backes, C., Deutscher, S., et al., 2013. A blood based 12-miRNA signature of Alzheimer disease patients. Genome Biol. 14 (7), R78.

Li, B., Yamamori, H., Tatebayashi, Y., et al., 2008. Failure of neuronal maturation in Alzheimer disease dentate gyrus. J. Neuropathol. Exp. Neurol. 67 (1), 78–84.

Lie, D.-C., Colamarino, S.A., Song, H.-J., et al., 2005. Wnt signalling regulates adult hippocampal neurogenesis. Nature 437 (7063), 1370–1375.

Lilja, A.M., Röjdner, J., Mustafiz, T., et al., 2013. Age-dependent neuroplasticity mechanisms in Alzheimer Tg2576 mice following modulation of brain amyloid-β levels. PLoS ONE 8 (3), e58752.

Lim, D.A., Huang, Y.-C., Swigut, T., et al., 2009. Chromatin remodelling factor Mll1 is essential for neurogenesis from postnatal neural stem cells. Nature 458 (7237), 529–533.

Lim, L.P., Lau, N.C., Garrett-Engele, P., et al., 2005. Microarray analysis shows that some microRNAs downregulate large numbers of target mRNAs. Nature 433 (7027), 769–773.

Liu, C., Teng, Z.-Q., Santistevan, N.J., et al., 2010. Epigenetic regulation of miR-184 by MBD1 governs neural stem cell proliferation and differentiation. Cell Stem Cell 6 (5), 433–444.

Lopez-Ramirez, M.A., Nicoli, S., 2014. Role of miRNAs and epigenetics in neural stem cell fate determination. Epigenetics 9 (1), 90–100.

Lu, M., Grove, E.A., Miller, R.J., 2002. Abnormal development of the hippocampal dentate gyrus in mice lacking the CXCR4 chemokine receptor. Proc. Natl. Acad. Sci. U. S. A. 99 (10), 7090–7095.

Lunyak, V.V., Burgess, R., Prefontaine, G.G., et al., 2002. Corepressor-dependent silencing of chromosomal regions encoding neuronal genes. Science 298 (5599), 1747–1752.

Ma, D.K., Marchetto, M.C., Guo, J.U., et al., 2010. Epigenetic choreographers of neurogenesis in the adult mammalian brain. Nat. Neurosci. 13 (11), 1338–1344.

Maccioni, R.B., Muñoz, J.P., Barbeito, L., 2001. The molecular bases of Alzheimer's disease and other neurodegenerative disorders. Arch. Med. Res. 32 (5), 367–381.

McKiernan, R.C., Jimenez-Mateos, E.M., Sano, T., et al., 2012. Expression profiling the microRNA response to epileptic preconditioning identifies miR-184 as a modulator of seizure-induced neuronal death. Exp. Neurol. 237 (2), 346–354.

Ming, G., Song, H., 2005. Adult neurogenesis in the mammalian central nervous system. Annu. Rev. Neurosci. 28, 223–250.

Ming, G.-L., Song, H., 2011. Adult neurogenesis in the mammalian brain: significant answers and significant questions. Neuron 70 (4), 687–702.

Moon, C., Yoo, J.-Y., Matarazzo, V., et al., 2002. Leukemia inhibitory factor inhibits neuronal terminal differentiation through STAT3 activation. Proc. Natl. Acad. Sci. U. S. A. 99 (13), 9015–9020.

Morgenstern, N.A., Giacomini, D., Lombardi, G., et al., 2013. Delayed dendritic development in newly generated dentate granule cells by cell-autonomous expression of the amyloid precursor protein. Mol. Cell. Neurosci. 56, 298–306.

Murphy, B.L., Hofacer, R.D., Faulkner, C.N., et al., 2012. Abnormalities of granule cell dendritic structure are a prominent feature of the intrahippocampal kainic acid model of epilepsy despite reduced postinjury neurogenesis. Epilepsia 53 (5), 908–921.

Murphy, B.L., Pun, R.Y.K., Yin, H., et al., 2011. Heterogeneous integration of adult-generated granule cells into the epileptic brain. J. Neurosci. 31 (1), 105–117.

Nakatomi, H., Kuriu, T., Okabe, S., et al., 2002. Regeneration of hippocampal pyramidal neurons after ischemic brain injury by recruitment of endogenous neural progenitors. Cell 110 (4), 429–441.

Packer, A.N., Xing, Y., Harper, S.Q., et al., 2008. The bifunctional microRNA miR-9/miR-9* regulates REST and CoREST and is downregulated in Huntington's disease. J. Neurosci. 28 (53), 14341–14346.

Palmer, T.D., Willhoite, A.R., Gage, F.H., 2000. Vascular niche for adult hippocampal neurogenesis. J. Comp. Neurol. 425 (4), 479−494.

Parent, J.M., Yu, T.W., Leibowitz, R.T., et al., 1997. Dentate granule cell neurogenesis is increased by seizures and contributes to aberrant network reorganization in the adult rat hippocampus. J. Neurosci. 17 (10), 3727−3738.

Peng, J., Omran, A., Ashhab, M.U., et al., 2013. Expression patterns of miR-124, miR-134, miR-132, and miR-21 in an immature rat model and children with mesial temporal lobe epilepsy. J. Mol. Neurosci. 50 (2), 291−297.

Perez-Moreno, M., Jamora, C., Fuchs, E., 2003. Sticky business: orchestrating cellular signals at adherens junctions. Cell 112 (4), 535−548.

Perry, E.K., Johnson, M., Ekonomou, A., et al., 2012. Neurogenic abnormalities in Alzheimer's disease differ between stages of neurogenesis and are partly related to cholinergic pathology. Neurobiol. Dis. 47 (2), 155−162.

Pichardo-Casas, I., Goff, L.A., Swerdel, M.R., et al., 2012. Expression profiling of synaptic microRNAs from the adult rat brain identifies regional differences and seizure-induced dynamic modulation. Brain Res. 1436, 20−33.

Pitkänen, A., Immonen, R.J., Gröhn, O.H.J., et al., 2009. From traumatic brain injury to posttraumatic epilepsy: what animal models tell us about the process and treatment options. Epilepsia 50 (Suppl 2), 21−29.

Pitkänen, A., Kharatishvili, I., Karhunen, H., et al., 2007. Epileptogenesis in experimental models. Epilepsia 48 (Suppl 2), 13−20.

Qu, Q., Sun, G., Li, W., et al., 2010. Orphan nuclear receptor TLX activates Wnt/beta-catenin signalling to stimulate neural stem cell proliferation and self-renewal. Nat. Cell Biol. 12 (1), 31−40, sup pp 1−9.

Sano, T., Reynolds, J.P., Jimenez-Mateos, E.M., et al., 2012. MicroRNA-34a upregulation during seizure-induced neuronal death. Cell Death Dis. 3, e287.

Santarelli, L., 2003. Requirement of hippocampal neurogenesis for the behavioral effects of antidepressants. Science 301 (5634), 805−809.

Sanuki, R., Onishi, A., Koike, C., et al., 2011. MiR-124a is required for hippocampal axogenesis and retinal cone survival through Lhx2 suppression. Nat. Neurosci. 14 (9), 1125−1134.

Scharfman, H.E., Goodman, J.H., Sollas, A.L., 2000. Granule-like neurons at the hilar/CA3 border after status epilepticus and their synchrony with area CA3 pyramidal cells: functional implications of seizure-induced neurogenesis. J. Neurosci. 20 (16), 6144−6158.

Scharfman, H.E., McCloskey, D.P., 2009. Postnatal neurogenesis as a therapeutic target in temporal lobe epilepsy. Epilepsy Res. 85 (2−3), 150−161.

Schonrock, N., Ke, Y.D., Humphreys, D., et al., 2010. Neuronal microRNA deregulation in response to Alzheimer's disease amyloid-beta. PLoS ONE 5 (6), e11070.

Schouten, M., Buijink, M.R., Lucassen, P.J., et al., 2012. New neurons in aging brains: molecular control by small non-coding RNAs. Front. Neurosci. 6, 25.

Schouten, M., Fratantoni, S.A., Hubens, C.J., et al., 2015. MicroRNA-124 and -137 cooperativity controls caspase-3 activity through BCL2L13 in hippocampal neural stem cells. Sci. Rep. 5, 12448.

Schratt, G.M., Tuebing, F., Nigh, E.A., et al., 2006. A brain-specific microRNA regulates dendritic spine development. Nature 439 (7074), 283−289.

Schwamborn, J.C., Berezikov, E., Knoblich, J.A., 2009. The TRIM-NHL protein TRIM32 activates microRNAs and prevents self-renewal in mouse neural progenitors. Cell 136 (5), 913−925.

Seidenfaden, R., Desoeuvre, A., Bosio, A., et al., 2006. Glial conversion of SVZ-derived committed neuronal precursors after ectopic grafting into the adult brain. Mol. Cell. Neurosci. 32 (1−2), 187−198.

Sethi, P., Lukiw, W.J., 2009. Micro-RNA abundance and stability in human brain: specific alterations in Alzheimer's disease temporal lobe neocortex. Neurosci. Lett. 459 (2), 100−104.

Shi, Y., Chichung Lie, D., Taupin, P., et al., 2004. Expression and function of orphan nuclear receptor TLX in adult neural stem cells. Nature 427 (6969), 78–83.

Shibata, M., Kurokawa, D., Nakao, H., et al., 2008. MicroRNA-9 modulates Cajal–Retzius cell differentiation by suppressing Foxg1 expression in mouse medial pallium. J. Neurosci. 28 (41), 10415–10421.

Shihabuddin, L.S., Horner, P.J., Ray, J., et al., 2000. Adult spinal cord stem cells generate neurons after transplantation in the adult dentate gyrus. J. Neurosci. 20 (23), 8727–8735.

Siegel, G., Obernosterer, G., Fiore, R., et al., 2009. A functional screen implicates microRNA-138-dependent regulation of the depalmitoylation enzyme APT1 in dendritic spine morphogenesis. Nat. Cell Biol. 11 (6), 705–716.

Smrt, R.D., Szulwach, K.E., Pfeiffer, R.L., et al., 2010. MicroRNA miR-137 regulates neuronal maturation by targeting ubiquitin ligase mind bomb-1. Stem Cells 28 (6), 1060–1070.

Snyder, J.S., Kee, N., Wojtowicz, J.M., 2001. Effects of adult neurogenesis on synaptic plasticity in the rat dentate gyrus. J. Neurophysiol. 85 (6), 2423–2431.

Song, Y.-J., Tian, X.-B., Zhang, S., et al., 2011. Temporal lobe epilepsy induces differential expression of hippocampal miRNAs including let-7e and miR-23a/b. Brain Res. 1387, 134–140.

Spalding, K.L., Bergmann, O., Alkass, K., et al., 2013. Dynamics of hippocampal neurogenesis in adult humans. Cell 153 (6), 1219–1227.

Stolt, C.C., Lommes, P., Sock, E., et al., 2003. The Sox9 transcription factor determines glial fate choice in the developing spinal cord. Genes Dev. 17 (13), 1677–1689.

Sun, G., Ye, P., Murai, K., et al., 2011. MiR-137 forms a regulatory loop with nuclear receptor TLX and LSD1 in neural stem cells. Nat. Commun. 2, 529.

Szulwach, K.E., Li, X., Smrt, R.D., et al., 2010. Cross talk between microRNA and epigenetic regulation in adult neurogenesis. J. Cell Biol. 189 (1), 127–141.

Takizawa, T., Nakashima, K., Namihira, M., et al., 2001. DNA methylation is a critical cell-intrinsic determinant of astrocyte differentiation in the fetal brain. Dev. Cell 1 (6), 749–758.

Van den Hove, D.L., Kompotis, K., Lardenoije, R., et al., 2014. Epigenetically regulated microRNAs in Alzheimer's disease. Neurobiol. Aging 35 (4), 731–745.

Wang, X., Liu, P., Zhu, H., et al., 2009. MiR-34a, a microRNA up-regulated in a double transgenic mouse model of Alzheimer's disease, inhibits bcl2 translation. Brain Res. Bull. 80 (4–5), 268–273.

Wood, J.C., Jackson, J.S., Jakubs, K., et al., 2011. Functional integration of new hippocampal neurons following insults to the adult brain is determined by characteristics of pathological environment. Exp. Neurol. 229 (2), 484–493.

Xin, H., Li, Y., Buller, B., et al., 2012. Exosome-mediated transfer of miR-133b from multipotent mesenchymal stromal cells to neural cells contributes to neurite outgrowth. Stem Cells 30 (7), 1556–1564.

Zhao, C., Sun, G., Li, S., et al., 2009. A feedback regulatory loop involving microRNA-9 and nuclear receptor TLX in neural stem cell fate determination. Nat. Struct. Mol. Biol. 16 (4), 365–371.

Zhao, C., Sun, G., Li, S., et al., 2010. MicroRNA let-7b regulates neural stem cell proliferation and differentiation by targeting nuclear receptor TLX signaling. Proc. Natl. Acad. Sci. U. S. A. 107 (5), 1876–1881.

Ziv, Y., Finkelstein, A., Geffen, Y., et al., 2007. A novel immune-based therapy for stroke induces neuroprotection and supports neurogenesis. Stroke 38 (2 Suppl), 774–782.

Zovoilis, A., Agbemenyah, H.Y., Agis-Balboa, R.C., et al., 2011. MicroRNA-34c is a novel target to treat dementias. EMBO J. 30 (20), 4299–4308.

TRANSCRIPTIONAL AND EPIGENETIC CONTROL OF ASTROGLIOGENESIS

10

Neha Tiwari and Benedikt Berninger

University Medical Center of the Johannes Gutenberg University Mainz, Mainz, Germany

INTRODUCTORY REMARKS

Astrocytes are a heterogeneous class of glial cells in the brain that fulfill an ever increasing list of functions important for the formation, maintenance, and plasticity of the brain (Bayraktar et al., 2014). In particular, astrocytes play key roles in regulating the neuronal microenvironment by contributing to ion and neurotransmitter homeostasis (Walz, 2000; Rothstein et al., 1996) and are important for neuronal energy supply (Choi et al., 2012). Moreover, astrocytes sense neuronal activity through neurotransmitter receptors on their surface and release so-called gliotransmitters, which in turn act on neurons, thereby contributing to synaptic processing and perhaps even plasticity (Henneberger et al., 2010; Jourdain et al., 2007; Pascual et al., 2005; Wang et al., 2006). In some instances, astrocytes serve as chemosensors, as shown for CO_2-sensitive astrocytes in respiratory control centers (Gourine et al., 2010). Furthermore, astrocytes, together with pericytes, play a fundamental role in the maintenance of the blood−brain-barrier by interacting with microvessels through their astrocytic endfeet (Abbott et al., 2006; Daneman et al., 2010). Finally, astrocytes are a critical component of the so-called neurovascular unit, which couples neural activity to local cerebral blood flow (Petzold and Murthy, 2011; Attwell et al., 2010). Given their important functions, astrocytes have recently been implicated in many neurological diseases such as amyotrophic lateral sclerosis, Alzheimer's disease, Huntington's disease, and Parkinson's disease (Maragakis and Rothstein, 2006; Molofsky et al., 2012). Although this remarkable plethora of specialized functions suggests a high degree of specification and molecular and functional heterogeneity, little is known regarding how these functions are controlled on a molecular level during the development of the nervous system. Therefore, it is important to understand the molecular mechanisms underlying the differentiation of astrocytes from neural stem cells (NSC), a process referred to herein as astrogliogenesis which commences largely after the end of neurogenesis (Kriegstein and Alvarez-Buylla, 2009). Throughout the last two decades, interest into the role of transcription factors (TFs) and epigenetic mechanisms in astrogliogenesis has markedly risen. The goal of this chapter is to provide an overview of our current knowledge regarding the transcriptional and epigenetic mechanisms underlying astrogliogenesis.

Essentials of Noncoding RNA in Neuroscience. DOI: http://dx.doi.org/10.1016/B978-0-12-804402-5.00010-8

TRANSCRIPTIONAL CONTROL OF ASTROGLIOGENESIS
STAT3 SIGNALING AND ASTROGLIOGENESIS

Twenty years ago, it was first shown by Mc Kay's group that the differentiation of multipotent stem cells from the embryonic and adult central nervous system (CNS) is influenced by extracellular signals that can instruct specific cell fates. For example, ciliary neurotrophic factor (CNTF) generates clones of astrocytes from NSC, whereas platelet-derived growth factor (PDGF-AA, -AB, and -BB) supports neuronal differentiation (Johe et al., 1996). The ligand-binding subunit of the CNTF receptor α is expressed in the embryonic cortical ventricular zone (VZ), where the fate of proliferating NSC is determined (Ip et al., 1993). CNTF promotes the differentiation of cortical neuroepithelial precursor cells into so-called type 1 astrocytes (Bonni et al., 1997). LIF (leukemia inhibitory factor) receptor (LIFR) and its co-receptor, gp130, both contribute to the CNTF-mediated induction of astrocyte differentiation in vitro. Cytokines such as LIF or CNTF induce dimerization of the LIFR with the co-receptor gp130, thereby leading to phosphorylation and activation of Janus kinases (JAKs). Receptor dimerization also creates docking sites for the TF STAT3 (Signal transducer and activator of transcription 3) which becomes dimerized after phosphorylation by JAKs and translocates to the nucleus, where it binds to the Glial fibrillary activated protein, (*Gfap*) promoter (Bonni et al., 1997) and regulates astroglial differentiation (Fig. 10.1). Genetic ablation of LIF, its receptors LIFR and gp130, or the signaling molecules STAT1/3 leads to impaired astroglial differentiation (Bugga et al., 1998; Koblar et al., 1998; Nakashima et al., 1999; Cao et al., 2010). During neurogenesis, the basic helix loop helix TF, Neurogenin-1 (*Neurog1*), inhibits the differentiation of NSC into astrocytes by sequestering the CBP/p300/SMAD1 transcription complex away from Stat3 and astrocytic genes and by directly suppressing the JAK/STAT pathway (Sun et al., 2001). In addition, blockage of other proneural genes such as Neurogenin-2 (*Neurog2*), and Achaete-scute homolog 1 (*Ascl1*), also induces astrogliogenesis at the expense of neurogenesis (Nieto et al., 2001; Tomita et al., 2000). Moreover, the onset of astrogliogenesis is controlled by a positive autoregulatory loop of signaling, whereby STAT1/3 directly induces the expression of various components of the cytokine-induced JAK/STAT pathway (He et al., 2005) (Fig. 10.1). During neurogenesis, when proneural genes are highly expressed, the autoregulatory loop is suppressed, whereas it is activated during the gliogenic phase once the level of pro-neural factors decreases (He et al., 2005) (Fig. 10.1).

JAK-STAT signaling is repressed by the suppressor of cytokine signaling (SOCS) family of proteins (Yoshimura et al., 2005), especially SOCS3. SOCS3 directly binds to the gp130 receptor via its SH2 domain (Nicholson et al., 2000; Schmitz et al., 2000) and inhibits signaling mediated by this receptor (Lang et al., 2003). Moreover, Cao et al. (2006) have shown that overexpression of SOCS3 blocks the differentiation of NSC into astrocytes and increases neurogenesis. Together, these findings demonstrate that STAT3 signaling plays a fundamental role in astrogliogenesis.

BMP SIGNALING AND ASTROGLIOGENESIS

Bone morphogenetic proteins, BMPs, are members of the transforming growth factor β superfamily, with the exception of BMP1, which is a metalloprotease (Nohe et al., 2004). These proteins are known for their multitude of functions in maintaining homeostasis in the body, for example, in regulating the anabolic and catabolic processes in bone (Sanchez-Duffhues et al., 2015). BMPs bind to

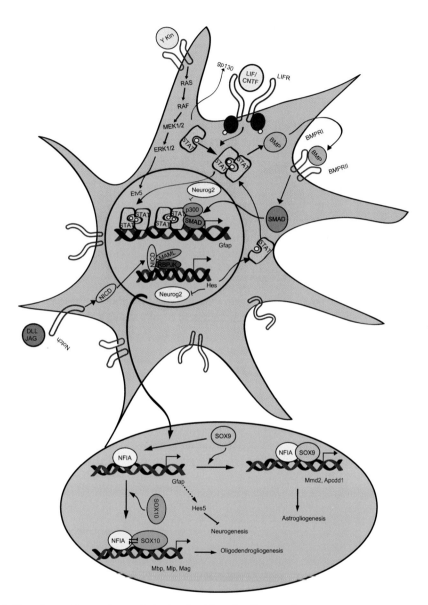

FIGURE 10.1 Signaling pathways known to be involved in astrogliogenesis

STAT and BMP signaling play a fundamental role in the induction of astrocyte fate. Cytokines (LIF or CNTF) induce dimerization of the LIF receptor (LIFR) with the co-receptor gp130, thereby leading to phosphorylation and activation of Janus kinases (JAKs). Receptor dimerization also creates docking sites for the transcription factor STAT3, which becomes dimerized after phosphorylation by JAKs and translocates into the nucleus, where it binds for example to the *Gfap* promoter. BMP dimers induce the tetramerization of BMPR-I and BMPR-II receptors. BMPR-II then

(Continued)

the BMP receptors type I (BMPR-I) and II (BMPR-II) and lead to the tetramerization of BMPR-I and BMPR-II. Subsequently, BMPR-II phosphorylates and activates BMPR-I, which in turn phosphorylates SMAD1. Phosphorylated SMAD1 interacts with SMAD4, and the complex translocates to the nucleus, recruiting co-factors such as p300 or corepressors such as TGIF to mediate transcriptional changes (Shi and Massague, 2003). BMP signaling is inhibited intracellularly by I-Smads and extracellularly by antagonists such as Noggin and Chordin (Rider and Mulloy, 2010) (Fig. 10.1). During early development, BMP is mainly expressed in the lateral edges of the neural plate and later in the dorsal midline of the neural tube, and it promotes dorsal-medial patterning and ventral forebrain development (Gamez et al., 2013; Mehler et al., 1997). During later development BMP has a broad spectrum of expression, especially in the hippocampus and cerebral cortex (Mehler et al., 1997; Gamez et al., 2013). At this stage BMP plays a major role in promoting astrogliogenesis and inhibiting oligodendrogliogenesis and neurogenesis (Gamez et al., 2013; Mehler et al., 1997). Findings from Goldman's group have demonstrated that the application of Noggin, a BMP inhibitor, and brain-derived neurotrophic factor (BDNF) to the ventricular walls of adult rats induces neurogenesis in the striatum (Chmielnicki et al., 2004). In adulthood BMP is required for the maintenance of the stem cell niche and for the survival of neuroblasts (Colak et al., 2008). Genetic ablation of SMAD4 or the introduction of Noggin alters the transit amplifying precursors' identity and results in generation of oligodendrocyte progenitor cells (OPCs) at the expense of neurons. This effect is rescued by *Olig2* suppression. These results suggest that during adulthood BMP signaling allows the progression toward the neurogenic lineage (Colak et al., 2008).

In the context of astrogliogenesis, during late embryonic and early postnatal periods, BMP signaling induces astrocyte differentiation from NSC (Gross et al., 1996; Mehler, 2002; Mehler et al., 1997). Oligodendroglial–astroglial progenitor cells derived from postnatal day 2 (P2) rat cortex differentiate into astrocytes after BMP treatment instead of oligodendrocytes (Mabie et al., 1997). Moreover, BMPR double knockout mice show a reduction in the number of mature astrocytes during spinal cord development (See et al., 2007). At the molecular level, these effects are mediated by the BMP-induced expression of Inhibitor of differentiation proteins *Id2* and *Id4* in neural progenitors, which sequester *Olig1* and *Olig2* away from their targets and prevent OPC differentiation (Samanta and Kessler, 2004), or by the expression of *Id1*, *Id3*, and Hairy-and-enhancer-of-Split5 *(Hes5)* in NSC, thus preventing neurogenesis (Nakashima et al., 2001). Although *Hes5* and *Ids* contribute to the acquisition of an astrocytic fate, to maintain this fate BMP relies on the continuous expression of

---

◄ phosphorylates and activates BMPR-I, which in turn phosphorylates SMAD1. Phosphorylated SMAD1 associates with SMAD4, and the complex translocates to the nucleus and recruits p300, resulting in its binding to the *Gfap* promoter. These signaling pathways are inhibited by *Neurog2*, which is thought to act by blocking STAT dimerization and sequestering p300. Notch receptors, after stimulation with ligands such as DLL and JAG, release the intracellular domain NICD, which also translocates to the nucleus and forms a complex with MAML and RBPJK at the promoters of Notch targets such as *Hes1* and *Hes5*. These effectors inhibit proneural genes such as *Neurog2* and promote astrogliogenesis. Tyrosine kinases also induce RAS/RAF/MEK/ERK signaling that results in the activation of *Etv5* in radial glial cells, thereby promoting astrogliogenesis. On the other hand, NFIA directly binds to the Gfap promoter to induce its expression. It also enhances *Hes5* expression that then blocks neurogenesis. NFIA also interacts with SOX9, a TF upstream of *Nfia*, to regulate different sets of genes that regulate astroglial progenitor migration (e.g., *Mmd2*), and astroglial metabolism (e.g., *Apcdd1*). During OPC generation, NFIA interacts with SOX10 and regulates oligodendrogliogenesis-related genes. Excess of NFIA blocks this interaction and induces astrogliogenesis.

the transcriptional repressor RE1 silencer of transcription/neuron-restrictive silencer factor to inhibit neurogenesis (Kohyama et al., 2010). Using SOCS3-deficient NSC derived from E14.5 telencephalon a study from Taga's group showed that STAT3 activation by LIF leads to an unexpected increase in *Smad1* expression and changes the morphology of astrocytes (Fukuda et al., 2007) (Fig. 10.1). This group also showed that STAT3 and SMAD1 work in synergy through direct interaction of p300 with STAT3 and SMAD1, which in turn activates astrocyte-specific promoters (Nakashima et al., 1999) (Fig. 10.1). Both signaling pathways generate different types of astrocytes. Astrocytes generated by STAT signaling are *Gfap* + and continue to express NSC markers, whereas BMP-induced astrocytes exit the cell cycle and are characterized as mature *Gfap* + astrocytes expressing no NSC markers (Bonaguidi et al., 2005). The blockage of STAT3 interaction with SMAD1 by HDACs leads to suppression of BMP-mediated astrogliogenesis (Shaked et al., 2008; Scholl et al., 2012). However, an alternative model has been proposed by Rajan et al. (2003), in which STAT activation is mediated by the serine-threonine kinase FKBP12/rapamycin-associated protein/mammalian target of rapamycin, which becomes activated after binding of BMP4 to its receptor and mediates the phosphorylation of STAT and astrogliogenesis. Moreover, another signaling molecule, sonic-hedgehog, promotes oligodendrogliogenesis by activating HDAC1/2 and compacting chromatin, whereas BMP4 favors histone acetylation, which prevents oligodendrogliogenesis and induces astrogliogenesis (Wu et al., 2012). Thus, BMP signaling, along with Stat3, plays a crucial role in conferring astrocyte identity (Fig. 10.1).

NOTCH SIGNALING AND ASTROGLIOGENESIS

Notch signaling is activated when transmembrane Notch ligands such as Delta or Jagged activate the Notch receptors on neighboring cells. After activation, NICD is cleaved by γ-secretase and translocates to the nucleus, where it activates target genes such as *Hes1* and *Hes5* by interacting with the DNA-binding protein RBPJK (Schroeter et al., 1998; Struhl and Adachi, 1998). The role of Notch signaling in astrogliogenesis was first demonstrated by Morrison and colleagues (Morrison et al., 2000) (Fig. 10.1). They showed that the activation of Notch signaling inhibits the differentiation of neural crest stem cells into neurons; instead, even in the presence of neurogenic factors, they give rise to astrocytes without affecting cell proliferation or stem cell self-renewal properties in the peripheral nervous system (Morrison et al., 2000). In the hippocampus *Fgf2*-dependent rat adult hippocampal progenitors (AHP) also differentiate into astrocytes in response to the Notch activators Notch1 and Notch3. This differentiation is independent of JAK-STAT signaling, and it appears that Notch signaling plays a critical role in cell commitment (Tanigaki et al., 2001). In addition, the constitutive activation of Notch1 in mouse forebrain progenitors at E9.5 or E14.5 leads to radial glial self-renewal before birth, whereas postnatally these cells become astrocytes (Gaiano et al., 2000, Chambers et al., 2001). To delineate the mechanism underlying Notch-mediated activation of astrogliogenesis, Ge et al. (2002) conducted studies showing that the downstream effector of Notch signaling RBPJK directly induces Gfap expression, and there is a delay in the differentiation of NSC into astrocytes in RBPJκ mutant embryonic stem (ES) cell-derived NSC (Fig. 10.1). Moreover, RBPJκ binding to the co-repressor NCoR leads to the repression of its binding to the Gfap promoter in NSC, whereas during the astrogliogenic phase, RBPJκ binds to the co-activator NICD and subsequently stimulates *Gfap* expression (Hermanson et al., 2002) (Fig. 10.1). Furthermore, there is crosstalk between STAT and NOTCH signaling as first shown by the Gotoh lab: the interaction of the Notch effectors *Hes1* and *Hes5* with STAT and JAK promoted STAT phosphorylation and subsequently astrogliogenesis (Kamakura et al., 2004) (Fig. 10.1). Moreover, STAT3 signaling has been found to induce *Hes1* and *Hes5* expression in cultures of neuroepithelial cells prepared from E12.5 mice, thus supporting the existence of a positive feedback loop (Yoshimatsu et al., 2006). Furthermore, *Hes1* and *Hes5* functionally suppress proneural genes and promote astrogliogenesis in E14.5 neural progenitor cells (Hirabayashi and Gotoh, 2005; Bertrand et al., 2002) (Fig. 10.1). A loss of function of these proneural genes has been shown to result in premature glial differentiation at the expense of neuronal differentiation (Nieto et al., 2001; Sun et al., 2001; Tomita et al., 2000).

NFIA, one of the central molecules in astrogliogenesis (see the section Cooperative roles of NFIA with SOX9 or SOX10 in Gliogenesis for detail), is a target of Notch signaling. As with Notch and STAT signaling, there is a feedback loop, although a negative one, that exists between NFIA and Notch. Accordingly, Notch activates *Nfia* expression, but *Nfia* represses *Hes1* expression in E18 cortical tissue and consequently prevents excessive generation of astrocytes (Piper et al., 2010). Given the importance of the Notch-NFIA axis in astrogliogenesis, Tole's group has demonstrated the importance of the Lim-homeodomain TF, LHX2 in the suppressing astrogliogenesis. They showed that upon LHX2 loss in the hippocampus there is precocious production of astrocytes in the neurogenic phase while neurogenesis is prolonged far into the astrogliogenic phase upon LHX2 overexpression. Overexpression of *Lhx2* along with *Notch* or *Nfia* restores normal neurogenesis, thus demonstrating the functional interplay between LHX2, Notch and NFIA (Subramanian et al., 2011). Degradation of the Notch receptors mediated by F-box- and WD repeat domain-containing protein 7, a component of a ubiquitin protein ligase complex, prevents neurogenesis during early development (Hoeck et al., 2010). Postnatally, when Notch1 is highly expressed, this degradation induces astrogliogenesis (Matsumoto et al., 2011). Thus, Notch plays a fundamental role in regulating neuron-glia-cell fate decisions.

MEK SIGNALING AND GLIOGENESIS

The MEK signaling cascade is well known for its important role in cell proliferation and differentiation (Zhang and Liu, 2002). For example, FGF4 is involved in the lineage commitment of ES cells from self-renewal to differentiation (Kunath et al., 2007; Ying et al., 2008), and FGF10 regulates the timing of transition of neural epithelium to radial progenitors Sahara and O'leary (2009). Moreover, FGF2 has previously been implicated in the process of gliogenesis: FGF2 induces generation of OPC from dorsally derived precursor cells from embryonic forebrain and spinal cord (Gabay et al., 2003; Chandran et al., 2003; Kessaris et al., 2004, Abematsu et al., 2006; Naruse et al., 2006; Furusho et al., 2011; Morrow et al., 2001). Recently, Snider's group has shown that MEK signaling is required for expression of gp130 by radial glia and MEK1/2 deletion in radial glia leads to failure of CNTF to induce STAT3 phosphorylation, which attenuates gliogenesis and prolongs neurogenesis (Fig. 10.1). Furthermore, the constitutively activation of MEK1 increases the number of astrocytes at the expense of neurons in the developing mouse cortex (Li et al., 2012). The results of this study are in line with a previous study in which NF1 (neurofibromin 1) loss in neural progenitors during early embryonic stages was found to be involved in gliogenesis by activating RAF/MEK/ERK signaling, and the effect was reversed by a MEK/ERK inhibitor (Hegedus et al., 2007). NF1 is a RAS GTPase that converts the GTP-bound active form of RAS proteins to the inactive, guanosine diphosphate-bound form (Scheffzek et al., 1997) and is affected in Neurofibromatosis type 1 disorder, in which individuals have numerous CNS abnormalities related to neuronal and glial cell lineage (Gutmann et al., 2012). NF1 also controls the generation of OPCs in the developing cortex (Bennett et al., 2003; Dasgupta and Gutmann, 2005). On the mechanistic level, Snider's group has provided evidence that *Etv5*, a member of the Ets family, is regulated by MEK signaling (Fig. 10.1). *Etv5* is expressed in the VZ at E14.5-E18.5. *Etv5* overexpression in MEK-deleted radial glia leads to the restoration of CNTF-induced astrogliogenesis (Li et al., 2012). These observations are in line with a recent publication showing that hyperactivation of Ras signaling in P2 VZ leads to depletion of the NSC pool and expansion of progenitors cells which subsequently yield glioma by upregulating *Etv5* (Breunig et al., 2015). Moreover FGF2 can also chromatin remodelling to allow access of the STAT/CBP complex to the STAT3 binding site at the Gfap promoter (Song and Ghosh, 2004). These results suggest a critical role of Mek signaling in astrogliogenesis and brain cancer.

COOPERATIVE ROLES OF NFIA WITH SOX9 OR SOX10 IN GLIOGENESIS

NFIA, a member of the Nf-1 family, contains 509 amino acids and binds to the palindromic DNA sequence 5′-TGGCANNNTGCCA-3′ (Gronostajski, 2000). The family comprises three more members, NFIB, NFIC, and NFIX (Gronostajski, 2000). *Nfia* plays a crucial role in the onset of gliogenesis in the developing spinal cord, where it is induced at E11.5 within the VZ to maintain the pool of Glast-positive gliogenic progenitors (Deneen et al., 2006) (Fig. 10.1). At later stages of gliogenesis, *Nfia* promotes the migration and differentiation of astrocytic progenitors, whereas *Nfia* function is suppressed in OPCs by *Olig2* (Deneen et al., 2006). Similarly, *Nfia* regulates *Gfap* expression by directly binding to its promoter in astrocytes (Cebolla and Vallejo, 2006). Moreover, NFIA knockout mice show defects in cortical glial development (Shu et al., 2003). In addition, in the hippocampus, NFIA induces gliogenesis by repressing the Notch effector *Hes1* (Piper et al., 2010).

NFIA induction is regulated upstream by the TF SOX9 (Kang et al., 2012) (Fig. 10.1). *Sox9* belongs to the SoxE family and binds to its CCTTGAG motif by the HMG box DNA-binding domain. SOX9 knockout mice show an extended period of neurogenesis at the expense of oligodendrogliogenesis (Stolt et al., 2003). Moreover, *Sox9* is required for the maintenance of the NSC pool in the embryonic and adult central nervous system (Cheng et al., 2009; Scott et al., 2010). *Sox9* is induced prior to *Nfia* in the VZ of the embryonic spinal cord, and *Sox9* expression is required for *Nfia* expression and its function in the initiation of gliogenesis (Kang et al., 2012). In later stages of gliogenesis (E12.5), SOX9 and NFIA interact physically and form a complex, which regulates a set of genes such as *Apcdd1* and *Mmd2* (Fig. 10.1). These genes restore gliogenesis in the absence of *Nfia* and perform a key role in the migration and proliferation of glial precursors during astrogliogenesis. Together, the SOX9 and NFIA-mediated transcriptional regulatory cascade is important for the initiation of gliogenesis and for regulating the migration and metabolic properties of astroglial progenitors during development (Kang et al., 2012) (Fig. 10.1). In addition, a recent study has shown that mouse embryonic fibroblasts can be reprogrammed into functional astrocytes through overexpression of *Nfia* and *Sox9* along with *Nfib*, another member of the Nf-1 family (Caiazzo et al., 2015).

SOX10, which also belongs to SoxE family of TFs, plays a significant role in myelin gene expression (Li et al., 2007b). A study of SOX10 knockout mice has shown that OPCs differentiate into astrocytes instead of oligodendrocytes in the absence of SOX10 (Glasgow et al., 2014). *Nfia* directly antagonizes the SOX10-mediated induction of *MBP*, *PLP1*, and *MAG* during oligodendrocyte development in the chick spinal cord by forming a complex with SOX10 similar to the complex with SOX9 (Fig. 10.1). Subsequent downregulation of *Nfia* during OPCs development leads to release of *Sox10* from repression to induce OPC-specific gene expression instead of *Nfia*-mediated astrocytic gene expression (Glasgow et al., 2014) (Fig. 10.1).

EPIGENETIC CONTROL OF ASTROGLIOGENESIS
DNA METHYLATION AND ASTROGLIOGENESIS

DNA methylation is an epigenetic mechanism used by cells to regulate gene expression. DNA methylation occurs on the CpG islands present near promoters and leads to transcriptional repression (Jaenisch and Bird, 2003). DNA methytransferases (DNMTs) are the enzymes that deposit the methyl group on cytosine residues and further help in maintaining the methylation pattern during replication (Jaenisch and Bird, 2003). While methylation of the Gfap promoter in rat brain cells and non-neuronal cells had been reported earlier (Teter et al., 1994), Taga and colleagues showed that the STAT3 binding site in the *Gfap* promoter is methylated in E11.5 neuroepithelial cells and becomes remethylated in post-mitotic neurons, even though STAT3 is active in these cells, but it is demethylated in late progenitors and in astrocytes, thus allowing for *Gfap* expression (Takizawa et al., 2001) (Fig. 10.2A). Treatment of E11.5 neuroepithelial cells with demethylating agents such as 5-aza-2-deoxycytidine (5-aza-CdR) leads to demethylation of the *Gfap* promoter, and the cells become responsive to LIF signaling. These results suggest a role of DNA methylation in silencing of *Gfap* in neuroepithelial cells.

Fan et al. (2005) have shown that DNA methylation regulates the timing and extent of astrogliogenesis by regulating JAK-STAT signaling and silencing glial gene expression including *Gfap* in the developing central nervous system, as shown by the conditional deletion of DNMT1 in NSC. The conditional knockout of DNMT1 leads to precocious astrogliogenesis and a decrease in neuron generation. Moreover, genes in the Stat3 signaling pathway also are repressed after DNMT1 loss. The authors of this study further demonstrated that methyl-CpG-binding proteins, which recruit histone deacetylases and histone methyltransferases to promote an inactive chromatin state, are closely associated with glial genes such as *Gfap*, *S100b*, and *Stat3* in early progenitors. However, this association is lost at later stages, and astroglial genes gain the association with STAT3, thus promoting astrogliogenesis (Fig. 10.2A). In addition, early progenitors acquire inactive chromatin modifications such as demethylation at lysine 9 of histone H3 (H3K9me2) and loss of active marks such as H3k4me2/3. In addition, Notch signaling activates *Nfia* expression in E11.5 progenitor cells, which leads to NFIA binding to the promoter of astrocyte-specific genes responsive to JAK-STAT signaling and subsequent dissociation of DNMT1 resulting in promoter demethylation (Namihira et al., 2009) (Fig. 10.2A). NSC derived from E14.5 Nfia deficient mice preserve methylation even in the presence of Notch signaling. These results indicate that NFIA is indispensable for Notch-induced demethylation of promoters of astrocyte-specific during brain development.

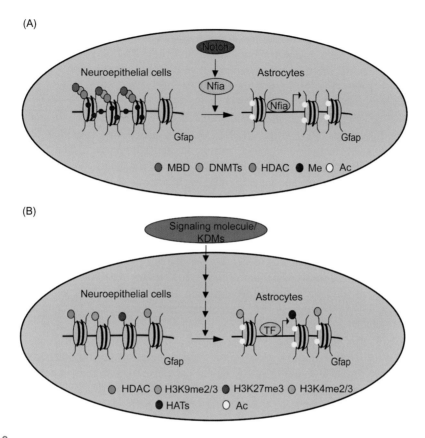

FIGURE 10.2

(A) In neuroepithelial cells, chromatin at the *Gfap* promoter is closed due to DNA methylation (Me) deposited by DNA methyltransferases (DNMTs) that recruit methyl-CpG-binding proteins (MBD) that in turn bring histone deacetylases (HDAC). At the later stage of development, when astrogliogenesis initiates, Notch signaling induces *Nfia* expression that targets the *Gfap* promoter, displaces DNMTs, and confers H3K27ac (Ac) to induce Gfap expression. (B) Similar to DNA methylation, chromatin of astroglial genes in neuroepithelial cells is enriched with inactive histone marks such as H3K9me2/3 and H3K27me3. Upon onset of astrogliogenesis, distinct signaling pathways induce histone demethylases that in turn remove the inactive marks. Concomitant to this these genes gain active histone modifications such as H3K4me2/3 and histone acetylation (Ac) deposited by histone methyltransferases and histone acetyltransferases (HATs), respectively.

HISTONE MODIFICATIONS AND ASTROGLIOGENESIS

Similar to DNA methylation, histone modification is another way to control the transcription. The globular domain and the tail of histones are subjected to various post-translational modifications such as methylation, acetylation, ubiquitination, ribosylation, sumoylation, and phosphorylation. These modifications either repress transcription by closing chromatin or activate transcription by opening it (Li et al., 2007a; Kouzarides, 2007). In context of astrogliogenesis, as discussed earlier in section BMP Signaling and Astrogliogenesis, a role for histone modifications was first implicated by Taga's group, who showed that the STAT-SMAD-p300/CBP complex is required for astrocyte-specific gene expression mediated by synergistic STAT and BMP signaling (Nakashima et al., 1999) (Fig. 10.1). Findings from Ghosh's group have demonstrated the role of another histone mark: FGF2 potentiates CNTF-induced astrogliogenesis by replacing the repressive mark H3K9me2 by the active mark H3K4me2 at the STAT3 binding site of the *Gfap* promoter (Song and Ghosh, 2004) (Fig. 10.2B). Gotoh's group has provided the first genetic evidence for a role of histone-modifying factors in astrogliogenesis. They showed that the Polycomb group (PcG) proteins ring 1B and EZH2 restrict the neurogenic competence of NSC via the repression of pro-neuronal genes, such as *Neurog1*, thereby triggering the onset of gliogenesis (Hirabayashi et al., 2009). In contrast, early cortex-specific deletion of the catalytic subunit of the PcG repressor complex 2 (PRC2) EZH2 causes premature astrocyte differentiation (Pereira et al., 2010). In addition, during the early phase of cortical development, chromodomain helicase DNA-binding protein-4 (CHD4) forms a complex with EZH2 and recruits PRC2 to astroglial genes such as *Gfap* (but not to neurogenic genes), thus resulting in gene silencing and the inhibition of astrogliogenesis (Sparmann et al., 2013) (Fig. 10.2B). Furthermore, expression of *Ezh2* in astrocytes induces their dedifferentiation toward NSC (Sher et al., 2011). On the acetylation level, HDACs suppress BMP-promoted astrogliogenesis by targeting its crucial partner STAT3 which interacts with BMP2-responsive TF SMAD1/5/8 (Scholl et al., 2012). Likewise, HDAC3 also controls the decision between oligodendrogliogenesis and astrogliogenesis (Zhang et al., 2016a). Although *Hdac3* is mainly detected in the oligodendrocyte lineage of the corpus callosum and white matter of the spinal cord at P14, it is only weakly present in the corpus callosum's astrocytes. The inhibition of HDAC3 specifically leads to decreased expression of *Olig2* in OPCs isolated from P2 rat brains (Zhang et al., 2016a). Moreover, conditional deletion of HDAC3 in the oligodendrocytic lineage leads to fewer OPCs and more astrocytes. Furthermore, these mice exhibit myelination defects (Zhang et al., 2016a). On the molecular level, HDAC3 competes with STAT3 for the co-activator protein p300 to induce *Olig2* expression. Furthermore, HDAC3 inhibits the STAT3 acetylation required for the activation of JAK-STAT signaling to prevent astrogliogenesis (Zhang et al., 2016a). In addition to PRC2 and HDACs, high-mobility group nucleosome-binding family proteins have been also shown to play a critical role in promoting astrocyte differentiation of neural precursor cells (Nagao et al., 2014). Finally, the histone methyltransferase ESET represses the premature onset of astrogliogenesis by directly binding to the *Gfap* promoter and repressing its expression via inducing histone 3 lysine 9 trimethlyation (H3K9me3) (Tan et al., 2012).

MiRNA AND ASTROGLIOGENESIS

microRNAs (miRNAs) are small noncoding RNAs that play critical roles in post-transcriptional regulation, for example, by inhibiting the translation of their target genes by destabilizing mRNA (Krol et al., 2010). Transcribed miRNAs are processed into pre-miRNAs by the RNA binding protein Dgcr8 and the RNaseIII enzyme Drosha, and the transcribed miRNAs are further processed by the enzyme Dicer into 18−21 nucleotide duplexes in the cytoplasm. One miRNA strand gets loaded into a silencing complex comprising Argonaute proteins. After loading, the 5′ end of the miRNA binds to mRNA by a seed sequence of 6−8 nucleotides and destabilizes the targeted mRNAs. miRNA targets form networks of genes that collaborate in promoting or inhibiting cell fate transitions (Judson et al., 2013). Significantly, miRNAs are essential for the neurogenic to gliogenic switch in the developing brain (Zheng et al., 2012, 2010; Kawase-Koga et al., 2009). Some of these miRNAs and their role in astrogliogenesis are described below:

miR-153: mir-153 is expressed throughout the CNS during early embryonic development (E9.5−10.5), including VZ. However, at midgestation the expression level of miR-153 in NSC is decreased. Overexpression of miR-153 leads to the inhibition of astrogliogenesis and maintenance of NSC in an undifferentiated state. Precocious astrogliogenesis occurs after miR-153 inhibition in early neurogenic NSC. This neurogenic-to-gliogenic switch is facilitated by increased NFI levels, which is a target of miR-153 (Tsuyama et al., 2015).

miR-9: miR-9 is expressed in the VZ and SVZ progenitors of the cerebral cortex and reaches a maximum level at E16, before the onset of astrogliogenesis. The proneural gene *Neurog1* binds upstream of the miR-9 transcription start site (TSS) at E15 (i.e., during the neurogenic phase) but is absent at P3. The overexpression of *Neurog1* enhances miR-9 expression, which promotes the blockage of an astroglial program. In addition, in the adult mouse brain miR-9 is specifically expressed in neurons, but not in astrocytes. On a mechanistic level, the 3′UTRs of three upstream components of the JAK-STAT pathway, LIFR, IL6, and JAK1, contain binding sites for miR-9, and transfection of mouse NSC with exogenous miR-9 duplex decreases the protein levels of LIFR, IL6, and JAK1, an outcome similar to *Neurog1* overexpression, and attenuates astrogliogenesis by abolishing Stat1/3 phosphorylation (Zhao et al., 2015).

Let7 and miR-125: The Let-7 family of miRNA is known for its role in the regulation of the timing of glial differentiation (Nishino et al., 2013; Patterson et al., 2014). Let-7 was first identified as a gene essential for regulating the timing of C. elegans' larval development (Pasquinelli et al., 2000). By decreasing the expression of the high-mobility group-AT-hook 2, a member of the high-mobility group A family, which encodes a small non-histone chromatin-associated protein that modulates transcription by altering chromatin structure, let-7b has been shown to be involved in age-dependent decline in NSC activity (Morrison et al., 2000). Let-7b expression also induces the cell cycle inhibitors p16[INK4a] and p19[ARF] and thus blocks NSC function with age (Tanigaki et al., 2001). Let-7b has further been implicated in the differentiation of NSC into neurons and glia through suppression of an orphan nuclear receptor *TLX* and cyclin D1 as well as insulin-like growth factor 2 mRNA binding protein 1 (*IGF2BP1*), whose loss decreases HMGA2 levels (Ge et al., 2002). Let-7 influences the fate of human NSC to differentiate into glial or neurons by targeting HMGA2 (Patterson et al., 2014). HMGA2 modulates chromatin, thereby promoting access of NICD to the *Hes5* promoter and subsequent induction of astrogliogenesis (Hirabayashi et al., 2009). Recently, Blelloch's group has shown the importance of Let7 and the miR-25 family of miRNA in astrogliogenesis by using DGCR8 conditional knockout mice (Shenoy et al., 2015). DGCR8, along with Drosha, is a RNA binding protein required for the processing of transcribed pri-miRNA to pre-miRNA. The conditional deletion of DGCR8 in ES-derived glial progenitor cells (GPC) leads to a defect in astrocyte differentiation, which is in part mediated by decreased JAK-STAT signaling. Following a candidate approach, this group has identified the miRNAs Let-7 and miR-125 as being able to rescue the phenotype in GPCs and increase astrogliogenesis independently of JAK-STAT signaling. Let-7 and miR-125 share many common targets, including *Plagl2* and *Igf2bp2*, which are actively repressed by let-7 and miR-125 in GPCs, and *Plagl2* and *Igf2bp2* upregulation blocks astrogliogenesis (Shenoy et al., 2015).

miR-17/106: miR-17 and 106a/b (miR-17/106) belong to the miR-17 family of miRNAs that share a common seed sequence. These miRs act downstream of chicken ovalbumin upstream promoter-TF I and II (COUP-TFI and COUP-TFII) and play a major role in regulating the neuro-to-gliogenic switch in neural stem/progenitor cell (NSPC). Sustained neurogenesis has been observed after the double knockdown (KD) of *Coup-tfI/II* in ESC-derived NSPCs. Moreover, this knockdown also increases the time window for the generation of specific types of neurons, which are normally born only at the early neurogenic stage. miR-17/106 is expressed in NSPCs specifically at the early neurogenic stage and is increased in response to the *Coup-tfI/II* KD in ESC-derived NSPCs and is subsequently downregulated in developing NSPCs. This time-dependent decrease in miR-17/106 expression in developing NSPCs leads to an increase in the expression of mitogen-activated protein kinase 14 (MAPK14), which is essential for the acquisition of gliogenic competence (Naka-Kaneda et al., 2014). The overexpression of miR-17 or loss of Mapk14 in gliogenic NSPCs leads to the restoration of neurogenic competence even after completion of the neurogenic phase. However, their expression does not affect the epigenetic status of the *Gfap* promoter (Naka-Kaneda et al., 2014).

CONCLUDING REMARKS

Recent years have revealed that astrogliogenesis involves precise rewiring of gene expression programs in neural stem cells and their progeny. However, despite these advances, we still know very little of the actual players guiding this process including TFs and epigenetic regulators. The existing knowledge on transcriptional control further gets thinner when it comes to distinct types of astrocytes present in the brain such as protoplasmic astrocytes, residing in gray matter (Sofroniew and Vinters, 2010), and fibrous astrocytes, residing in white matter (Sofroniew and Vinters, 2010). Similarly, transcriptional changes following ischemic stroke and neuroinflammation-induced-reactive astrocytes as well as differences between immature and mature parenchymal astrocytes are not well understood. Gotoh's lab has established two factors, Hmgn and Zbtb20, in regulating the gene expression program underlying astrocyte generation (Nagao et al., 2014, 2016). Moreover, recent findings from Murai's group showed that sonic-hedgehog secreted from mature neurons can dictate the identity and molecular profile of distinct astrocytes present in cerebellum, cortex, and hippocampus (Farmer et al., 2016). It is possible that there are many more secreted factors from neurons that influence astrocyte heterogeneity. There are few publically available datasets where researchers have compared astrocytes with other two lineages derived from NSCs, i.e., neurons and OPCs as well as different subtypes of astrocytes in mouse as well as human (Cahoy et al., 2008; Zhang et al., 2014, 2016b; Ziller et al., 2015; Lovatt et al., 2007; Doyle et al., 2008; Zamanian et al., 2012; Michelucci et al., 2016). With the recent developments in computational and system biology, it will eventually become possible to delineate the gene regulatory networks underlying astrocyte generation (Okawa et al., 2016).

REFERENCES

Abbott, N.J., Ronnback, L., Hansson, E., 2006. Astrocyte-endothelial interactions at the blood-brain barrier. Nat. Rev. Neurosci. 7, 41–53.

Abematsu, M., Kagawa, T., Fukuda, S., Inoue, T., Takebayashi, H., Komiya, S., et al., 2006. Basic fibroblast growth factor endows dorsal telencephalic neural progenitors with the ability to differentiate into oligodendrocytes but not gamma-aminobutyric acidergic neurons. J. Neurosci. Res. 83, 731–743.

Attwell, D., Buchan, A.M., Charpak, S., Lauritzen, M., Macvicar, B.A., Newman, E.A., 2010. Glial and neuronal control of brain blood flow. Nature 468, 232–243.

Bayraktar, O.A., Fuentealba, L.C., Alvarez-Buylla, A., Rowitch, D.H., 2014. Astrocyte development and heterogeneity. Cold Spring Harb. Perspect. Biol. 7, a020362.

Bennett, M.R., Rizvi, T.A., Karyala, S., Mckinnon, R.D., Ratner, N., 2003. Aberrant growth and differentiation of oligodendrocyte progenitors in neurofibromatosis type 1 mutants. J. Neurosci. 23, 7207–7217.

Bertrand, N., Castro, D.S., Guillemot, F., 2002. Proneural genes and the specification of neural cell types. Nat. Rev. Neurosci. 3, 517–530.

Bonaguidi, M.A., Mcguire, T., Hu, M., Kan, L., Samanta, J., Kessler, J.A., 2005. LIF and BMP signaling generate separate and discrete types of GFAP-expressing cells. Development 132, 5503–5514.

Bonni, A., Sun, Y., Nadal-Vicens, M., Bhatt, A., Frank, D.A., Rozovsky, I., et al., 1997. Regulation of gliogenesis in the central nervous system by the JAK-STAT signaling pathway. Science 278, 477–483.

Breunig, J.J., Levy, R., Antonuk, C.D., Molina, J., Dutra-Clarke, M., Park, H., et al., 2015. Ets factors regulate neural stem cell depletion and gliogenesis in Ras pathway glioma. Cell Rep. 12, 258–271.

Bugga, L., Gadient, R.A., Kwan, K., Stewart, C.L., Patterson, P.H., 1998. Analysis of neuronal and glial phenotypes in brains of mice deficient in leukemia inhibitory factor. J. Neurobiol. 36, 509–524.

Cahoy, J.D., Emery, B., Kaushal, A., Foo, L.C., Zamanian, J.L., Christopherson, K.S., et al., 2008. A transcriptome database for astrocytes, neurons, and oligodendrocytes: a new resource for understanding brain development and function, J. Neurosci. 28, pp. 264–278.

Caiazzo, M., Giannelli, S., Valente, P., Lignani, G., Carissimo, A., Sessa, A., et al., 2015. Direct conversion of fibroblasts into functional astrocytes by defined transcription factors. Stem Cell Rep. 4, 25–36.

Cao, F., Hata, R., Zhu, P., Ma, Y.J., Tanaka, J., Hanakawa, Y., et al., 2006. Overexpression of SOCS3 inhibits astrogliogenesis and promotes maintenance of neural stem cells. J. Neurochem. 98, 459–470.

Cao, F., Hata, R., Zhu, P., Nakashiro, K., Sakanaka, M., 2010. Conditional deletion of Stat3 promotes neurogenesis and inhibits astrogliogenesis in neural stem cells. Biochem. Biophys. Res. Commun. 394, 843–847.

Cebolla, B., Vallejo, M., 2006. Nuclear factor-I regulates glial fibrillary acidic protein gene expression in astrocytes differentiated from cortical precursor cells. J. Neurochem. 97, 1057–1070.

Chambers, C.B., Peng, Y., Nguyen, H., Gaiano, N., Fishell, G., Nye, J.S., 2001. Spatiotemporal selectivity of response to Notch1 signals in mammalian forebrain precursors. Development 128, 689–702.

Chandran, S., Kato, H., Gerreli, D., Compston, A., Svendsen, C.N., Allen, N.D., 2003. FGF-dependent generation of oligodendrocytes by a hedgehog independent pathway. Development 130, 6599–6609.

Cheng, L.C., Pastrana, E., Tavazoie, M., Doetsch, F., 2009. miR-124 regulates adult neurogenesis in the subventricular zone stem cell niche. Nat. Neurosci. 12, 399–408.

Chmielnicki, E., Benraiss, A., Economides, A.N., Goldman, S.A., 2004. Adenovirally expressed noggin and brain-derived neurotrophic factor cooperate to induce new medium spiny neurons from resident progenitor cells in the adult striatal ventricular zone. J. Neurosci. 24, 2133–2142.

Choi, H.B., Gordon, G.R., Zhou, N., Tai, C., Rungta, R.L., Martinez, J., et al., 2012. Metabolic communication between astrocytes and neurons via bicarbonate-responsive soluble adenylyl cyclase. Neuron 75, 1094–1104.

Colak, D., Mori, T., Brill, M.S., Pfeifer, A., Falk, S., Deng, C., et al., 2008. Adult neurogenesis requires Smad4-mediated bone morphogenic protein signaling in stem cells. J. Neurosci. 28, 434–446.

Daneman, R., Zhou, L., Kebede, A.A., Barres, B.A., 2010. Pericytes are required for blood-brain barrier integrity during embryogenesis. Nature 468, 562−566.

Dasgupta, B., Gutmann, D.H., 2005. Neurofibromin regulates neural stem cell proliferation, survival, and astroglial differentiation in vitro and in vivo. J. Neurosci. 25, 5584−5594.

Deneen, B., Ho, R., Lukaszewicz, A., Hochstim, C.J., Gronostajski, R.M., Anderson, D.J., 2006. The transcription factor NFIA controls the onset of gliogenesis in the developing spinal cord. Neuron 52, 953−968.

Doyle, J.P., Dougherty, J.D., Heiman, M., Schmidt, E.F., Stevens, T.R., Ma, G., et al., 2008. Application of a translational profiling approach for the comparative analysis of CNS cell types. Cell 135, 749−762.

Fan, G., Martinowich, K., Chin, M.H., He, F., Fouse, S.D., Hutnick, L., et al., 2005. DNA methylation controls the timing of astrogliogenesis through regulation of JAK-STAT signaling. Development 132, 3345−3356.

Farmer, W.T., Abrahamsson, T., Chierzi, S., Lui, C., Zaelzer, C., Jones, E.V., et al., 2016. Neurons diversify astrocytes in the adult brain through sonic hedgehog signaling. Science 351, 849−854.

Fukuda, S., Abematsu, M., Mori, H., Yanagisawa, M., Kagawa, T., Nakashima, K., et al., 2007. Potentiation of astrogliogenesis by STAT3-mediated activation of bone morphogenetic protein-Smad signaling in neural stem cells. Mol. Cell Biol. 27, 4931−4937.

Furusho, M., Kaga, Y., Ishii, A., Hebert, J.M., Bansal, R., 2011. Fibroblast growth factor signaling is required for the generation of oligodendrocyte progenitors from the embryonic forebrain. J. Neurosci. 31, 5055−5066.

Gabay, L., Lowell, S., Rubin, L.L., Anderson, D.J., 2003. Deregulation of dorsoventral patterning by FGF confers trilineage differentiation capacity on CNS stem cells in vitro. Neuron 40, 485−499.

Gaiano, N., Nye, J.S., Fishell, G., 2000. Radial glial identity is promoted by Notch1 signaling in the murine forebrain. Neuron 26, 395−404.

Gamez, B., Rodriguez-Carballo, E., Ventura, F., 2013. BMP signaling in telencephalic neural cell specification and maturation. Front. Cell. Neurosci. 7, 87.

Ge, W., Martinowich, K., Wu, X., He, F., Miyamoto, A., Fan, G., et al., 2002. Notch signaling promotes astrogliogenesis via direct CSL-mediated glial gene activation. J. Neurosci. Res. 69, 848−860.

Glasgow, S.M., Zhu, W., Stolt, C.C., Huang, T.W., Chen, F., Loturco, J.J., et al., 2014. Mutual antagonism between Sox10 and NFIA regulates diversification of glial lineages and glioma subtypes. Nat. Neurosci. 17, 1322−1329.

Gourine, A.V., Kasymov, V., Marina, N., Tang, F., Figueiredo, M.F., Lane, S., et al., 2010. Astrocytes control breathing through pH-dependent release of ATP. Science 329, 571−575.

Gronostajski, R.M., 2000. Roles of the NFI/CTF gene family in transcription and development. Gene 249, 31−45.

Gross, R.E., Mehler, M.F., Mabie, P.C., Zang, Z., Santschi, L., Kessler, J.A., 1996. Bone morphogenetic proteins promote astroglial lineage commitment by mammalian subventricular zone progenitor cells. Neuron 17, 595−606.

Gutmann, D.H., Parada, L.F., Silva, A.J., Ratner, N., 2012. Neurofibromatosis type 1: modeling CNS dysfunction. J. Neurosci. 32, 14087−14093.

He, F., Ge, W., Martinowich, K., Becker-Catania, S., Coskun, V., Zhu, W., et al., 2005. A positive autoregulatory loop of Jak-STAT signaling controls the onset of astrogliogenesis. Nat. Neurosci. 8, 616−625.

Hegedus, B., Dasgupta, B., Shin, J.E., Emnett, R.J., Hart-Mahon, E.K., Elghazi, L., et al., 2007. Neurofibromatosis-1 regulates neuronal and glial cell differentiation from neuroglial progenitors in vivo by both cAMP- and Ras-dependent mechanisms. Cell Stem Cell 1, 443−457.

Henneberger, C., Papouin, T., Oliet, S.H., Rusakov, D.A., 2010. Long-term potentiation depends on release of D-serine from astrocytes. Nature 463, 232−236.

Hermanson, O., Jepsen, K., Rosenfeld, M.G., 2002. N-CoR controls differentiation of neural stem cells into astrocytes. Nature 419, 934−939.

Hirabayashi, Y., Gotoh, Y., 2005. Stage-dependent fate determination of neural precursor cells in mouse forebrain. Neurosci. Res. 51, 331−336.

Hirabayashi, Y., Suzki, N., Tsuboi, M., Endo, T.A., Toyoda, T., Shinga, J., et al., 2009. Polycomb limits the neurogenic competence of neural precursor cells to promote astrogenic fate transition. Neuron 63, 600−613.

Hoeck, J.D., Jandke, A., Blake, S.M., Nye, E., Spencer-Dene, B., Brandner, S., et al., 2010. Fbw7 controls neural stem cell differentiation and progenitor apoptosis via Notch and c-Jun. Nat. Neurosci. 13, 1365−1372.

Ip, N.Y., Mcclain, J., Barrezueta, N.X., Aldrich, T.H., Pan, L., Li, Y., et al., 1993. The alpha component of the CNTF receptor is required for signaling and defines potential CNTF targets in the adult and during development. Neuron 10, 89−102.

Jaenisch, R., Bird, A., 2003. Epigenetic regulation of gene expression: how the genome integrates intrinsic and environmental signals. Nat. Genet. 33 (Suppl), 245−254.

Johe, K.K., Hazel, T.G., Muller, T., Dugich-Djordjevic, M.M., Mckay, R.D., 1996. Single factors direct the differentiation of stem cells from the fetal and adult central nervous system. Genes Dev. 10, 3129−3140.

Jourdain, P., Bergersen, L.H., Bhaukaurally, K., Bezzi, P., Santello, M., Domercq, M., et al., 2007. Glutamate exocytosis from astrocytes controls synaptic strength. Nat. Neurosci. 10, 331−339.

Judson, R.L., Greve, T.S., Parchem, R.J., Blelloch, R., 2013. MicroRNA-based discovery of barriers to dedifferentiation of fibroblasts to pluripotent stem cells. Nat. Struct. Mol. Biol. 20, 1227−1235.

Kamakura, S., Oishi, K., Yoshimatsu, T., Nakafuku, M., Masuyama, N., Gotoh, Y., 2004. Hes binding to STAT3 mediates crosstalk between Notch and JAK-STAT signalling. Nat. Cell Biol. 6, 547−554.

Kang, P., Lee, H.K., Glasgow, S.M., Finley, M., Donti, T., Gaber, Z.B., et al., 2012. Sox9 and NFIA coordinate a transcriptional regulatory cascade during the initiation of gliogenesis. Neuron 74, 79−94.

Kawase-Koga, Y., Otaegi, G., Sun, T., 2009. Different timings of Dicer deletion affect neurogenesis and gliogenesis in the developing mouse central nervous system. Dev. Dyn. 238, 2800−2812.

Kessaris, N., Jamen, F., Rubin, L.L., Richardson, W.D., 2004. Cooperation between sonic hedgehog and fibroblast growth factor/MAPK signalling pathways in neocortical precursors. Development 131, 1289−1298.

Koblar, S.A., Turnley, A.M., Classon, B.J., Reid, K.L., Ware, C.B., Cheema, S.S., et al., 1998. Neural precursor differentiation into astrocytes requires signaling through the leukemia inhibitory factor receptor. Proc. Natl. Acad. Sci. U. S. A. 95, 3178−3181.

Kohyama, J., Sanosaka, T., Tokunaga, A., Takatsuka, E., Tsujimura, K., Okano, H., et al., 2010. BMP-induced REST regulates the establishment and maintenance of astrocytic identity. J. Cell Biol. 189, 159−170.

Kouzarides, T., 2007. Chromatin modifications and their function. Cell 128, 693−705.

Kriegstein, A., Alvarez-Buylla, A., 2009. The glial nature of embryonic and adult neural stem cells. Annu. Rev. Neurosci. 32, 149−184.

Krol, J., Loedige, I., Filipowicz, W., 2010. The widespread regulation of microRNA biogenesis, function and decay. Nat. Rev. Genet. 11, 597−610.

Kunath, T., Saba-El-Leil, M.K., Almousailleakh, M., Wray, J., Meloche, S., Smith, A., 2007. FGF stimulation of the Erk1/2 signalling cascade triggers transition of pluripotent embryonic stem cells from self-renewal to lineage commitment. Development 134, 2895−2902.

Lang, R., Pauleau, A.L., Parganas, E., Takahashi, Y., Mages, J., Ihle, J.N., et al., 2003. SOCS3 regulates the plasticity of gp130 signaling. Nat. Immunol. 4, 546−550.

Li, B., Carey, M., Workman, J.L., 2007a. The role of chromatin during transcription. Cell 128, 707−719.

Li, H., Lu, Y., Smith, H.K., Richardson, W.D., 2007b. Olig1 and Sox10 interact synergistically to drive myelin basic protein transcription in oligodendrocytes. J. Neurosci. 27, 14375−14382.

Li, X., Newbern, J.M., Wu, Y., Morgan-Smith, M., Zhong, J., Charron, J., et al., 2012. MEK is a key regulator of gliogenesis in the developing brain. Neuron 75, 1035−1050.

Lovatt, D., Sonnewald, U., Waagepetersen, H.S., Schousboe, A., He, W., Lin, J.H., et al., 2007. The transcriptome and metabolic gene signature of protoplasmic astrocytes in the adult murine cortex. J. Neurosci. 27, 12255−12266.

Mabie, P.C., Mehler, M.F., Marmur, R., Papavasiliou, A., Song, Q., Kessler, J.A., 1997. Bone morphogenetic proteins induce astroglial differentiation of oligodendroglial-astroglial progenitor cells. J. Neurosci. 17, 4112−4120.

Maragakis, N.J., Rothstein, J.D., 2006. Mechanisms of disease: astrocytes in neurodegenerative disease. Nat. Clin. Pract. Neurol. 2, 679–689.

Matsumoto, A., Onoyama, I., Sunabori, T., Kageyama, R., Okano, H., Nakayama, K.I., 2011. Fbxw7-dependent degradation of Notch is required for control of "stemness" and neuronal-glial differentiation in neural stem cells. J. Biol. Chem. 286, 13754–13764.

Mehler, M.F., 2002. Mechanisms regulating lineage diversity during mammalian cerebral cortical neurogenesis and gliogenesis. Results Probl. Cell Differ. 39, 27–52.

Mehler, M.F., Mabie, P.C., Zhang, D., Kessler, J.A., 1997. Bone morphogenetic proteins in the nervous system. Trends Neurosci. 20, 309–317.

Michelucci, A., Bithell, A., Burney, M.J., Johnston, C.E., Wong, K.Y., Teng, S.W., et al., 2016. The Neurogenic Potential of Astrocytes Is Regulated by Inflammatory Signals. Mol. Neurobiol. 53, 3724–3739.

Molofsky, A.V., Krencik, R., Ullian, E.M., Tsai, H.H., Deneen, B., Richardson, W.D., et al., 2012. Astrocytes and disease: a neurodevelopmental perspective. Genes Dev. 26, 891–907.

Morrison, S.J., Perez, S.E., Qiao, Z., Verdi, J.M., Hicks, C., Weinmaster, G., et al., 2000. Transient Notch activation initiates an irreversible switch from neurogenesis to gliogenesis by neural crest stem cells. Cell 101, 499–510.

Morrow, T., Song, M.R., Ghosh, A., 2001. Sequential specification of neurons and glia by developmentally regulated extracellular factors. Development 128, 3585–3594.

Nagao, M., Lanjakornsiripan, D., Itoh, Y., Kishi, Y., Ogata, T., Gotoh, Y., 2014. High mobility group nucleosome-binding family proteins promote astrocyte differentiation of neural precursor cells. Stem Cells 32, 2983–2997.

Nagao, M., Ogata, T., Sawada, Y., Gotoh, Y., 2016. Zbtb20 promotes astrocytogenesis during neocortical development. Nat. Commun. 7, 11102.

Naka-Kaneda, H., Nakamura, S., Igarashi, M., Aoi, H., Kanki, H., Tsuyama, J., et al., 2014. The miR-17/106-p38 axis is a key regulator of the neurogenic-to-gliogenic transition in developing neural stem/progenitor cells. Proc. Natl. Acad. Sci. U. S. A. 111, 1604–1609.

Nakashima, K., Takizawa, T., Ochiai, W., Yanagisawa, M., Hisatsune, T., Nakafuku, M., et al., 2001. BMP2-mediated alteration in the developmental pathway of fetal mouse brain cells from neurogenesis to astrocytogenesis. Proc. Natl. Acad. Sci. U. S. A. 98, 5868–5873.

Nakashima, K., Yanagisawa, M., Arakawa, H., Kimura, N., Hisatsune, T., Kawabata, M., et al., 1999. Synergistic signaling in fetal brain by STAT3-Smad1 complex bridged by p300. Science 284, 479–482.

Namihira, M., Kohyama, J., Semi, K., Sanosaka, T., Deneen, B., Taga, T., et al., 2009. Committed neuronal precursors confer astrocytic potential on residual neural precursor cells. Dev. Cell 16, 245–255.

Naruse, M., Nakahira, E., Miyata, T., Hitoshi, S., Ikenaka, K., Bansal, R., 2006. Induction of oligodendrocyte progenitors in dorsal forebrain by intraventricular microinjection of FGF-2. Dev. Biol. 297, 262–273.

Nicholson, S.E., De Souza, D., Fabri, L.J., Corbin, J., Willson, T.A., et al., 2000. Suppressor of cytokine signaling-3 preferentially binds to the SHP-2-binding site on the shared cytokine receptor subunit gp130. Proc. Natl. Acad. Sci. U. S. A. 97, 6493–6498.

Nieto, M., Schuurmans, C., Britz, O., Guillemot, F., 2001. Neural bHLH genes control the neuronal versus glial fate decision in cortical progenitors. Neuron 29, 401–413.

Nishino, J., Kim, S., Zhu, Y., Zhu, H., Morrison, S.J., 2013. A network of heterochronic genes including Imp1 regulates temporal changes in stem cell properties. Elife 2, e00924.

Nohe, A., Keating, E., Knaus, P., Petersen, N.O., 2004. Signal transduction of bone morphogenetic protein receptors. Cell. Signal. 16, 291–299.

Okawa, S., Nicklas, S., Zickenrott, S., Schwamborn, J.C., Del Sol, A., 2016. A generalized gene-regulatory network model of stem cell differentiation for predicting lineage specifiers. Stem Cell Rep. 7, 307–315.

Pascual, O., Casper, K.B., Kubera, C., Zhang, J., Revilla-Sanchez, R., Sul, J.Y., et al., 2005. Astrocytic purinergic signaling coordinates synaptic networks. Science 310, 113−116.

Pasquinelli, A.E., Reinhart, B.J., Slack, F., Martindale, M.Q., Kuroda, M.I., Maller, B., et al., 2000. Conservation of the sequence and temporal expression of let-7 heterochronic regulatory RNA. Nature 408, 86−89.

Patterson, M., Gaeta, X., Loo, K., Edwards, M., Smale, S., Cinkornpumin, J., et al., 2014. let-7 miRNAs can act through notch to regulate human gliogenesis. Stem Cell Rep. 3, 758−773.

Pereira, J.D., Sansom, S.N., Smith, J., Dobenecker, M.W., Tarakhovsky, A., Livesey, F.J., 2010. Ezh2, the histone methyltransferase of PRC2, regulates the balance between self-renewal and differentiation in the cerebral cortex. Proc. Natl. Acad. Sci. U. S. A. 107, 15957−15962.

Petzold, G.C., Murthy, V.N., 2011. Role of astrocytes in neurovascular coupling. Neuron 71, 782−797.

Piper, M., Barry, G., Hawkins, J., Mason, S., Lindwall, C., Little, E., et al., 2010. NFIA controls telencephalic progenitor cell differentiation through repression of the Notch effector Hes1. J. Neurosci. 30, 9127−9139.

Rajan, P., Panchision, D.M., Newell, L.F., Mckay, R.D., 2003. BMPs signal alternately through a SMAD or FRAP-STAT pathway to regulate fate choice in CNS stem cells. J. Cell Biol. 161, 911−921.

Rider, C.C., Mulloy, B., 2010. Bone morphogenetic protein and growth differentiation factor cytokine families and their protein antagonists. Biochem. J. 429, 1−12.

Rothstein, J.D., Dykes-Hoberg, M., Pardo, C.A., Bristol, L.A., Jin, L., Kuncl, R.W., et al., 1996. Knockout of glutamate transporters reveals a major role for astroglial transport in excitotoxicity and clearance of glutamate. Neuron 16, 675−686.

Sahara, S., O'leary, D.D., 2009. Fgf10 regulates transition period of cortical stem cell differentiation to radial glia controlling generation of neurons and basal progenitors. Neuron 63, 48−62.

Samanta, J., Kessler, J.A., 2004. Interactions between ID and OLIG proteins mediate the inhibitory effects of BMP4 on oligodendroglial differentiation. Development 131, 4131−4142.

Sanchez-Duffhues, G., Hiepen, C., Knaus, P., Ten Dijke, P., 2015. Bone morphogenetic protein signaling in bone homeostasis. Bone 80, 43−59.

Scheffzek, K., Ahmadian, M.R., Kabsch, W., Wiesmüller, L., Lautwein, A., Schmitz, F., et al., 1997. The Ras-RasGap complex: structural basis for GTPase activation and its loss in oncogenic Ras mutants. Science 277, 333−339.

Schmitz, J., Weissenbach, M., Haan, S., Heinrich, P.C., Schaper, F., 2000. SOCS3 exerts its inhibitory function on interleukin-6 signal transduction through the SHP2 recruitment site of gp130. J. Biol. Chem. 275, 12848−12856.

Scholl, C., Weibetamuller, K., Holenya, P., Shaked-Rabi, M., Tucker, K.L., Wolfl, S., 2012. Distinct and overlapping gene regulatory networks in BMP- and HDAC-controlled cell fate determination in the embryonic forebrain. BMC Genomics 13, 298.

Schroeter, E.H., Kisslinger, J.A., Kopan, R., 1998. Notch-1 signalling requires ligand-induced proteolytic release of intracellular domain. Nature 393, 382−386.

Scott, C.E., Wynn, S.L., Sesay, A., Cruz, C., Cheung, M., Gomez Gaviro, M.V., et al., 2010. SOX9 induces and maintains neural stem cells. Nat. Neurosci. 13, 1181−1189.

See, J., Mamontov, P., Ahn, K., Wine-Lee, L., Crenshaw 3rd, E.B., Grinspan, J.B., 2007. BMP signaling mutant mice exhibit glial cell maturation defects. Mol. Cell. Neurosci. 35, 171−182.

Shaked, M., Weissmuller, K., Svoboda, H., Hortschansky, P., Nishino, N., Wolfl, S., et al., 2008. Histone deacetylases control neurogenesis in embryonic brain by inhibition of BMP2/4 signaling. PLoS ONE 3, e2668.

Shenoy, A., Danial, M., Blelloch, R.H., 2015. Let-7 and miR-125 cooperate to prime progenitors for astrogliogenesis. EMBO J. 34, 1180−1194.

Sher, F., Boddeke, E., Copray, S., 2011. Ezh2 expression in astrocytes induces their dedifferentiation toward neural stem cells. Cell Rep. 13, 1−6.

Shi, Y., Massague, J., 2003. Mechanisms of TGF-beta signaling from cell membrane to the nucleus. Cell 113, 685−700.

Shu, T., Butz, K.G., Plachez, C., Gronostajski, R.M., Richards, L.J., 2003. Abnormal development of forebrain midline glia and commissural projections in Nfia knock-out mice. J. Neurosci. 23, 203−212.

Sofroniew, M.V., Vinters, H.V., 2010. Astrocytes: biology and pathology. Acta Neuropathol. 119, 7−35.

Song, M.R., Ghosh, A., 2004. FGF2-induced chromatin remodeling regulates CNTF-mediated gene expression and astrocyte differentiation. Nat. Neurosci. 7, 229−235.

Sparmann, A., Xie, Y., Verhoeven, E., Vermeulen, M., Lancini, C., Gargiulo, G., et al., 2013. The chromodomain helicase Chd4 is required for Polycomb-mediated inhibition of astroglial differentiation. EMBO J. 32, 1598−1612.

Stolt, C.C., Lommes, P., Sock, E., Chaboissier, M.C., Schedl, A., Wegner, M., 2003. The Sox9 transcription factor determines glial fate choice in the developing spinal cord. Genes Dev. 17, 1677−1689.

Struhl, G., Adachi, A., 1998. Nuclear access and action of notch in vivo. Cell 93, 649−660.

Subramanian, L., Sarkar, A., Shetty, A.S., Muralidharan, B., Padmanabhan, H., Piper, M., et al., 2011. Transcription factor Lhx2 is necessary and sufficient to suppress astrogliogenesis and promote neurogenesis in the developing hippocampus. Proc. Natl. Acad. Sci. U. S. A. 108, E265−E274.

Sun, Y., Nadal-Vicens, M., Misono, S., Lin, M.Z., Zubiaga, A., Hua, X., et al., 2001. Neurogenin promotes neurogenesis and inhibits glial differentiation by independent mechanisms. Cell 104, 365−376.

Takizawa, T., Nakashima, K., Namihira, M., Ochiai, W., Uemura, A., Yanagisawa, M., et al., 2001. DNA methylation is a critical cell-intrinsic determinant of astrocyte differentiation in the fetal brain. Dev. Cell 1, 749−758.

Tan, S.L., Nishi, M., Ohtsuka, T., Matsui, T., Takemoto, K., Kamio-Miura, A., et al., 2012. Essential roles of the histone methyltransferase ESET in the epigenetic control of neural progenitor cells during development. Development 139, 3806−3816.

Tanigaki, K., Nogaki, F., Takahashi, J., Tashiro, K., Kurooka, H., Honjo, T., 2001. Notch1 and Notch3 instructively restrict bFGF-responsive multipotent neural progenitor cells to an astroglial fate. Neuron 29, 45−55.

Teter, B., Osterburg, H.H., Anderson, C.P., Finch, C.E., 1994. Methylation of the rat glial fibrillary acidic protein gene shows tissue-specific domains. J. Neurosci. Res. 39, 680−693.

Tomita, K., Moriyoshi, K., Nakanishi, S., Guillemot, F., Kageyama, R., 2000. Mammalian achaete-scute and atonal homologs regulate neuronal versus glial fate determination in the central nervous system. EMBO J. 19, 5460−5472.

Tsuyama, J., Bunt, J., Richards, L.J., Iwanari, H., Mochizuki, Y., Hamakubo, T., et al., 2015. MicroRNA-153 regulates the acquisition of gliogenic competence by neural stem cells. Stem Cell Rep. 5, 365−377.

Walz, W., 2000. Role of astrocytes in the clearance of excess extracellular potassium. Neurochem. Int. 36, 291−300.

Wang, X., Lou, N., Xu, Q., Tian, G.F., Peng, W.G., Han, X., et al., 2006. Astrocytic Ca2 + signaling evoked by sensory stimulation in vivo. Nat. Neurosci. 9, 816−823.

Wu, M., Hernandez, M., Shen, S., Sabo, J.K., Kelkar, D., Wang, J., et al., 2012. Differential modulation of the oligodendrocyte transcriptome by sonic hedgehog and bone morphogenetic protein 4 via opposing effects on histone acetylation. J. Neurosci. 32, 6651−6664.

Ying, Q.L., Wray, J., Nichols, J., Batlle-Morera, L., Doble, B., Woodgett, J., et al., 2008. The ground state of embryonic stem cell self-renewal. Nature 453, 519−523.

Yoshimatsu, T., Kawaguchi, D., Oishi, K., Takeda, K., Akira, S., Masuyama, N., et al., 2006. Non-cell-autonomous action of STAT3 in maintenance of neural precursor cells in the mouse neocortex. Development 133, 2553−2563.

Yoshimura, A., Nishinakamura, H., Matsumura, Y., Hanada, T., 2005. Negative regulation of cytokine signaling and immune responses by SOCS proteins. Arthritis Res. Ther. 7, 100−110.

Zamanian, J.L., Xu, L., Foo, L.C., Nouri, N., Zhou, L., Giffard, R.G., et al., 2012. Genomic analysis of reactive astrogliosis. J. Neurosci. 32, 6391−6410.

Zhang, L., He, X., Liu, L., Jiang, M., Zhao, C., Wang, H., et al., 2016a. Hdac3 Interaction with p300 histone acetyltransferase regulates the oligodendrocyte and astrocyte lineage fate switch. Dev. Cell 36, 316−330.

Zhang, W., Liu, H.T., 2002. MAPK signal pathways in the regulation of cell proliferation in mammalian cells. Cell Res. 12, 9−18.

Zhang, Y., Chen, K., Sloan, S.A., Bennett, M.L., Scholze, A.R., O'keeffe, S., et al., 2014. An RNA-sequencing transcriptome and splicing database of glia, neurons, and vascular cells of the cerebral cortex. J. Neurosci. 34, 11929−11947.

Zhang, Y., Sloan, S.A., Clarke, L.E., Caneda, C., Plaza, C.A., Blumenthal, P.D., et al., 2016b. Purification and characterization of progenitor and mature human astrocytes reveals transcriptional and functional differences with mouse. Neuron 89, 37−53.

Zhao, J., Lin, Q., Kim, K.J., Dardashti, F.D., Kim, J., He, F., et al., 2015. Ngn1 inhibits astrogliogenesis through induction of miR-9 during neuronal fate specification. Elife 4, e06885.

Zheng, K., Li, H., Huang, H., Qiu, M., 2012. MicroRNAs and glial cell development. Neuroscientist 18, 114−118.

Zheng, K., Li, H., Zhu, Y., Zhu, Q., Qiu, M., 2010. MicroRNAs are essential for the developmental switch from neurogenesis to gliogenesis in the developing spinal cord. J. Neurosci. 30, 8245−8250.

Ziller, M.J., Edri, R., Yaffe, Y., Donaghey, J., Pop, R., Mallard, W., et al., 2015. Dissecting neural differentiation regulatory networks through epigenetic footprinting. Nature 518, 335−359.

MicroRNAs IN OLIGODENDROCYTE MYELINATION AND REPAIR IN THE CENTRAL NERVOUS SYSTEM

11

Xianghui Zhao[1,2], Haibo Wang[2] and Qing Richard Lu[2,3]

[1]*Fourth Military Medical University, Xi'an, China* [2]*Cincinnati Children's Hospital Medical Center, Cincinnati, OH, United States* [3]*University of Cincinnati, Cincinnati, OH, United States*

INTRODUCTION

Myelin is required to enable saltatory conduction of action potentials as well as to provide long-term trophic and metabolic support for axonal integrity (Nave and Werner, 2014). Disruptions in myelination or failure of remyelination contribute to a broad spectrum of debilitating neurological diseases including multiple sclerosis (MS) and leukodystrophies (Franklin and Gallo, 2014; Nave and Werner, 2014). In the central nervous system (CNS), myelin membrane is synthesized by oligodendrocytes (OLs). During CNS development, OLs arise after successive stages of lineage progression from OL precursor cells (OPCs) to immature OLs and finally to mature myelinating OLs. This complex process initiates in embryonic stage and continues into young adult life and is coordinately regulated by extracellular signals and a network of intrinsic factors in a spatially and temporally specific manner (He and Lu, 2013; Zuchero and Barres, 2013).

MicroRNAs (miRNAs) are central regulators of gene expression in many biological and disease processes (Beavers et al., 2015; Stefani and Slack, 2008), and recent studies have revealed their critical functions in regulation of OL development (Barca-Mayo and Lu, 2012; Dugas and Notterpek, 2011; He et al., 2012). Primary transcripts of miRNAs are mainly transcribed by RNA polymerase II. These long transcripts are cleaved by a nuclear "microprocessor" RNase III Drosha/DGCR8 to release short hairpin intermediates (pre-miRNAs). Pre-miRNAs are transported to the cytoplasm where they are further processed by the RNase III Dicer into mature miRNAs, which are duplexes of approximately 22 base pairs (Carthew and Sontheimer, 2009; Desvignes et al., 2015). One strand of the duplex produced by Dicer is selected and loaded onto an Argonaute ribonucleoprotein (AGO) to form the miRNA-induced silencing complex, which recognizes and represses expression of targeted messenger RNA (mRNAs). In mammals, there are four AGO proteins, AGO1−4; only AGO2 can mediate specific endonucleolytic cleavage of a target mRNA in the context of perfect or nearly perfect pairing with miRNA (Frohn et al., 2012; Liu et al., 2004; Meister

Essentials of Noncoding RNA in Neuroscience. DOI: http://dx.doi.org/10.1016/B978-0-12-804402-5.00011-X

et al., 2004). miRNAs may also trigger translational repression and degradation without inducing cleavage of the mRNA. Recent studies indicate that miRNA-mediated translational repression is a prerequisite for mRNA target degradation (Valinezhad Orang et al., 2014; Wilczynska and Bushell, 2015).

Control of gene expression mediated by miRNAs is clearly critical in normal brain function, and dysregulation of miRNA-mediated gene expression regulation is associated with brain dysfunction, neurodegenerative diseases, and brain cancer (Barbano et al., 2014; Caputo et al., 2015; Luo et al., 2015; Persengiev et al., 2012; Szafranski et al., 2015). In this chapter, we summarize and discuss recent evidence for the role of miRNAs in regulating the development of myelinating OLs and in myelin repair in the CNS.

MiRNAs IN OL DEVELOPMENT
DICER1-MEDIATED miRNA PROCESSING IS REQUIRED FOR OL DEVELOPMENT

A crucial approach for assessment of the overall role of miRNAs in brain development is to disable the miRNA biogenesis pathway. This has been accomplished by the conditional deletion of the gene encoding Dicer, which results in a failure to generate mature Dicer-dependent miRNAs. Cre-mediated deletion of *Dicer1* at different stages in the OL lineage has been carried out (Barca-Mayo and Lu, 2012; Dugas and Notterpek, 2011; He et al., 2012). Knockout of *Dicer1* in the uncommitted neural precursors by *Nestin* promoter-driven Cre (Nestin-Cre) leads to a reduction in overall OL number (including both OPCs and mature OLs) in the spinal cord (Kawase-Koga et al., 2009). Since no obvious cell death is detected in the Nestin-Cre: *Dicer* spinal cord, the OL defect is likely caused by a decreased expansion of the OPC pool.

Similarly, when *Dicer1* was selectively ablated in Olig1-expressing cells using the Olig1-Cre system, which results in *Dicer1* deletion in early progenitors of OL lineage cells or primitive OPCs, generation of OPCs is nearly completely blocked (Zhao et al., 2010). Intriguingly, this inhibition of oligodendrogenesis is not associated with the expansion of motor neuron formation, suggesting that neurogenesis and gliogenesis are uncoupled and controlled by distinct molecular mechanisms.

OL differentiation and myelin formation are severely impaired when *Dicer1* is deleted in OL lineage cells directed by Olig1-Cre, Olig2-Cre-Tva, or cyclic nucleotide 3′ phosphohydrolase (CNP)-Cre. OPCs isolated from the cortices of these *Dicer1*-mutant animals fail to differentiate normally in vitro (Dugas et al., 2010; Zhao et al., 2010). Although mice with the *Dicer1* deletion mediated by Olig1-Cre or Olig2-Cre-Tva exhibit similar dysmyelination, the former mutants lack the majority of myelin normally observed in the CNS and die around postnatal week three. In contrast, the latter show a developmental delay but ultimate recovery of myelination in adulthood. A possible explanation is that floxed *Dicer1* allele is not completely removed by Olig2-Cre-Tva in a population of immature OL precursors. Those OPCs that escape Cre excision may expand and eventually restore near-normal levels of myelination in adulthood (Dugas et al., 2010; He et al., 2012).

Finally, *Dicer1* has been disrupted specifically in mature OLs by driving tamoxifen-inducible Cre under an OL-expressing proteolipid protein (PLP) promoter (PLP-CreERT). Ablation of *Dicer1* induced by tamoxifen treatment leads to degeneration of fully formed CNS myelin (Shin et al., 2009).

Collectively, these studies indicate that miRNAs are required at various stages of OL development including the initial production of fate-committed OPCs, the differentiation of OPCs and myelination during development, and the maintenance of myelin sheath integrity in adult animals.

DYNAMIC EXPRESSION OF miRNAs DURING OL DEVELOPMENT

MiRNAs are dynamically regulated during OL development (Dugas et al., 2010; Lau et al., 2008; Letzen et al., 2010; Zhao et al., 2010). Expression of a cohort of miRNAs is altered during the transition from A2B5$^+$ OPCs to premyelinating GalC$^+$ cells (Lau et al., 2008) and when rat OPCs are induced to differentiate in vitro (Dugas et al., 2010). Among them, miR-219, miR-138, and miR-338 are strongly upregulated in differentiating OLs (Dugas et al., 2010). Expression of a subset of miRNAs is increased during the differentiation process of OLs derived from human embryonic stem cells (Letzen et al., 2010). Although a few miRNAs are in similarly expressed in rodent and human cells, the majority do not overlap at the same developmental stage, which might be due to differences between species or to differences in the assigned developmental stages between rat and human. The miRNAs expressed at the highest expression in rat OPCs, such as miR-130a, miR-7, miR-16, miR-17, and miR-20a, are also observed in human oligodendroglial progenitors. Similarly, miR-17, miR-20a, miR-21, miR-16, miR-103, and miR-107 identified in rat OPCs are also expressed in the human progenitors (Lau et al., 2008; Letzen et al., 2010).

    miRNA microarray profiling of the CNS tissues of wild-type and myelin-deficient *Dicer1*-knockout mice has also identified miRNAs that are preferentially enriched in mature OLs or, conversely, downregulated in *Dicer1* mutants. miR-219 and miR-338 are substantially increased at the onset of OL myelination and in mature OLs (Shin et al., 2009; Zhao et al., 2010). Intriguingly, three miRNAs (miR-23, miR-26, and miR-29) that are found in OL lineage cells (Lau et al., 2008) are also enriched in astrocytes (Smirnova et al., 2005), suggesting that a population of miRNAs are shared between OLs and astrocytes. In addition, OLs and neurons share high levels of expression of miR-34c, miR-137, miR-146, miR-186, miR-218, and miR-449 (Tsang et al., 2007). Neuronally expressed miRNAs, such as miR-124, may negatively regulate gliogenesis and expression of myelin genes such as *Mobp* (Letzen et al., 2010) while promoting neurogenesis (Cheng et al., 2009). These data thus suggest that different neural cell types may share a set of miRNAs and that a balance of neurogenic and gliogenic miRNA expression may be crucial for the binary fate choice of neural progenitor cells or for proper cellular functions in specific contexts.

ROLE OF miRNAs IN OL DEVELOPMENT

MiRNAs in Glial Fate Specification and OPC Proliferation

miR-7a is highly enriched in OPCs (Lau et al., 2008; Zhao et al., 2012), and it shares a similar temporal expression pattern with the miR-17−92 cluster (Miska et al., 2004). miR-7a promotes the generation of OPCs from neural progenitors in vitro and in vitro during brain development; blocking the function of miR-7a in differentiating neural progenitors leads to a reduction of OPCs and an expansion of neuronal progenitors (Zhao et al., 2012). miR-7a exerts these effects by directly repressing expression of genes encoding proneuronal differentiation factors, including Pax6 and NeuroD4, and of genes involved in OL maturation such as *CNP* and *Sp1* (Zhao et al., 2012). These results suggest that miR-7a regulates OPC fate commitment while suppressing neurogenesis and OL maturation.

miR-9 is highly expressed in rat OPCs, and its level is inversely correlated with that of *peripheral myelin protein 22* (*PMP22*) mRNA (Lau et al., 2008) (Fig. 11.1A). This interrelated expression pattern may indicate that miR-9 is expressed to silence the "leaky" expression of genes that should not be produced in OL lineage cells. In addition, there is evidence that miR-9 is important for suppressing neural stem cell (NSC) proliferation and for accelerating differentiation through inhibition of expression of Tailless (TLX), a nuclear receptor that is essential for NSC self-renewal and for maintaining NSCs in an undifferentiated state. Therefore, miR-9 may regulate the switch from NSC proliferation to differentiation through a feedback loop with TLX (Zhao et al., 2009). Overexpression of miR-9 together with miR-124a inhibits glial fibrillary protein positive astrocyte formation from embryonic stem cell-derived neural progenitors (Krichevsky et al., 2006). In addition to its inhibition of PMP22 production, miR-9 may repress expression of nonlineage-related proteins in OLs and function as a guardian to maintain oligodendroglial identity (Fig. 11.1A).

The oncogenic miRNAs produced from a common precursor transcript synthesized from the miR-17/92 cluster, miR-17, miR-18a, miR-19a, miR-20a, miR-19b, and miR-92a (also known as Oncomir-1) are highly enriched in OL lineage cells (Budde et al., 2010; Lau et al., 2008). In vivo inactivation of the miR-17−92 cluster leads to a reduction in the number of OLs (Fig. 11.1A). Expression of Oncomir-1 in OPC cultures promotes cell proliferation; this miRNA targets *Pten* resulting in activation of Akt signaling (Budde et al., 2010), suggesting that Oncomir-1 promotes OPC expansion.

MiRNAs IN OL DIFFERENTIATION

OL differentiation is tightly linked to the cessation of OPC proliferation and cell cycle withdrawal (Raff et al., 1998). Deletion of *Dicer1* in OL lineage cells results in an increase of OPC proliferation and blocks OPC differentiation and myelination (Dugas et al., 2010; Zhao et al., 2010), suggesting that miRNAs are required for normal OPC cell cycle exit and differentiation. These two studies indicate that miR-219, and miR-338 target OPC-proliferation-related genes such as *platelet-derived growth factor receptor alpha (PDGFRa)*, which is required for OPC expansion and survival (Fig. 11.1A). miR-138 appears to target Sox4 to enhance the OPC differentiation process (Dugas et al., 2010).

miR-219 and miR-338 levels are substantially increased at the onset of OL differentiation, and these miRNAs promote OL maturation and the myelination processes. Expression of miRNA mimics of miR-338-3p and either miR-219 or miR-338-5p together promote OPC differentiation and partially rescue the differentiation deficit of *Dicer1*-deleted OPCs in vitro (Dugas et al., 2010; Zhao et al., 2010). In the developing chick and mouse CNS, overexpression of miR-219 or miR-338 is sufficient to promote precocious expression of OLs from neural progenitors as indicated by the presence of lineage markers including PDGFRα and myelin basic protein (MBP) (Zhao et al., 2010). Consistently, inhibition of expression of these two miRNAs in cultured OPCs (Dugas et al., 2010; Zhao et al., 2010) or of miR-219 in zebrafish embryos (Zhao et al., 2010) inhibits OPC differentiation. Thus, miR-219 appears to be both necessary and sufficient to initiate OL differentiation. miR-138 is transiently expressed in early postmitotic OLs. Expression of miR-138 stimulates early differentiation, but inhibits OL terminal differentiation (Dugas et al., 2010).

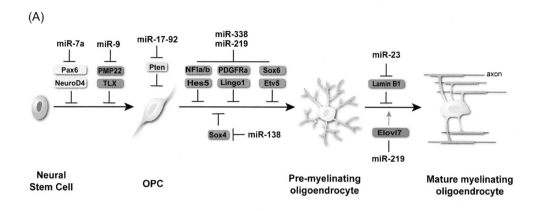

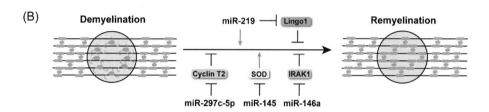

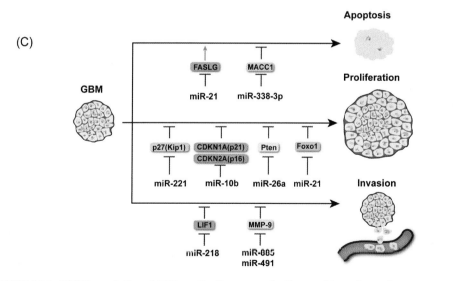

FIGURE 11.1 MiRNA regulation of CNS myelination, remyelination, and tumorigenesis

(A) Diagram depicting stage-specific miRNAs involved in regulating the neural precursor to OPC, OPC differentiation, and OL myelination as well as myelin maintenance by targeting the documented and predicted targets as indicated. (B) Candidate miRNAs and their potential targets in OL remyelination in demyelinating diseases such as MS. (C) Diagram showing candidate miRNAs and their potential targets in regulation of apoptosis, proliferation, and invasion of glioma cells.

miR-219 and miR-338 target a number of genes that have previously been shown to be involved in inhibiting OPC differentiation and in maintaining OPCs in their proliferative state including *PDGFRα, Hes5*, and *Sox6* (Dugas et al., 2010; Zhao et al., 2010). Significantly, miR-219 may also target the mRNAs encoding other transcription factors, such as Zfp238, FoxJ3, NeuroD1, Isl1, and Otx2 that are potentially involved in promoting neurogenesis (Fig. 11.1A). Since overexpression of these proneural factors inhibits OPC differentiation, miR-219 likely inhibits production of the factors that restrict differentiation of neural progenitors to the OL lineage.

Unbiased biotinylated miRNA-mRNA pull-down assays revealed that miR-219 targets neurogenic genes or OL differentiation inhibitors such as *Nfia, Nfib*, and *Lingo1* in OPCs and *Etv5* in differentiating OLs. These observations suggest that miR-219 targets distinct sets of genes at different stages to promote OL lineage progression (Wang et al., 2017). Furthermore, the studies of miR-219 knockout and over-expression in transgenic mice demonstrate that miR-219 is both necessary and sufficient for OL differentiation (Wang et al., 2017). In addition, miR-219 and miR-338 cooperate to regulate the full extent of OL maturation (Wang et al., 2017).

Current evidence thus indicates that a network of miRNAs regulates distinct stages of OL lineage progression and safeguards against the expression of neuronal and other cell lineage genes. In addition to regulating the OL differentiation program, miR-219 also regulates lipid metabolism by targeting the mRNA encoding fatty acid elongase ELOVL7, which synthesizes very long chain fatty acids. Accumulation of fatty acids is observed in PLP-CreERT *Dicer*-floxed mice (Shin et al., 2009), suggesting that miRNAs such as miR-219 also play a role in myelin lipid homeostasis. Collectively, miRNAs function at multiple stages during OL lineage development to direct and fine-tune the OL differentiation processes (Fig. 11.1A).

Recently, Diao et al. (2015) have developed a nanofiber-mediated miRNA delivery method to control OPC differentiation through a combination of fiber topography and gene silencing. Using poly(ε-caprolactone) nanofibers integrated with miR-219 and miR-338, efficient inhibition of expression of OL differentiation inhibitors such as *PDGFRα, Sox6, Hes5, FoxJ3*, and *ZFP238* were achieved. As compared with two-dimensional culture, nanofiber 3D topography in the presence of miRNAs enhanced OPC differentiation and maturation by increasing the number of RIP$^+$ and MBP$^+$ mature OLs. Such scaffolds carrying myelination promoting miRNAs such as miR-219 may help to promote OL differentiation and remyelination for treatment of demyelinating diseases and pathological insults.

MiRNAs AND MYELIN MAINTENANCE

Failure to maintain myelin sheaths or OL death may trigger demyelinating diseases including those of autoimmune origin (Traka et al., 2016). miRNAs are necessary for maintenance of proper myelin structure. For instance, miR-23 and its downstream target *SLC45A3* regulate myelin maintenance by modulating glucose and lipid metabolism and myelin protein expression in OLs (Shin et al., 2012). miR-20a, a component of miR-17−92 cluster that controls OL cell number, regulates the expression of the mRNA that codes for major myelin PLP (Wang and Cambi, 2012). miR-23 is a negative regulator of *lamin B1* expression (Lin and Fu, 2009). Excess lamin

B1 causes severe CNS myelin loss in adult-onset autosomal dominant leukodystrophy patients by repressing production of myelin proteins such as MBP, PLP, and myelin OL glycoprotein myelin oligodendrocyte glycoprotein (MOG). Transgenic mice engineered to overexpress miR-23a have increased myelin thickness, providing in vivo evidence that miR-23a enhances both OL differentiation and myelin synthesis (Lin et al., 2013). These studies suggest that miRNAs such as miR-23 play important roles in regulation of myelin homeostasis (Fig. 11.1A).

TRANSCRIPTIONAL REGULATION OF miRNA NETWORK DURING OL DEVELOPMENT

Although miRNAs regulate OL differentiation through action on various targeted mRNAs, how miRNA expression is regulated is not fully understood. Among the transcription factors expressed in OLs, Sox10, Olig1, Olig2, and Myrf may regulate expression of a network of miRNAs critical during OL lineage progression. Olig2 is critical for OL lineage specification and is required for OPC formation (Ligon et al., 2006; Lu et al., 2002; Takebayashi et al., 2002). Sox10 is essential for OPC differentiation into mature OLs (Stolt et al., 2002). *Olig1*-null mice have normal OPCs and even generate premyelinating OLs expressing O4 and CNP but fail to generate mature MBP-expressing OLs (Xin et al., 2005). Myrf is highly enriched in differentiated OLs and required for OL differentiation and maturation (Emery et al., 2009).

The miRNAs dependent on particular transcription factors were identified by analyzing CNS tissues of mice deficient in particular factors and by analyzing OLs in which expression of particular transcription factors were inhibited. The microarray analysis revealed significant reductions in miR-338 and miR-219 levels in the spinal cords of *Olig1*-null mice (Zhao et al., 2010). Other miRNAs, such as miR-322, miR-20b, miR-138, and the miR-17/92 cluster miRNAs were downregulated in *Olig1*-null mice (Zhao et al., 2010). Expression of miR-338 was downregulated in Sox10-deficient CG4 OL cells, and the miR-17−92 cluster expression is dependent on Sox10 in Schwann cells (Gokey et al., 2012). Interestingly, the miR-219-2 gene is located within the intron of a long non-coding RNA, *lncOL4* regulates myelin gene expression and its function in OL differentiation is likely channeled through the control of miR-219 (He et al., 2017).

Chromatin immunoprecipitation (ChIP) followed by sequencing in CNS tissues or purified OLs provides evidence that the transcriptional regulators such as Olig2, Sox10, Myrf, and Tcf7l2/Tcf4 target the promoters of miRNAs such as miR-219 (Bujalka et al., 2013; Yu et al., 2013; Zhao et al., 2016). In addition, ChIP analysis of spinal cords identified a strong Sox10 binding site upstream of the miR-17/92 cluster and of miR-338 (Svaren, 2014). These data suggest that miRNA expression is transcriptionally regulated by OL differentiation-promoting factors.

MiRNAs IN DEMYELINATING DISEASES AND MYELIN REPAIR

MiRNAs have not only been identified as key regulators of development but have also been implicated in disease progression. Identification of OL-specific miRNAs and their critical role in OL development and myelination in the CNS suggest that certain miRNAs have potential as biomarkers for demyelinating diseases or as therapeutic targets (Fitzpatrick et al., 2015).

MS is one of the most devastating autoimmune demyelinating diseases. Failure of OL remyelination or OL cell death in the CNS due to pathological insults or injuries is the primary event in MS pathogenesis. OL differentiation and maturation defects are the major cause of poor remyelination and axonal degeneration in MS lesions, and a differentiation block in OPCs is a major determinant of remyelination failure in chronic MS lesions (Kuhlmann et al., 2008). Although substantial efforts have centered on suppression of the immune response that attacks myelin, it is becoming increasingly clear that in addition to immune modulation, targeted therapy that promotes myelin repair is essential for better management of the disease (Chang et al., 2002; Franklin and Ffrench-Constant, 2008). Enhancing remyelination through treatment with miRNA mimics may provide an important potential therapeutic intervention to promote myelin repair. miRNA profiles from active and inactive demyelinating lesions (Junker et al., 2009) and peripheral blood cells (Du et al., 2009; Keller et al., 2009; Otaegui et al., 2009) indicate that there are alterations of miRNA expression in MS patients compared with healthy controls. Strikingly, the miRNAs enriched in mature OLs (e.g., miR-219 and miR-338) are present at very low levels or are not detectable in chronic MS lesions (Junker et al., 2009), suggesting that these miRNAs may also function in human OL maturation and myelin repair (Fig. 11.1B). Our recent studies indicated that miR-219 targets and attenuates a remyelination inhibitor Lingo1 as well as OL differentiation inhibitors Etv5 and NFIa/b. In addition, augmentation of miR-219 enhances myelin restoration in different animal models of multiple sclerosis including toxin-induced demyelination and experimental autoimmune encephalomyelitis (EAE). These suggesting a therapeutic role for miR-219 in promoting myelin repair in demyelinating diseases in the CNS (Wang et al., 2017).

In a cuprizone-induced demyelination model, miR-297c-5p was identified as a novel regulator of OPC function (Kuypers et al., 2016). miR-297c-5p increased during mouse OPC differentiation in vitro and during callosal development in vivo. miR-297c-5p exhibits dual functions as a negative regulator of OPC proliferation, via targeting *cyclin T2*, and as a positive regulator of OL maturation (Fig. 11.1B). These data indicate that miR-297c-5p likely facilitates myelin repair in pathological conditions.

Because of substantial gliosis and immune cell infiltration in demyelinating lesions, highly upregulated miRNAs in tissue lesions are likely expressed in astrocytes, T cells, and monocytes (Junker et al., 2009). Several miRNAs upregulated in active MS lesions, including miR-34a, miR-155, and miR-326 (Junker et al., 2009; Waschbisch et al., 2011), target the mRNA encoding CD47, which functions as an inhibiting "don't eat me" signal for macrophage activity (Oldenborg et al., 2000). In EAE, larger amounts of miR-326 are expressed by interleukin 17-producing T helper cells (T_H-17 cells) than by naïve $CD4^+$ T cells (Waschbisch et al., 2011). miR-326 positively regulates T_H-17 cell differentiation and cell number through targeting endogenous Ets-1 protein, a negative regulator of T_H-17 differentiation. Inhibition of expression of miR-326 alleviates EAE symptoms. Thus, miRNAs may permit macrophages to execute phagocytosis of myelin debris in MS lesions and may act as enhancers of the immune response in MS.

Intriguingly, levels of miR-338, miR-155, and miR-491 are upregulated in patients with MS and in EAE brains. miR-338 and miR-155 repress expression of neurosteroidogenic enzymes and allopregnanolone, suggesting that miRNAs can also impair neurosteroid biogenesis in MS and EAE brain (Noorbakhsh et al., 2011).

miR-17 and miR-20a, two members of the miR-17−92 cluster, are present at significantly lower concentrations in the whole blood of MS patients than in blood samples from healthy controls. The downregulation of these miRNAs facilitates astrocyte differentiation and also promotes

MS-associated T-cell activation via the PI3K/Akt/mTOR pathway, contributing to immune-mediated OL cell death in the disease (Cox et al., 2010).

miRNA profiling has also been used to identify potential biomarkers for disease progression and therapies for MS. Using human miRNA microarray and the Geniom Real Time Analyzer platform, 165 miRNAs were identified that are significantly deregulated in the blood cells of patients with relapsing−remitting MS as compared with healthy controls (Keller et al., 2009). A single miRNA, hsa-miR-145, that is increased during the transition from OPC to OL (Lau et al., 2008; Letzen et al., 2010), discriminates MS from controls with high specificity and sensitivity (Keller et al., 2009). Another study revealed that miR-18b and miR-599 may be relevant to the relapse status of MS patients, whereas miR-96 may be associated with remission (Otaegui et al., 2009). In addition, quantities of miR-326 and miR-26a appear to discriminate between relapsing and remitting MS (Honardoost et al., 2014).

Ischemic stroke is also linked to functional myelin loss due to the vulnerability of OLs to ischemic injury. OL regeneration and remyelination is critical for functional recovery after ischemic injury, therefore miRNA profiles have been developed as biomarkers for diagnosis and prognosis of cerebral ischemic stroke (Fitzpatrick et al., 2015; Li and Yao, 2012). Although their activities in stroke recovery are not fully elucidated, a cluster of miRNAs are upregulated in an ischemic stroke animal model. In particular, miR-145, which is upregulated during the transition from OPC to OL (Lau et al., 2008; Letzen et al., 2010), appears to have a role in ischemic stroke etiology and in the pathology of MS (Dharap and Vemuganti, 2010). The mRNA encoding superoxide dismutase (SOD) is the downstream target of miR-145, and SOD1/2 overexpression enhances proliferation of OPCs and accelerates their differentiation to mature OLs in vitro (Veiga et al., 2011). Downregulation of miR-145 leads to increased levels of SOD in the postischemic brain (Dharap and Vemuganti, 2010) (Fig. 11.1B). Therefore, together with other miRNAs, strategies that decrease miR-145 levels have potential as treatments for brain injury. Further miRNA profiling and verification in different demyelinating diseases may point to new biomarkers and potential therapies for MS.

In a middle cerebral artery occlusion-induced stroke model, it was found that miR-146a increases substantially in the corpus callosum and subventricular zone of the lateral ventricle of the ischemic hemisphere (Liu et al., 2016). In vitro overexpression of miR-146a in neural progenitor cells significantly promotes the generation of $O4^+$ OPCs and expression of myelin proteins. miR-146a negatively regulates *IRAK1* expression in OPCs; IRAK1 is a negative regulator of myelin proteins and OPC survival (Fig. 11.1B). These data suggest that miR-146a might mediate stroke-induced OL regeneration.

MiRNAs IN BRAIN TUMORIGENESIS

Glioma represents a heterogeneous group of CNS malignancies that arise from neural progenitors or glial precursors. Glioblastoma (GBM) is the most common and malignant brain tumor (Louis et al., 2007). As in other diseases, miRNAs are appealing therapeutic targets and potential biomarkers for GBMs (Hummel et al., 2011). Deregulation of miRNAs that impact cell proliferation, cell cycle regulation, and angiogenesis underlies the pathogenesis of GBM (Novakova et al., 2009). Early work that assessed miRNA expression via microarrays in tumor tissues obtained from GBM patients or GBM cell lines revealed the elevation of miR-221 levels compared with controls

(Ciafre et al., 2005). The mRNA encoding tumor suppressor p27(Kip1), which is reduced in GBMs, is a direct target of miR-221 (Gillies and Lorimer, 2007).

Levels of two other tumor suppressors, CDKN1A (p21) and CDKN2A (p16), are regulated miR-10b, which is significantly upregulated in malignant gliomas (Gabriely et al., 2011) (Fig. 11.1C). In addition, miR-21 and miR-26a, which are also highly expressed in primary GBM tumors, promote tumor cell proliferation by multiple modes of actions including targeting of *Pten* (Kim et al., 2010) and *Foxo1* (Lei et al., 2014) and inhibiting expression of proteins involved in cell death pathways such as Fas ligand (Shang et al., 2015) (Fig. 11.1C). GBM cell invasion is also mediated by miRNAs such as miR-218 and miR-491 that target invasion-related *LEF1* and *MMP-9* (Liu et al., 2012) (Fig. 11.1C).

Since primary GBM tumor tissues are enriched with miRNAs, miRNAs can also be released from tumor cells into various body fluids, thus positioning these molecules as circulating biomarkers of malignancy (Westphal and Lamszus, 2015). Upregulation of miR-21 in plasma samples (Ilhan-Mutlu et al., 2012) and in extracellular vesicles isolated from cerebrospinal fluid of GBM patients (Akers et al., 2013) further supports the diagnostic relevance of miR-21. Overall, these studies reveal that miRNAs have potential as noninvasive biomarkers for GBM diagnosis.

FUTURE PERSPECTIVES

A number of studies have probed the functions of miRNAs in myelinating glia. Most have focused on miRNA profiling, target identification, and elucidation of miRNA function through use of Dicer mutants in OLs. Future studies analyzing targeted ablation of individual miRNAs in mice or the genome editing technology CRISPR-Cas9 (Dominguez et al., 2016; Doudna and Charpentier, 2014; Shalem et al., 2015) are highly anticipated. miRNAs clearly regulate gliogenesis, cell proliferation, and differentiation of OLs and have roles in pathogenesis and brain tumorigenesis; however, a challenge remains to understand the individual components of the miRNA networks and their functional interplay with regulatory circuitry that controls the myelination program during development and myelin repair.

Understanding miRNA signatures during development and in disease states will be valuable both for diagnostic and therapeutic purposes. miRNAs have an advantage over mRNAs as they are more stable and do not undergo significant decay during the tissue sample processing (Jung et al., 2010). The robustness of profiling data collected to date suggests that miRNAs are potential biomarkers for diagnosis or monitoring of demyelinating diseases. Further study of miRNAs will illuminate important aspects of their influence on the myelination processes. New knowledge about these small regulatory molecules will suggest novel therapeutic interventions by which disease-related miRNAs can be antagonized or functionally restored to induce myelin repair and to shrink tumors. A future challenge will be to translate this knowledge into improved outcome for patients with demyelinating diseases and brain tumors.

ACKNOWLEDGMENTS

The authors would like to thank Xianyao Zhou for diagram drawing. This study was funded in part by grants from the US National Institutes of Health R01NS072427 and R01NS075243 and the National Multiple Sclerosis Society (NMSS-4727) to QRL.

REFERENCES

Akers, J.C., Ramakrishnan, V., Kim, R., Skog, J., Nakano, I., Pingle, S., et al., 2013. MiR-21 in the extracellular vesicles (EVs) of cerebrospinal fluid (CSF): a platform for glioblastoma biomarker development. PLoS ONE 8, e78115.

Barbano, R., Palumbo, O., Pasculli, B., Galasso, M., Volinia, S., D'Angelo, V., et al., 2014. A miRNA signature for defining aggressive phenotype and prognosis in gliomas. PLoS ONE 9, e108950.

Barca-Mayo, O., Lu, Q.R., 2012. Fine-tuning oligodendrocyte development by microRNAs. Front. Neurosci. 6, 13.

Beavers, K.R., Nelson, C.E., Duvall, C.L., 2015. MiRNA inhibition in tissue engineering and regenerative medicine. Adv. Drug Deliv. Rev. 88, 123−137.

Budde, H., Schmitt, S., Fitzner, D., Opitz, L., Salinas-Riester, G., Simons, M., 2010. Control of oligodendroglial cell number by the miR-17-92 cluster. Development 137, 2127−2132.

Bujalka, H., Koenning, M., Jackson, S., Perreau, V.M., Pope, B., Hay, C.M., et al., 2013. MYRF is a membrane-associated transcription factor that autoproteolytically cleaves to directly activate myelin genes. PLoS Biol. 11, e1001625.

Caputo, V., Ciolfi, A., Macri, S., Pizzuti, A., 2015. The emerging role of microRNA in schizophrenia. CNS Neurol. Disord. Drug Targets 14, 208−221.

Carthew, R.W., Sontheimer, E.J., 2009. Origins and mechanisms of miRNAs and siRNAs. Cell 136, 642−655.

Chang, A., Tourtellotte, W.W., Rudick, R., Trapp, B.D., 2002. Premyelinating oligodendrocytes in chronic lesions of multiple sclerosis. N. Engl. J. Med. 346, 165−173.

Cheng, L.C., Pastrana, E., Tavazoie, M., Doetsch, F., 2009. miR-124 regulates adult neurogenesis in the subventricular zone stem cell niche. Nat. Neurosci. 12, 399−408.

Ciafre, S.A., Galardi, S., Mangiola, A., Ferracin, M., Liu, C.G., Sabatino, G., et al., 2005. Extensive modulation of a set of microRNAs in primary glioblastoma. Biochem. Biophys. Res. Commun. 334, 1351−1358.

Cox, M.B., Cairns, M.J., Gandhi, K.S., Carroll, A.P., Moscovis, S., Stewart, G.J., et al., 2010. MicroRNAs miR-17 and miR-20a inhibit T cell activation genes and are under-expressed in MS whole blood. PLoS ONE 5, e12132.

Desvignes, T., Batzel, P., Berezikov, E., Eilbeck, K., Eppig, J.T., McAndrews, M.S., et al., 2015. miRNA nomenclature: a view incorporating genetic origins, biosynthetic pathways, and sequence variants. Trends Genet. 31, 613−626.

Dharap, A., Vemuganti, R., 2010. Ischemic pre-conditioning alters cerebral microRNAs that are upstream to neuroprotective signaling pathways. J. Neurochem. 113, 1685−1691.

Diao, H.J., Low, W.C., Lu, Q.R., Chew, S.Y., 2015. Topographical effects on fiber-mediated microRNA delivery to control oligodendroglial precursor cells development. Biomaterials 70, 105−114.

Dominguez, A.A., Lim, W.A., Qi, L.S., 2016. Beyond editing: repurposing CRISPR-Cas9 for precision genome regulation and interrogation. Nat. Rev. Mol. Cell Biol. 17, 5−15.

Doudna, J.A., Charpentier, E., 2014. Genome editing. The new frontier of genome engineering with CRISPR-Cas9. Science 346, 1258096.

Du, C., Liu, C., Kang, J., Zhao, G., Ye, Z., Huang, S., et al., 2009. MicroRNA miR-326 regulates TH-17 differentiation and is associated with the pathogenesis of multiple sclerosis. Nat. Immunol. 10, 1252−1259.

Dugas, J.C., Notterpek, L., 2011. MicroRNAs in oligodendrocyte and Schwann cell differentiation. Dev. Neurosci. 33, 14−20.

Dugas, J.C., Cuellar, T.L., Scholze, A., Ason, B., Ibrahim, A., Emery, B., et al., 2010. Dicer1 and miR-219 are required for normal oligodendrocyte differentiation and myelination. Neuron 65, 597−611.

Emery, B., Agalliu, D., Cahoy, J.D., Watkins, T.A., Dugas, J.C., Mulinyawe, S.B., et al., 2009. Myelin gene regulatory factor is a critical transcriptional regulator required for CNS myelination. Cell 138, 172−185.

Fitzpatrick, J.M., Anderson, R.C., McDermott, K.W., 2015. MicroRNA: key regulators of oligodendrocyte development and pathobiology. Int. J. Biochem. Cell Biol. 65, 134−138.

Franklin, R.J., Ffrench-Constant, C., 2008. Remyelination in the CNS: from biology to therapy. Nat. Rev. Neurosci. 9, 839−855.

Franklin, R.J., Gallo, V., 2014. The translational biology of remyelination: past, present, and future. Glia 62, 1905–1915.

Frohn, A., Eberl, H.C., Stohr, J., Glasmacher, E., Rudel, S., Heissmeyer, V., et al., 2012. Dicer-dependent and -independent Argonaute2 protein interaction networks in mammalian cells. Mol. Cell. Proteomics 11, 1442–1456.

Gabriely, G., Yi, M., Narayan, R.S., Niers, J.M., Wurdinger, T., Imitola, J., et al., 2011. Human glioma growth is controlled by microRNA-10b. Cancer Res. 71, 3563–3572.

Gillies, J.K., Lorimer, I.A., 2007. Regulation of p27Kip1 by miRNA 221/222 in glioblastoma. Cell Cycle 6, 2005–2009.

Gokey, N.G., Srinivasan, R., Lopez-Anido, C., Krueger, C., Svaren, J., 2012. Developmental regulation of microRNA expression in Schwann cells. Mol. Cell. Biol. 32, 558–568.

He, D., Wang, J., Lu, Y., Deng, Y., Zhao, C., Xu, L., et al., 2017. lncRNA functional networks in oligodendrocytes reveal stage-specific myelination control by an lncOL1/Suz12 complex in the CNS. Neuron 93, 362–378.

He, L., Lu, Q.R., 2013. Coordinated control of oligodendrocyte development by extrinsic and intrinsic signaling cues. Neurosci. Bull. 29, 129–143.

He, X., Yu, Y., Awatramani, R., Lu, Q.R., 2012. Unwrapping myelination by microRNAs. Neuroscientist 18, 45–55.

Honardoost, M.A., Kiani-Esfahani, A., Ghaedi, K., Etemadifar, M., Salehi, M., 2014. miR-326 and miR-26a, two potential markers for diagnosis of relapse and remission phases in patient with relapsing-remitting multiple sclerosis. Gene 544, 128–133.

Hummel, R., Maurer, J., Haier, J., 2011. MicroRNAs in brain tumors: a new diagnostic and therapeutic perspective? Mol. Neurobiol. 44, 223–234.

Ilhan-Mutlu, A., Wagner, L., Wohrer, A., Furtner, J., Widhalm, G., Marosi, C., et al., 2012. Plasma microRNA-21 concentration may be a useful biomarker in glioblastoma patients. Cancer Invest. 30, 615–621.

Jung, M., Schaefer, A., Steiner, I., Kempkensteffen, C., Stephan, C., Erbersdobler, A., et al., 2010. Robust microRNA stability in degraded RNA preparations from human tissue and cell samples. Clin. Chem. 56, 998–1006.

Junker, A., Krumbholz, M., Eisele, S., Mohan, H., Augstein, F., Bittner, R., et al., 2009. MicroRNA profiling of multiple sclerosis lesions identifies modulators of the regulatory protein CD47. Brain 132, 3342–3352.

Kawase-Koga, Y., Otaegi, G., Sun, T., 2009. Different timings of Dicer deletion affect neurogenesis and gliogenesis in the developing mouse central nervous system. Dev. Dyn. 238, 2800–2812.

Keller, A., Leidinger, P., Lange, J., Borries, A., Schroers, H., Scheffler, M., et al., 2009. Multiple sclerosis: microRNA expression profiles accurately differentiate patients with relapsing-remitting disease from healthy controls. PLoS ONE 4, e7440.

Kim, H., Huang, W., Jiang, X., Pennicooke, B., Park, P.J., Johnson, M.D., 2010. Integrative genome analysis reveals an oncomir/oncogene cluster regulating glioblastoma survivorship. Proc. Natl. Acad. Sci. U. S. A. 107, 2183–2188.

Krichevsky, A.M., Sonntag, K.C., Isacson, O., Kosik, K.S., 2006. Specific microRNAs modulate embryonic stem cell-derived neurogenesis. Stem Cells 24, 857–864.

Kuhlmann, T., Miron, V., Cui, Q., Wegner, C., Antel, J., Bruck, W., 2008. Differentiation block of oligodendroglial progenitor cells as a cause for remyelination failure in chronic multiple sclerosis. Brain 131, 1749–1758.

Kuypers, N.J., Bankston, A.N., Howard, R.M., Beare, J.E., Whittemore, S.R., 2016. Remyelinating oligodendrocyte precursor cell miRNAs from the Sfmbt2 cluster promote cell cycle arrest and differentiation. J. Neurosci. 36, 1698–1710.

Lau, P., Verrier, J.D., Nielsen, J.A., Johnson, K.R., Notterpek, L., Hudson, L.D., 2008. Identification of dynamically regulated microRNA and mRNA networks in developing oligodendrocytes. J. Neurosci. 28, 11720–11730.

Lei, B.X., Liu, Z.H., Li, Z.J., Li, C., Deng, Y.F., 2014. miR-21 induces cell proliferation and suppresses the chemosensitivity in glioblastoma cells via downregulation of FOXO1. Int. J. Clin. Exp. Med. 7, 2060−2066.

Letzen, B.S., Liu, C., Thakor, N.V., Gearhart, J.D., All, A.H., Kerr, C.L., 2010. MicroRNA expression profiling of oligodendrocyte differentiation from human embryonic stem cells. PLoS ONE 5, e10480.

Li, J.S., Yao, Z.X., 2012. MicroRNAs: novel regulators of oligodendrocyte differentiation and potential therapeutic targets in demyelination-related diseases. Mol. Neurobiol. 45, 200−212.

Ligon, K.L., Kesari, S., Kitada, M., Sun, T., Arnett, H.A., Alberta, J.A., et al., 2006. Development of NG2 neural progenitor cells requires Olig gene function. Proc. Natl. Acad. Sci. U. S. A. 103, 7853−7858.

Lin, S.T., Fu, Y.H., 2009. miR-23 regulation of lamin B1 is crucial for oligodendrocyte development and myelination. Dis. Model. Mech. 2, 178−188.

Lin, S.T., Huang, Y., Zhang, L., Heng, M.Y., Ptacek, L.J., Fu, Y.H., 2013. MicroRNA-23a promotes myelination in the central nervous system. Proc. Natl. Acad. Sci. U. S. A. 110, 17468−17473.

Liu, J., Carmell, M.A., Rivas, F.V., Marsden, C.G., Thomson, J.M., Song, J.J., et al., 2004. Argonaute2 is the catalytic engine of mammalian RNAi. Science 305, 1437−1441.

Liu, X.S., Chopp, M., Pan, W.L., Wang, X.L., Fan, B.Y., Zhang, Y., et al., 2016. MicroRNA-146a promotes oligodendrogenesis in stroke. Mol. Neurobiol.

Liu, Y., Yan, W., Zhang, W., Chen, L., You, G., Bao, Z., et al., 2012. MiR-218 reverses high invasiveness of glioblastoma cells by targeting the oncogenic transcription factor LEF1. Oncol. Rep. 28, 1013−1021.

Louis, D.N., Ohgaki, H., Wiestler, O.D., Cavenee, W.K., Burger, P.C., Jouvet, A., et al., 2007. The 2007 WHO classification of tumours of the central nervous system. Acta Neuropathol. 114, 97−109.

Lu, Q.R., Sun, T., Zhu, Z., Ma, N., Garcia, M., Stiles, C.D., et al., 2002. Common developmental requirement for Olig function indicates a motor neuron/oligodendrocyte connection. Cell 109, 75−86.

Luo, J.W., Wang, X., Yang, Y., Mao, Q., 2015. Role of micro-RNA (miRNA) in pathogenesis of glioblastoma. Eur. Rev. Med. Pharmacol. Sci. 19, 1630−1639.

Meister, G., Landthaler, M., Patkaniowska, A., Dorsett, Y., Teng, G., Tuschl, T., 2004. Human Argonaute2 mediates RNA cleavage targeted by miRNAs and siRNAs. Mol. Cell 15, 185−197.

Miska, E.A., Alvarez-Saavedra, E., Townsend, M., Yoshii, A., Sestan, N., Rakic, P., et al., 2004. Microarray analysis of microRNA expression in the developing mammalian brain. Genome Biol. 5, R68.

Nave, K.A., Werner, H.B., 2014. Myelination of the nervous system: mechanisms and functions. Annu. Rev. Cell Dev. Biol. 30, 503−533.

Noorbakhsh, F., Ellestad, K.K., Maingat, F., Warren, K.G., Han, M.H., Steinman, L., et al., 2011. Impaired neurosteroid synthesis in multiple sclerosis. Brain 134, 2703−2721.

Novakova, J., Slaby, O., Vyzula, R., Michalek, J., 2009. MicroRNA involvement in glioblastoma pathogenesis. Biochem. Biophys. Res. Commun. 386, 1−5.

Oldenborg, P.A., Zheleznyak, A., Fang, Y.F., Lagenaur, C.F., Gresham, H.D., Lindberg, F.P., 2000. Role of CD47 as a marker of self on red blood cells. Science 288, 2051−2054.

Otaegui, D., Baranzini, S.F., Armananzas, R., Calvo, B., Munoz-Culla, M., Khankhanian, P., et al., 2009. Differential micro RNA expression in PBMC from multiple sclerosis patients. PLoS ONE 4, e6309.

Persengiev, S.P., Kondova, I.I., Bontrop, R.E., 2012. The impact of MicroRNAs on brain aging and neurodegeneration. Curr. Gerontol. Geriatr. Res. 2012, 359369.

Raff, M.C., Durand, B., Gao, F.B., 1998. Cell number control and timing in animal development: the oligodendrocyte cell lineage. Int. J. Dev. Biol. 42, 263−267.

Shalem, O., Sanjana, N.E., Zhang, F., 2015. High-throughput functional genomics using CRISPR-Cas9. Nat. Rev. Genet. 16, 299−311.

Shang, C., Guo, Y., Hong, Y., Liu, Y.H., Xue, Y.X., 2015. MiR-21 up-regulation mediates glioblastoma cancer stem cells apoptosis and proliferation by targeting FASLG. Mol. Biol. Rep. 42, 721−727.

Shin, D., Howng, S.Y., Ptacek, L.J., Fu, Y.H., 2012. miR-32 and its target SLC45A3 regulate the lipid metabolism of oligodendrocytes and myelin. Neuroscience 213, 29−37.

Shin, D., Shin, J.Y., McManus, M.T., Ptacek, L.J., Fu, Y.H., 2009. Dicer ablation in oligodendrocytes provokes neuronal impairment in mice. Ann. Neurol. 66, 843–857.

Smirnova, L., Grafe, A., Seiler, A., Schumacher, S., Nitsch, R., Wulczyn, F.G., 2005. Regulation of miRNA expression during neural cell specification. Eur. J. Neurosci. 21, 1469–1477.

Stefani, G., Slack, F.J., 2008. Small non-coding RNAs in animal development. Nat. Rev. Mol. Cell Biol. 9, 219–230.

Stolt, C.C., Rehberg, S., Ader, M., Lommes, P., Riethmacher, D., Schachner, M., et al., 2002. Terminal differentiation of myelin-forming oligodendrocytes depends on the transcription factor Sox10. Genes Dev. 16, 165–170.

Svaren, J., 2014. MicroRNA and transcriptional crosstalk in myelinating glia. Neurochem. Int. 77, 50–57.

Szafranski, K., Abraham, K.J., Mekhail, K., 2015. Non-coding RNA in neural function, disease, and aging. Front. Genet. 6, 87.

Takebayashi, H., Nabeshima, Y., Yoshida, S., Chisaka, O., Ikenaka, K., Nabeshima, Y., 2002. The basic helix–loop–helix factor olig2 is essential for the development of motoneuron and oligodendrocyte lineages. Curr. Biol. 12, 1157–1163.

Traka, M., Podojil, J.R., McCarthy, D.P., Miller, S.D., Popko, B., 2016. Oligodendrocyte death results in immune-mediated CNS demyelination. Nat. Neurosci. 19, 65–74.

Tsang, J., Zhu, J., van Oudenaarden, A., 2007. MicroRNA-mediated feedback and feedforward loops are recurrent network motifs in mammals. Mol. Cell 26, 753–767.

Valinezhad Orang, A., Safaralizadeh, R., Kazemzadeh-Bavili, M., 2014. Mechanisms of miRNA-mediated gene regulation from common downregulation to mRNA-specific upregulation. Int. J. Genomics 2014, 970607.

Veiga, S., Ly, J., Chan, P.H., Bresnahan, J.C., Beattie, M.S., 2011. SOD1 overexpression improves features of the oligodendrocyte precursor response in vitro. Neurosci. Lett. 503, 10–14.

Wang, E., Cambi, F., 2012. MicroRNA expression in mouse oligodendrocytes and regulation of proteolipid protein gene expression. J. Neurosci. Res. 90, 1701–1712.

Wang, H., Moyano, A.L., Ma, Z., Deng, Y., Lin, Y., Zhao, C., et al., 2017. miR-219 cooperates with miR-338 in myelination and promotes myelin repair in the CNS. Dev. Cell in press.

Waschbisch, A., Atiya, M., Linker, R.A., Potapov, S., Schwab, S., Derfuss, T., 2011. Glatiramer acetate treatment normalizes deregulated microRNA expression in relapsing remitting multiple sclerosis. PLoS ONE 6, e24604.

Westphal, M., Lamszus, K., 2015. Circulating biomarkers for gliomas. Nat. Rev. Neurol. 11, 556–566.

Wilczynska, A., Bushell, M., 2015. The complexity of miRNA-mediated repression. Cell Death Differ. 22, 22–33.

Xin, M., Yue, T., Ma, Z., Wu, F.F., Gow, A., Lu, Q.R., 2005. Myelinogenesis and axonal recognition by oligodendrocytes in brain are uncoupled in Olig1-null mice. J. Neurosci. 25, 1354–1365.

Yu, Y., Chen, Y., Kim, B., Wang, H., Zhao, C., He, X., et al., 2013. Olig2 targets chromatin remodelers to enhancers to initiate oligodendrocyte differentiation. Cell 152, 248–261.

Zhao, C., Deng, Y., Liu, L., Yu, K., Zhang, L., Wang, H., et al., 2016. Dual regulatory switch through interactions of Tcf7l2/Tcf4 with stage-specific partners propels oligodendroglial maturation. Nat. Commun. 7, 10883.

Zhao, C., Sun, G., Li, S., Shi, Y., 2009. A feedback regulatory loop involving microRNA-9 and nuclear receptor TLX in neural stem cell fate determination. Nat. Struct. Mol. Biol. 16, 365–371.

Zhao, X., He, X., Han, X., Yu, Y., Ye, F., Chen, Y., et al., 2010. MicroRNA-mediated control of oligodendrocyte differentiation. Neuron 65, 612–626.

Zhao, X., Wu, J., Zheng, M., Gao, F., Ju, G., 2012. Specification and maintenance of oligodendrocyte precursor cells from neural progenitor cells: involvement of microRNA-7a. Mol. Biol. Cell 23, 2867–2878.

Zuchero, J.B., Barres, B.A., 2013. Intrinsic and extrinsic control of oligodendrocyte development. Curr. Opin. Neurobiol. 23, 914–920.

miRNA IN NEURONAL NETWORKS MATURATION AND PLASTICITY

12

Debora Napoli[1] and Tommaso Pizzorusso[2,3]

[1]*Scuola Normale Superiore, Pisa, Italy* [2]*Institute of Neuroscience, Pisa, Italy* [3]*University of Florence, Florence, Italy*

INTRODUCTION

miRNAs are a class of noncoding RNA involved in the control of many biological processes such as cell proliferation, differentiation, and apoptosis (Ameres and Phillip, 2013; Jonas and Izaurralde, 2015). Their expression is ubiquitous with about 50% of them expressed in the brain (Lagos-Quintana et al., 2002; Landgraf et al., 2007); this abundance could be explained by the heterogenous composition of brain that implicates different miRNA profiling by area and cell populations (He et al., 2012; Ziats and Rennert, 2014). Even in the same neuron, the population of miRNAs can change among subcellular compartments (soma or dendrites) (Kye et al., 2007; Lugli et al., 2008; Schratt, 2011) where they play a crucial role in different brain functions. Studies using paradigms such as learning and memory, deprivation of sensory experience and drug abuse revealed a role for miRNAs in neuroplasticity, i.e., brain ability to modify its connections in response to internal (e.g., gut microbiota, Zeng et al., 2016) and external stimuli (e.g., sensory experiences, Tognini et al., 2011). Neuronal miRNAs, in cooperation with other transcriptional and posttranscriptional regulatory factors, modulate the final outcome of gene expression according to genetic instructions and in response to neuronal activity; in this way, they contribute to the functional organization of connections, in term of structural and functional synaptic plasticity. Structural plasticity consists in experience-dependent changes of morphology such as axonal sprouting, new dendritic spines, and synaptic contacts. Functional synaptic plasticity concerns the modulation of synaptic efficacy in response to neuronal activation and it is studied in models like long-term potentiation (LTP), long-term depression, and homeostatic plasticity. It is worthwhile to point out that, in many cases, functional and structural changes are just two faces of the same plasticity process, and the distinction between functional and structural events is more due to the level of analysis of the researcher than to the presence of distinct biological processes. Thus, structural and functional plasticity are collectively referred as synaptic plasticity. Since the first studies in *Aplysia*, it is widely accepted that plasticity involves transcriptional and translational regulation carried out by a variety of molecular effectors (Kandel, 2012). Two phases are distinguished in the mechanisms of long-term plasticity: an early phase and a late phase characterized by its dependence upon new local protein synthesis and gene transcription. miRNA and pre-miRNA have been found to be enriched in synaptic fractions as compared to total homogenate levels (Kye et al., 2007; Lugli et al., 2008; Pichardo-Casas et al., 2012) and components of their maturation mechanisms are found at postsynaptic density (Ashraf et al., 2006; Banerjee et al., 2009; Störchel et al., 2015). Thus, they could be part of the toolbox controlling translation in a mRNA and synapse-specific manner (Edbauer et al., 2010). In particular, it has been proposed that synaptic miRNAs could maintain their synaptic mRNA targets in a dormant state until their protein product is required for synaptic plasticity. miRNAs could be summoned to synapses by neuronal activity who directs also their transcription and stability (Krol et al., 2010); in turn, neuronal activity is modulated by miRNAs action (Edbauer et al., 2010). Given this contribution of miRNAs to neuronal function, it is not a surprise that they are enriched in the brain where regulation of gene expression by external stimuli is more dynamic than other tissues. Brain is continuously called upon to flexibly interface with the environmental variability; this implies that neuronal systems must have mechanisms that store past experiences in an editable format, but at the same time, these mechanisms should be robust enough to avoid noisy representations. In this sophisticated scenario, it has been proposed that miRNAs could be like biochemical shock absorbers making the system modifiable but not easy to be damaged, by their buffering action on the gene outcome.

The suggestion that miRNAs play a role in the regulation of complex neuronal networks comes also from evolutionary studies. It was found that mammalian genome conserves and transcribes thousands of noncoding RNA, including miRNAs (Olive et al., 2015). Moreover, Somel et al. (2011) performed a comparison of the expression of mRNAs and miRNAs among different primate species and found that the prefrontal cortex shows developmental profiles of miRNAs, as well as of their target genes, with rates of human-specific evolutionary change significantly faster than other genes, including transcription factors previously thought to be the leaders in the evolution process. A total of 19 of these developmentally regulated miRNAs are 24-fold more divergent in humans than in chimpanzee suggesting that miRNAs may have a significant role in the evolutionary improvements of higher cognition performances (further discussion on these topic can be also found in Chapters 15 and 16 of this book).

THE miRNA BIOGENESIS PATHWAY

Structural and maturation features of neuronal miRNAs are similar to that of other tissues and has been already discussed in Chapter 1, Making and Maintaining MicroRNAs in Animals of this book by Pasquinelli and colleagues. miRNAs are transcribed by RNA polymerase III or RNA polymerase II, their coding sequences are located in autonomous genes or included in introns or exons of other genes. In the first case, they have their own regulatory elements; in the second case, they are controlled by the transcription of their host genes. Furthermore, several miRNA genes are clustered and expressed together as polycistronic transcript. In the canonical pathway, the transcription produces a primary transcript (pri-miRNA) with a typical hairpin-stem loop structure processed, in the nucleus, by the RNAse III enzyme Drosha and the dsRNA-binding protein DGCR8. The resulting precursor miRNA (pre-miRNA, about 55–70 nt) is exported to the cytosol through the Exportin-5. Once in the cytoplasm, the RNAse III enzyme Dicer cleaves it forming a duplex miRNA. Argonaute proteins are recruited to Dicer and they unwind the duplex miRNA. Only the strand with the less thermodynamically stable pairing at the 5′ region is finally incorporate in the RNA-induced silencing complex (RISC). At this stage, the miRNA is ready for its regulatory function: It can bind the target mRNAs in the 3′ untranslated region (3′ UTR) by its seed sequence. Three alternative mechanisms are reported to occur: site-specific cleavage, enhanced mRNA decay, and translation inhibition. The choice depends on the complementarity between the miRNA and the mRNA targets. This pathway of miRNA biogenesis is considered the canonical one. But other pathways independent of the Drosha/DGCR8 processing are identified in mammals: the mirtrons are directly spliced from introns of protein-coding mRNA and exported to the cytoplasm where they are processed through the classical miRNA machinery (Ha and Kim, 2014).

ROLE OF THE miRNA SYSTEM IN THE POSTNATAL BRAIN

A popular manipulation to understand whether the miRNA system as a whole is involved in a biological process is genetic manipulation or interference with key enzymes of miRNA biosynthesis. In particular, since Dicer has proven to catalyze a key step in miRNA maturation it has become a preferential target for many studies on miRNA system role in the brain. For example, zebrafish Dicer KO showed abnormal brain morphology, while Dicer-null mice arrest their development at embryonic day 7.5 (before neurulation) because of stem cells lack (Giraldez, 2005). To overcome the problem caused by precocious embryonic mortality, and to better understand the role of miRNA system in later pre and postnatal developmental stages, many groups produced and analyzed different spatial and temporal conditional Dicer KO mice. In 2007, a study demonstrated that ablation of Dicer from Purkinje cells cause cell death, progressive cerebellar degeneration and development of ataxia first suggesting that miRNA were necessary at least for neuronal survival, and their misregulation could be implicated in neurodegenerative diseases such as Parkinson and Alzheimer (Schaefer et al., 2007). In the same year, a cKO mice with deletion of Dicer from midbrain dopamine neurons, the neuronal population that is lost in Parkinson disease, was generated (Kim et al., 2007), and its analysis revealed a progressive loss of dopaminergic neurons from 2 to 6 weeks of age, incriminating miR-133b downregulation as the major responsible. Indeed, miR-133b was found to be reduced in samples from Parkinson patients and in two additional animal models of dopaminergic neurons deficiency. This miRNA is responsible in particular for the downregulation of Pituitary homeobox 3 (Pix3), a transcription factor necessary for the expression of two dopaminergic markers: tyrosine hydroxylase and dopamine transporter. Pix3 also regulates miR-133b creating a negative feedback circuit whose misregulation could be responsible for the progressive neurodegeneration. In agreement with the hypothesis that misregulation of miRNA system could be responsible for neurodegeneration in neurodegenerative diseases, it was demonstrated that loss of Dicer restricted to postnatal forebrain neurons (using Cre recombinase expression driven by the CamKII promoter) could trigger neurodegeneration through mechanisms related to tau hyperphosphorylation (Hébert et al., 2010): In this case, the effect was proposed to be mediated by the impaired expression of miR-15 family members resulting in upregulation of extracellular signal-regulated kinase 1 (ERK1), a kinase that in turn could phosphorylate tau. The inactivation of Dicer has been recently performed using a tamoxifen Cre-inducible system to understand how perturbation of miRNA system impacts the physiology of the fully developed mouse brain (Fiorenza et al., 2016). The deletion of Dicer in adult forebrain neurons using the same CaMKII promoter of the Hebert et al. paper, but temporally regulated by tamoxifen administration, confirmed the presence of slow neurodegeneration. Interestingly, the data also reported a neuronal hyper-reactivity to external stimuli that made the mice more susceptible to epilepsy but also able to produce enhanced and more robust fear memory suggesting that miRNAs are key stabilizers of neuronal function in the adult brain. This role is in agreement with previous data showing a similar memory enhancement in mice with deletion of Dicer in the adult forebrain (Konopka et al., 2010). Although most of the works demonstrate that the inactivation of Dicer in neurons promotes neurodegeneration, it was also reported that the deletion of Dicer from dopaminoceptive striatal neurons achieved by the promoter sequences of dopaminergic receptor 1 driving Cre recombinase expression crossed with floxed Dicer mice, cause a phenotype characterized by hindlimb clasping, ataxia, reduction in brain size and neuronal cell size, astrogliosis, and reduced lifespan that was not due to neurodegeneration (Cuellar et al., 2008). Another study using viral transduction with Crerecombinase of the amygdala of floxed Dicer mice did not report neuronal death although the injected mice were followed for only 8 weeks (Haramati et al., 2011).

Until now, we have discussed the effects of miRNA system perturbations during postnatal development and during adulthood, but other studies have been done to elucidate also the role during earlier stages of development by using Cre-drivers like Wnt1-, Emx1-, and Nestin-Cre mice promoting the early deletion of Dicer in specific neural populations. The loss of Dicer achieved by Wnt-1 promoter (Huang et al., 2010; Zehir et al., 2010) leads to miRNA deficiency in neural crest cells resulting in a dramatic malformation of the tectum and cerebellum, and altered development of dorsal root ganglia, enteric nervous system, and sympathetic ganglia. Dicer inactivation controlled by promoter sequences of the transcription factor Emx1 (De Pietri Tonelli et al., 2008; Li et al., 2011), known to control the dorsal telencephalic development, promotes abnormal massive hypotrophy of the postnatal cortex due to neuronal apoptosis and defective cortical layering caused by dramatic impairments of differentiation. These mice present also abnormal hippocampal morphology, altered number of hippocampal progenitors and death shortly after weaning. Finally, when Dicer is deleted by Nestin promoter drive Cre recombinase, the migration of late-born neurons in the cortex and oligodendrocyte precursor expansion and differentiation in the spinal cord are greatly affected and the size of hippocampi is reduced by increased neuronal apoptosis (Kawase-Koga et al., 2009). Altogether these studies suggest that the miRNA system is crucial for correct neuronal development and maintenance of normal brain functions at all ages tested by regulating different sets of mRNAs depending on the age and on the brain region. These effects have been interpreted as a general role of the miRNA system in conferring stability to gene expression, but they could also involve the action of individual miRNAs. Indeed, several studies reported regulation of plasticity by manipulation of single miRNAs.

ROLE OF SELECTED miRNAs IN PLASTICITY

Activity-dependent mechanisms are obvious candidates as mechanisms controlling plasticity. A wealth of studies documented activity-dependent transcriptional changes mediated by regulation of transcription factors or factors involved in chromatin remodeling and revealed a role for these mechanisms in neural plasticity. Thus, it is not surprising that many labs investigated activity-dependent regulation of miRNAs.

A well-known activity-regulated miRNA is miR-132. It has been discovered in a genome-wide screening of genes regulated by cAMP response element-binding protein (CREB), a central transcription factor for the activation of plasticity genes. In vitro studies revealed that stimulation with KCl, which causes depolarization, and neurotrophins induces high levels of miR-132 expression. Activity-dependent regulation of miR-132 was confirmed in vivo by pharmacological induction of neuronal activation with pilocarpine and cocaine injection, applying odorant molecules, and after fear conditioning (Nudelman et al., 2010).

A series of studies analyzing the mechanisms of visual cortex plasticity revealed that miR-132 levels are quickly modulated by physiological visual stimulation in the visual cortex (Tognini et al., 2011; Tognini et al., 2015). Indeed, a brief period (3 days) of dark rearing during development downregulated miR-132 cortical levels. Reexposure to light induced a fivefold overshoot in miR-132 levels in only 105 minutes. Interestingly, the visual regulation of miR-132 is crucial to shape the developmental trajectory of the expression of this miRNA. MiR-132 in the visual cortex is very low before eye opening (P14−15) rising its expression dramatically between P20 and P30. Mice reared in darkness from birth did not show any developmental increase of miR-132, suggesting that the developmental trajectory of miR-132 is not genetically predetermined but is brought about by experience. The visual cortex is a classical model for experience-dependent refinement of cortical circuits, therefore it has been easy to move to functional studies on miR-132 role in developmental plasticity using the paradigm of monocular deprivation in juvenile mice (Tognini et al., 2011). It was found that miR-132 reduction due to the reduced visual input is necessary for the ocular dominance shift induced by monocular deprivation. Complete lack of miR-132 also resulted in defective ocular dominance shift (Mellios et al., 2011) indicating that intermediate levels of miR-132 are permissive, but lower or higher levels do not support ocular dominance plasticity. The analysis on visual cortex also shed light on mechanisms underlying experience-dependent regulation of miR-132. This miRNA derives from a bicistronic precursor that also includes miR-212. At genome level, CRE sites close to the transcriptional start site of the *miR-132/212* gene have been described. When mutated, these CRE sites dramatically reduce the induction of miR-132/miR-212 (Remenyi et al., 2010). The studies on the visual cortex found that these sites undergo dramatic changes in histone posttranslational modifications and DNA methylation in response to monocular deprivation and visual stimulation (Tognini et al., 2015). Moreover, blocking DNA methylation restores normal levels of miR-132 suggesting that epigenetic modifications play an important role in experience-dependent regulation of miR-132 transcription. The complexity of miR-132 regulation is further outlined by the observation that in adult mice visual experience exerts a much reduced regulation of miR-132 levels and of epigenetic modifications on CRE promoters (Tognini et al., 2011, 2015). This result could be linked to the different levels of plasticity present in juvenile and adult cortex. At neuronal level, previous work identified miR-132 as a regulator of dendritic spine stabilization and functional maturation of excitatory transmission in the cerebral cortex. Excitatory neurons seem to be a preferred target of miR-132. Indeed, miR-132 family has been reported to be preferentially expressed in cortical excitatory cells (He et al., 2012); *miR-132/212* deletion affects excitatory synaptic transmission (Remenyi et al., 2013), while miR-132 overexpression promotes functional and structural maturation of excitatory synapses (Edbauer et al., 2010; Hansen et al., 2010); and miR-132 mimic administration promotes dendritic spine maturation of excitatory pyramidal neurons (Tognini et al., 2011).

Several questions remain to be answered, for instance neither miR-132 nor other miRNAs have ever been investigated in the context of physiological maturation of cortical functions. The quantitatively relevant and temporally restricted rise in miR-132 cortical levels hints at a specific role of miR-132 in visual cortical development. In particular, by comparing the miR-132 developmental regulation with a timescale of the functional development of cortical neurons (Espinosa and Stryker, 2012) (Fig. 12.1) makes likely that late maturation of receptive field properties such as binocular matching of orientation preference, that are susceptible to manipulation of visual experience, could be linked to molecular changes driven by miR-132 increase in the cortex. Analyzing cortical development in mice with altered expression of miR-132 could answer this question and could represent a paradigmatic approach to understand the role of those miRNA that show steep age-dependent regulation. It is also unknown how the dramatic developmental rise in miR-132 impacts on the development of the molecular composition of cortical neurons. A genome-wide analysis performed at different ages after manipulation of miR-132 levels could answer this point.

It is also unknown if miR-132 could play similar roles in experience-dependent development of other sensory or nonsensory areas. Several papers analyzed the role of miR-132 in nonvisual areas: miR-132 contributes to the correct integration of newborn neurons in hippocampus and also to dendritic spine morphology, synaptic integration, and survival of neurons in olfactory bulb (Impey et al., 2012). These studies also investigated some molecular targets of miR-132. For instance, p250-GAP is a brain-enriched GTPase-activating protein for Rho family GTPases inversely correlated with spinogenesis and interacting with NMDA receptor and PSD-95 at the synapse. Indeed, KD of p250-GAP increases spine density and spine size such as mEPSC frequency and the positive GluR1 spines whereas inhibition of miR-132 has opposite effects. Finally, miR-132/p250-GAP regulates Rac1 activity and spine formation by modulating synapse-specific Kalirin7-Rac1-Pak signaling (Impey et al., 2012). A different pathway could also use miR-132 to regulates hippocampal function: Leptin, known to be a powerful regulator of energy homeostasis, promotes in hippocampus the formation of stable dendritic spines and functional synapses. This effect was mediated by

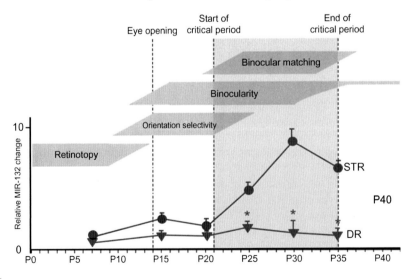

FIGURE 12.1

Developmental expression of miR-132 in the visual cortex in standard rearing (STR) or in dark rearing (DR) have been replotted (data from Tognini et al., 2011) on the principal milestones of visual cortical development (redrawn from Espinosa and Stryker, 2012).

inducing CREB transcription through the ERK pathway resulting in turn ins an increase in miRNA-132-mediated suppression of p250-GAP activity (Dhar et al., 2014).

Several studies found that miR-132 is important in learning and memory. An initial study reported that a fivefold overexpression of miR-132 in transgenic mice increased spine density, but caused deficits in novel object recognition memory (Hansen et al., 2010). However, experiments with a milder overexpression of miR-132 in the hippocampus resulted in enhanced cognitive capacity (Hansen et al., 2013). Recent work showed that mice with deletion of the *miR-132/−212* gene show learning impairments (Hansen et al., 2016), thus, like in the visual cortex, these findings suggest that miR-132 expression levels are tightly regulated with baseline levels that could be more or less permissive for plasticity depending on each individual circuit and developmental stage. However, the highly regulated expression of miR-132 would endow this miRNA with the ability to dynamically modulate plasticity levels.

Considering the role of miR-132 in neurodevelopment and in the adult, many papers have implicated this miRNA in the etiology of different brain disorders. Alterations of miR-132 are present in patients of disorders like Alzheimer's disease (Lau et al., 2013; Smith et al., 2011; Cogswell et al., 2008) and Huntington's disease (Lee et al., 2011; Johnson et al., 2008). In agreement with its role in neurodevelopment, miR-132 has been also involved in neurodevelopmental disorders. MeCP2, a methylated DNA binding protein that when mutated causes Rett syndrome is a target of miR-132 (Klein et al., 2007). Intriguingly, a highly precise regulation of MeCP2 seem to be necessary for normal brain function; indeed, symptoms are present both with MeCP2 duplication or in response to MeCP2 reduction with the severity of the symptoms being proportional to the extent of MeCP2 dysregulation (Chao and Zoghbi, 2012). It is tempting to speculate that miR-132 regulation contributes to this fine regulation, although the scenario could be complicated by the presence of a feedback regulatory loop of MeCP2 on miR-132 transcription (Klein et al., 2007). Indeed, reduced miR-132 levels were found in the cerebellum of MeCP2 null mice (Wu et al., 2010).

Another activity-dependent miRNA is miR-134; it has been discovered as dendritic miRNA of hippocampal neurons in 2006 (Schratt et al., 2006), and a recent paper demonstrates the loop sequence of its pre-miRNA form it necessary for the interaction with DHX36 and the accumulation to dendrites (Bicker et al., 2013). At the synapse, miR-134 regulates spine morphology by local inhibition of Limk1 mRNA, a cytoskeleton regulator. Stimulation with BDNF, usually secreted by neuronal activity, release Limk1 from miR-134 inhibition promoting Limk1 protein translation. Indeed, overexpression in vitro of miR-134 reduces dendritic spine volume and synaptic strength of hippocampal neurons, while miR-134 silencing has opposite effects (Schratt et al., 2006). Additional in vivo analysis demonstrated that miR-134 targets also the transcription factor CREB (Gao et al., 2010); in turn, miR-134 resulted to be negatively regulated by sirtuin 1 (SIRT1) in cooperation with the transcription factor YY1 during learning and memory. SIRT1 KO mice, that miss this mechanism, have LTP impairments in CA1 neurons and deficits in memory performances due to the reduction of CREB and BDNF. Recently, the role of miR-134 has also been studied in the context of disease mechanisms since it has been demonstrated that miR-134 is upregulated in experimental and human epilepsy (Gao et al., 2010; Jimenez-Mateos et al., 2012). Interestingly, miR-134 downregulation by antagomiR intrahippocampal injection protected against seizure induction and decreased spine density of hippocampal CA3 neurons caused by kainate, probably acting via derepression of Lmk1.

Because of its synaptic actions, miR-134 has been also studied in homeostatic plasticity. MiR-134 resulted to be necessary for both synapse elimination and the structural rearrangements leading to synaptic downscaling (Fiore et al., 2014). MiR-134 downregulation of Pumilio-2, a RBP involved in miRNA transport and translation inhibition, is required for miR-134-mediated homeostatic synaptic depression, in response to chronic activity. Using compartmentalized culture system, the authors demonstrated that this inhibition occurs specifically at dendritic compartment. Given that AMPA internalization is a necessary step of homeostatic plasticity, they investigated the possible contribution of miR-134 in this process identifying polo-like kinase 2 as a novel target of Pumilio-2 involved in the control of GluA2 surface expression. This a novel pathway of homeostatic plasticity that stabilizes neuronal circuits in response to increased network activity and future studies could be interesting to investigate how this pathway contribute to miR-134 neuroprotective role in epilepsy, searching for a possible therapeutic use of antagomirs.

Pumilio-2, Limk1, and CREB are not the only demonstrated targets of miR-134: a developmental in vivo study found that also Chordin-like 1 (Chrdl-1) and Doublecortin (Dcx) are targeted by this miRNA having stage-specific effects on cortical progenitors, migratory neurons, and differentiated neurons (Gaughwin et al., 2011). In neural progenitors, miR-134 promotes cell proliferation and counteracts Chrdl-1-induced apoptosis and Dcx-induced differentiation in vitro. In neurons, miR-134 reduces cell migration in vitro and in vivo in a Dcx-dependent manner. In differentiating neurons, miR-134 modulates process outgrowth in response to exogenous BMP-4 in a noggin-reversible manner (Gaughwin et al., 2011). MiR-134 is a good example of miRNAs versatility; their actions could be different depending on developmental stage, brain localization and cell compartment.

MiR-124 is a highly conserved and exclusively neuronal miRNA that accounts for 25%−48% of total adult brain miRNA (Lagos-Quintana et al., 2002). Its expression increases over 13-fold from E12 to E21, then remains stable in neuronal cells (Krichevsky et al., 2003). In mammalian neurons, miR-124 suppresses the levels of hundreds of nonneural genes, which contributes to the acquisition and maintenance of neuronal identity (Lim et al., 2005). For example, miR-124 inhibits the neuronal transcription regulator complex REST/SCP1 (Visvanathan et al., 2007). This complex represses the transcription of neuronal genes in nonneuronal cells by binding the conserved repressor sequence RE1, in cooperation with histone deacetylase, MeCP2, and SCP1 (small C-terminal domain phosphatase 1). During the transition from progenitors to postmitotic neurons, the inactivation of this complex is crucial. Indeed, miR-124 inhibition in the neural tube results in proliferating cells in the lateral postmitotic zone, but with a reduction of neuronal markers (Visvanathan et al., 2007). miR-124 acts during development to promote neuronal differentiation also by inhibition of PTBP1 (Makeyev et al., 2007), a repressor of neuron-specific splicing. miR-124 also contributes to the control of neurite outgrowth during neuronal differentiation possibly by cytoskeleton regulation (Yu et al., 2008) and affects dendritic differentiation by regulating RhoG (Franke et al., 2012). MiR-124 maintains its role of neuronal promoter also in the adult neurogenesis, a neural plasticity process that is also regulated by physical activity and environmental enrichment (van Praag et al., 1999; Brown et al., 2003). Indeed, in another study, it has been found that miR-124 increases the transition from amplifying cells to neuroblasts in the subventricular zone (SVZ), and further increases when neuroblasts exit the cell cycle (Chen et al., 2006). Knockdown of endogenous miR-124 of FACS purified SVZ stem cells astrocytes results in major numbers of proliferating cells, while knockdown of miR-124 in neuroblasts does not affect their survival or differentiation. In contrast to miR-124 knockdown, which maintained SVZ cells as dividing precursors, miR-124 overexpression caused a significant decrease in the overall proportion of dividing cells and an increase in postmitotic neurons. Furthermore, blocking miR-124 function in vivo during regeneration leads to hyperplasias. The authors demonstrated that the upregulation of miR-124 during cell differentiation in SVZ is necessary to repress the transcription factor Sox9, which controls the expression of glial genes and contribute to glial fate specification. Given that stem cells in SVZ are a pool of astrocytes, its downregulation is necessary to the progression of the lineage to mature neurons (Chen et al., 2006). Although miR-124 results to be crucial in particular in neuronal differentiation, its expression remains high in mature neurons raising the question about its functions in mature neurons. To answer this question, miR-124 role in maladaptive plasticity induced by cocaine addiction and chronic stress has been investigated. It has been demonstrated that cocaine causes the activation of the transcriptional repressor REST complex in nucleus accumbens; in turn, REST downregulates miR-124 expression resulting in the deregulation of gene expression in mesolimbic dopaminergic system. Although miR-124 is not the only reported misregulated-cocaine miRNA, it contributes to the perturbation of system by upregulation of important related plasticity genes such as BDNF, integrin β1, actin binding protein like NAC1, and axon guidance molecules like SEMA6A (Chandrasekar and Dreyer, 2009). Indeed, lentiviral overexpression of miR-124 in nucleus accumbens attenuates cocaine-induced conditioned place preference (Chandrasekar et al., 2011). miR-124 downregulation is found also in the hippocampus of mice exposed to chronic ultra-mild stress, a paradigm to induce depression-like behaviors and study mood disorders. Although this downregulation is not the cause of behavioral alterations, because perturbation of miR-124 expression in non-stressed mice does not promote depression-like behaviors, overexpression of miR-124 in stressed mice increases their resilience and downregulation of miR-124 increases susceptibility to a milder stress. These behavioral effects may be mediated by upregulation of two miR-124 targets: histone deacetylase HDAC4/5 and glycogen synthase kinase 3β. Indeed, inhibitors of these enzymes have antidepressant-like actions on behavior (Higuchi et al., 2016). In hippocampus, miR-124 could be involved also in the regulation of LTP and memory formation by targeting protein scaffold IQGAP1, that seems to be a key regulator of dendritic spine number in the hippocampus and lateral amygdala and that contributes to NR1/NR2A trafficking and NR2A-mediated ERK signaling (Gao et al., 2011; Yang et al., 2014).

Further studies have studied the role of miR-124 analyzing plasticity in a defined functional circuit from marine snail *Aplysia* containing a sensory neuron (SN) and a motor neuron (MN). In this model, a robust decrease of miR-124 was induced by serotonin stimulation in the presynaptic compartments. Serotonin is a neuromodulator critical for the induction of synaptic plasticity and memory for sensitization in *Aplysia*. While overexpression of miR-124 by miRNA mimics decreases the strength of connection between SNs and MNs (called long-term facilitation—LTF—when persisting for 48 hours), downregulation of miR-124 promotes an increase of LTF. The authors demonstrated that miR-124 acts by targeting the transcription factor CREB1. Indeed, knockdown of miR-124 induces an increase not only of CREB1 but of its controlled genes necessary (UCH, C/EBP, and KHC) for LTF (Rajasethupathy et al., 2009). Some years later, the authors used the same model of sensitization in *Aplysia* to demonstrate the role of piRNA in the control of synaptic plasticity. First they demonstrated that piRNAs, a class of 26−32 nucleotide small noncoding RNAs, are not exclusively expressed in germline and associated somatic tissues, but some of them are enriched in CNS and in particular in the nuclear compartment of neurons. Stimulation with serotonin of sensory−motor synapse induced an upregulation of piRNAs while the knockdown of their associated protein Piwi significantly impairs LTF increasing the expression of CREB2, a major inhibitory of LTF. Then, knowing that piRNAs control the silencing of transposable elements in germline by

DNA methylation and that CREB2 promotor contains a CpG island fully methylated after serotonin stimulation. The authors decided to block the DNMTs activity and obtained the same results of Piwi knockdown. Furthermore, they found that among the piRNAs upregulated by serotonin four are complementary to the CREB2 promoter and the knockdown of one of these (aca-piR-F) induces upregulation of CREB2. Although the authors did not directly demonstrate the demethylation of CREB2 promoter following aca-piR-F downregulation, the paper demonstrates for the first time the enrichment of piRNAs in the CNS and gives strong evidences of their role in synaptic plasticity.

We have reviewed here the evidences describing the role of three miRNAs with an established role in plasticity and that could be representative for a wealth of ongoing studies on other miRNAs. We are confident that these studies will enlarge the number of miRNAs involved in plasticity of neuronal connections.

CONCLUSIONS

miRNAs are perfect candidates to mediate the spatially and temporally accurate regulation of gene expression required for synaptic plasticity. Evidence supporting this theory is accumulating together with studies linking miRNAs to many neurological diseases. On the other hand, we still have a lot to understand about the biology of the miRNA system as a whole and of individual miRNAs. Novel "omic" approaches for the analysis of miRNA production, actively translating mRNAs and mRNA stability, cell-specific mRNA, and miRNA regulation will help to clarify outstanding questions with the broad perspective important for mechanistic studies at the system biology level. In vivo studies analyzing the outcome of specific alterations of miRNA expression patterns and methods to visualize miRNA action in space and time are dramatically required to obtain information on miRNA action in plasticity and disease. The cell-specific posttranscriptional control of miRNA biogenesis and activity is another aspect that seems to be little understood. Finally, we have to underscore that miRNAs represent also pharmacological targets that could be inhibited by miRNA sponges, decoys or antagomiRs, whereas upregulation could be achieved with miRNA mimics. For all these reasons, future studies on brain miRNA biology are expected to expand our knowledge on the complex molecular mechanisms underlying higher cognitive functions and possibly lead us closer to therapies for untreatable brain disorders.

REFERENCES

Ameres, S.L., Phillip, D.Z., 2013. Diversifying microRNA sequence and function. Nat. Rev. Mol. Cell Biol. 14 (8), 475−488.

Ashraf, S.I., Anna, L.M., Sclarsic, S.M., Kunes, S., 2006. Synaptic protein synthesis associated with memory is regulated by the RISC pathway in drosophila. Cell 124 (1), 191−205.

Banerjee, S., Neveu, P., Kosik, K.S., 2009. A coordinated local translational control point at the synapse involving relief from silencing and MOV10 degradation. Neuron 64 (6), 871−884.

Bicker, S., Khudayberdiev, S., Weiß, K., Zocher, K., Baumeister, S., Schratt, G., 2013. The DEAH-box helicase DHX36 mediates dendritic localization of the neuronal precursor-microRNA-134. Genes Dev. 27 (9), 991−996.

Brown, J., Christiana, M.C.-K., Gerd Kempermann, H.V., Praag, J., Winkler, F.H., Gage, Georg Kuhn, H., 2003. Enriched environment and physical activity stimulate hippocampal but not olfactory bulb neurogenesis. Eur. J. Neurosci. 17 (10), 2042−2046.

Chandrasekar, V., Dreyer, J.-L., 2009. microRNAs miR-124, let-7d and miR-181a regulate cocaine-induced plasticity. Mol. Cell. Neurosci. 42 (4), 350−362.

Chandrasekar, V., Chandrasekar, V., Dreyer, J.-L., 2011. Regulation of MiR-124, let-7d, and MiR-181a in the accumbens affects the expression, extinction, and reinstatement of cocaine-induced conditioned place preference. Neuropsychopharmacology 36 (6), 1149−1164.

Chao, H.-T., Zoghbi, H.Y., 2012. MeCP2: only 100% will do. Nat. Neurosci. 15 (2), 176−177.

Chen, A.I., de Nooij, J.C., Jessell, T.M., 2006. Graded activity of transcription factor Runx3 specifies the laminar termination pattern of sensory axons in the developing spinal cord. Neuron 49 (3), 395−408.

Cogswell, J.P., Ward, J., Taylor, I.A., Waters, M., Shi, Y., et al., 2008. Identification of miRNA changes in Alzheimer's disease brain and CSF yields putative biomarkers and insights into disease pathways. J. Alzheimers Dis.: JAD 14 (1), 27−41.

Cuellar, T.L., Tigwa, H.D., Nelson, P.T., Loeb, G.B., Harfe, B.D., et al., 2008. Dicer loss in striatal neurons produces behavioral and neuroanatomical phenotypes in the absence of neurodegeneration. Proc. Natl. Acad. Sci. U.S.A. 105 (14), 5614−5619.

De Pietri Tonelli., D., Pulvers, J.N., Haffner, C., Murchison, E.P., Hannon, G.J., Huttner, W.B., 2008. miRNAs are essential for survival and differentiation of newborn neurons but not for expansion of neural progenitors during early neurogenesis in the mouse embryonic neocortex. Development 135 (23), 3911−3921.

Dhar, M., Zhu, M., Impey, S., Lambert, T.J., Bland, T., Ilia, N., et al., 2014. Leptin induces hippocampal synaptogenesis via CREB-regulated microRNA-132 suppression of p250GAP. Mol. Endocrinol. 28 (7), 1073−1087.

Edbauer, D., Joel, R.N., Foster, K.A., Wang, C.-F., Seeburg, D.P., Matthew, N., et al., 2010. Regulation of synaptic structure and function by FMRP-associated microRNAs miR-125b and miR-132. Neuron 65 (3), 373−384.

Espinosa, J.S., Stryker, M.P., 2012. Development and plasticity of the primary visual cortex. Neuron 75 (2), 230−249.

Fiore, R., Rajman, M., Schwale, C., Bicker, S., Antoniou, A., et al., 2014. MiR-134-dependent regulation of pumilio-2 is necessary for homeostatic synaptic depression. EMBO J. 33 (19), 2231−2246.

Fiorenza, A., Jose, P.L.-A., Rovira, V., Scandaglia, M., Geijo-Barrientos, E., Barco, A., 2016. Blocking miRNA biogenesis in adult forebrain neurons enhances seizure susceptibility, fear memory, and food intake by increasing neuronal responsiveness. Cereb. Cortex 26 (4), 1619−1633.

Franke, K., Kristin, F., Wolfgang, O., Sascha, J., Jan, B., Nitsch, R., et al., 2012. miR-124-regulated RhoG reduces neuronal process complexity via ELMO/Dock180/Rac1 and Cdc42 signalling. EMBO J. 31 (13), 2908−2921.

Gao, C., Shanti, F.F., Guedea, A.L., Tronson, N.C., Jovasevic, V., et al., 2011. IQGAP1 regulates NR2A signaling, spine density, and cognitive processes. J. Neurosci. 31 (23), 8533−8542, Society for Neuroscience.

Gao, J., Jun, G., Wen-Yuan, W., Ying-Wei, M., Johannes, G., et al., 2010. A novel pathway regulates memory and plasticity via SIRT1 and miR-134. Nature 466 (7310), 1105−1109.

Gaughwin, P., Ciesla, M., Yang, H., Lim, B., Brundin, P., 2011. Stage-specific modulation of cortical neuronal development by mmu-miR-134. Cereb. Cortex 21 (8), 1857−1869.

Giraldez, A.J., 2005. MicroRNAs regulate brain morphogenesis in zebrafish. Science 308 (5723), 833−838.

Ha, M., Kim, V.N., 2014. Regulation of microRNA biogenesis. Nat. Rev. Mol. Cell Biol. 15 (8), 509−524.

Hansen, K.F., Karelina, K., Sakamoto, K., Wayman, G.A., Impey, S., Obrietan, K., 2013. miRNA-132: a dynamic regulator of cognitive capacity. Brain Struct. Funct. 218 (3), 817−831.

Hansen, K.F., Sakamoto, K., Wayman, G.A., Impey, S., Obrietan, K., 2010. Transgenic miR132 alters neuronal spine density and impairs novel object recognition memory. PLoS ONE 5 (11), e15497.

Hansen, K.F., Sakamoto, K., Aten, S., Snider, K.H., Loeser, J., Andrea, M., et al., 2016. Targeted deletion of miR-132/-212 impairs memory and alters the hippocampal transcriptome. Learn. Memory 23 (2), 61−71.

Haramati, S., Navon, I., Issler, O., Ezra-Nevo, G., Gil, S., et al., 2011. MicroRNA as Repressors of stress-induced anxiety: the case of amygdalar miR-34. J. Neurosci. 31 (40), 14191−14203.

He, M., Miao, H., Yu, L., Xiaowo, W., Zhang, M.Q., Gregory, J., et al., 2012. Cell-type-based analysis of MicroRNA profiles in the mouse brain. Neuron 73 (3), 620.

Hébert, S.S., Aikaterini, S.P., Smith, P., Galas, M.-C., Planel, E., Asli, N., et al., 2010. Genetic ablation of dicer in adult forebrain neurons results in abnormal tau hyperphosphorylation and neurodegeneration. Hum. Mol. Genet. 19 (20), 3959−3969.

Higuchi, F., Uchida, S., Yamagata, H., Abe-Higuchi, N., Hobara, T., et al., 2016. Hippocampal microRNA-124 enhances chronic stress resilience in mice. J. Neurosc. 36 (27), 7253−7267.

Huang, T., Liu, Y., Huang, M., Zhao, X., Cheng, L., 2010. Wnt1-Cre-mediated conditional loss of dicer results in malformation of the midbrain and cerebellum and failure of neural crest and dopaminergic differentiation in mice. J. Mol. Cell Biol. 2 (3), 152−163.

Impey, S., Soren, I., Monika, D., Adam, L., Dale, F., et al., 2012. Corrigendum to 'an activity-induced microRNA controls dendritic spine formation by regulating Rac1-PAK signaling' [YMCNE (2010) 146−156]. Mol. Cell. Neurosci. 49 (2), 250.

Jimenez-Mateos, E.M., Engel, T., Merino-Serrais, P., McKiernan, R.C., Tanaka, K., et al., 2012. Silencing microRNA-134 produces neuroprotective and prolonged seizure-suppressive effects. Nat. Med. 18 (7), 1087−1094.

Johnson, R., Zuccato, C., Belyaev, N.D., Guest, D.J., Cattaneo, E., Buckley, N.J., 2008. A microRNA-based gene dysregulation pathway in Huntington's disease. Neurobiol. Dis. 29 (3), 438−445.

Jonas, S., Izaurralde, E., 2015. Towards a molecular understanding of microRNA-mediated gene silencing. Nat. Rev. Genet. 16 (7), 421−433.

Kandel, E.R., 2012. The molecular biology of memory: cAMP, PKA, CRE, CREB-1, CREB-2, and CPEB. Mol. Brain 5 (May), 14.

Kawase-Koga, Y., Otaegi, G., Sun, T., 2009. Different timings of dicer deletion affect neurogenesis and gliogenesis in the developing mouse central nervous system. Dev. Dyn. 238 (11), 2800−2812.

Kim, J., Inoue, K., Ishii, J., Vanti, W.B., Voronov, S.V., et al., 2007. A microRNA feedback circuit in midbrain dopamine neurons. Science 317 (5842), 1220−1224.

Klein, M.E., Daniel, T.L., Ma, L., Impey, S., Mandel, G., Goodman, R.H., 2007. Homeostatic regulation of MeCP2 expression by a CREB-induced microRNA. Nat. Neurosci. 10 (12), 1513−1514.

Konopka, W., Kiryk, A., Novak, M., Herwerth, M., Parkitna, J.R., et al., 2010. MicroRNA loss enhances learning and memory in mice. J. Neurosci. 30 (44), 14835−14842.

Krichevsky, A.M., Kevin, S.K., Donahue, C.P., Khrapko, K., Kosik, K.S., 2003. A microRNA array reveals extensive regulation of microRNAs during brain development. RNA 9 (10), 1274−1281.

Krol, J., Busskamp, V., Markiewicz, I., Stadler, M.B., Ribi, S., et al., 2010. Characterizing light-regulated retinal microRNAs reveals rapid turnover as a common property of neuronal microRNAs. Cell 141 (4), 618−631.

Kye, M.-J., Liu, T., Levy, S.F., Xu, N.L., Groves, B.B., et al., 2007. Somatodendritic microRNAs identified by laser capture and multiplex RT-PCR. RNA 13 (8), 1224−1234.

Lagos-Quintana, M., Rauhut, R., Yalcin, A., Meyer, J., Lendeckel, W., Tuschl, T., 2002. Identification of tissue-specific microRNAs from mouse. Curr. Biol. 12 (9), 735−739.

Landgraf, P., Rusu, M., Sheridan, R., Sewer, A., Iovino, N., et al., 2007. A mammalian microRNA expression atlas based on small RNA library sequencing. Cell 129 (7), 1401−1414.

Lau, P., Koen Bossers, R.'sJ., Salta, E., Sala Frigerio, C., Barbash, S., Rothman, R., et al., 2013. Alteration of the microRNA network during the progression of Alzheimer's disease. EMBO Mol. Med. 5 (10), 1613−1634.

Lee, S.-T., Soon-Tae, L., Kon, C., Woo-Seok, I., Hye-Jin, Y., et al., 2011. Altered microRNA regulation in Huntington's disease models. Exp. Neurol. 227 (1), 172−179.

Li, Q., Bian, S., Hong, J., Kawase-Koga, Y., Zhu, E., et al., 2011. Timing Specific requirement of microRNA function is essential for embryonic and postnatal hippocampal development. PLoS ONE 6 (10), e26000.

Lim, L.P., Nelson, C.L., Garrett-Engele, P., Grimson, A., Schelter, J.M., et al., 2005. Microarray analysis shows that some microRNAs downregulate large numbers of target mRNAs. Nature 433 (7027), 769−773.

Lugli, G., Vetle, I.T., Larson, J., Smalheiser, N.R., 2008. Expression of microRNAs and their precursors in synaptic fractions of adult mouse forebrain. J. Neurochem. 106 (2), 650−661.

Makeyev, E.V., Zhang, J., Carrasco, M.A., Maniatis, T., 2007. The microRNA miR-124 promotes neuronal differentiation by triggering brain-specific alternative pre-mRNA splicing. Mol. Cell 27 (3), 435−448.

Mellios, N., Sugihara, H., Castro, J., Banerjee, A., Le, C., et al., 2011. miR-132, an experience-dependent microRNA, is essential for visual cortex plasticity. Nat. Neurosci. 14 (10), 1240−1242.

Nudelman, A.S., Derek, P.D., Lambert, T.J., Garelick, M.G., Le, J., Neil, M., et al., 2010. Neuronal activity rapidly induces transcription of the CREB-regulated microRNA-132, in vivo. Hippocampus 20 (4), 492−498.

Olive, V., Alex, C.M., He, L., 2015. Outside the coding genome, mammalian microRNAs confer structural and functional complexity. Sci. Signaling 8 (368), re2.

Pichardo-Casas, I., Goff, L.A., Swerdel, M.R., Athie, A., Davila, J., et al., 2012. Expression profiling of synaptic microRNAs from the adult rat brain identifies regional differences and seizure-induced dynamic modulation. Brain Res. 1436 (February), 20−33.

Rajasethupathy, P., Priyamvada, R., Ferdinando, F., Robert, S., Doron, B., Sathyanarayanan, V., et al., 2009. Characterization of small RNAs in *Aplysia* reveals a role for miR-124 in constraining synaptic plasticity through CREB. Neuron 63 (6), 803−817.

Remenyi, J., Christopher, J.H., Cole, C., Ando, H., Impey, S., Monk, C.E., et al., 2010. Regulation of the miR-212/132 locus by MSK1 and CREB in response to neurotrophins. Biochem. J. 428 (2), 281−291.

Remenyi, J., Remenyi, J., van den Bosch, M.W.M., Oleg, P., Mistry, R.B., McKenzie, C., et al., 2013. miR-132/212 knockout mice reveal roles for these miRNAs in regulating cortical synaptic transmission and plasticity. PLoS ONE 8 (4), e62509.

Schaefer, A., O'Carroll, D., Tan, C.L., Hillman, D., Sugimori, M., et al., 2007. Cerebellar neurodegeneration in the absence of microRNAs. J. Exp. Med. 204 (7), 1553−1558.

Schratt, G.M., Tuebing, F., Nigh, E.A., Kane, C.G., Sabatini, M.E., et al., 2006. A brain-specific microRNA regulates dendritic spine development. Nature 439 (7074), 283−289.

Schratt, G., 2011. microRNAs at the synapse. Nat. Rev. Neurosci. . Available from: http://dx.doi.org/10.1038/nrn3010.

Smith, P.Y., Delay, C., Girard, J., Papon, M.-A., Planel, E., et al., 2011. MicroRNA-132 loss is associated with tau exon 10 inclusion in progressive supranuclear palsy. Hum. Mol. Genet. 20 (20), 4016−4024.

Somel, M., Liu, X., Tang, L., Yan, Z., Hu, H., et al., 2011. MicroRNA-driven developmental remodeling in the brain distinguishes humans from other primates. PLoS Biol. 9 (12), e1001214.

Störchel, P.H., Thümmler, J., Siegel, G., Aksoy-Aksel, A., Zampa, F., et al., 2015. A large-scale functional screen identifies noval and Ncoa3 as regulators of neuronal miRNA function. EMBO J. 34 (17), 2237−2254.

Tognini, P., Napoli, D., Tola, J., Silingardi, D., Ragione, F.D., et al., 2015. Experience-dependent DNA methylation regulates plasticity in the developing visual cortex. Nat. Neurosci. 18 (7), 956−958.

Tognini, P., Putignano, E., Coatti, A., Pizzorusso, T., 2011. Experience-dependent expression of miR-132 regulates ocular dominance plasticity. Nat. Neurosci. 14 (10), 1237−1239.

van Praag, H., Kempermann, G., Gage, F.H., 1999. Running increases cell proliferation and neurogenesis in the adult mouse dentate gyrus. Nat. Neurosci. 2 (3), 266−270.

Visvanathan, J., Lee, S., Lee, B., Lee, J.W., Lee, S.-K., 2007. The microRNA miR-124 antagonizes the antineural REST/SCP1 pathway during embryonic CNS development. Genes Dev. 21 (7), 744−749.

Wu, H., Tao, J., Chen, P.J., Shahab, A., Ge, W., Hart, R.P., et al., 2010. Genome-wide analysis reveals methyl-CpG-binding protein 2-dependent regulation of microRNAs in a mouse model of Rett syndrome. Proc. Natl. Acad. Sci. U. S. A 107 (42), 18161−18166.

Yang, L., Zhang, R., Li, M., Wu, X., Wang, J., et al., 2014. A functional MiR-124 binding-site polymorphism in IQGAP1 affects human cognitive performance. PLoS ONE 9 (9), e107065, Public Library of Science.

Yu, J.-Y., Chung, K.-H., Deo, M., Thompson, R.C., Turner, D.L., 2008. MicroRNA miR-124 regulates neurite outgrowth during neuronal differentiation. Exp. Cell Res. 314 (14), 2618−2633.

Zehir, A., Lisa, L.H., Maska, E.L., Morikawa, Y., Cserjesi, P., 2010. Dicer is required for survival of differentiating neural crest cells. Dev. Biol. 340 (2), 459−467.

Zeng, L., Li, Z., Benhua, Z., Haiyang, W., Bo, L., et al., 2016. Microbiota modulates behavior and protein kinase C mediated cAMP response element-binding protein signaling. Sci. Rep. 6, 29998.

Ziats, M.N., Rennert, O.M., 2014. Identification of differentially expressed microRNAs across the developing human brain. Mol. Psychiatry 19 (7), 848−852.

SMALL RNA DYSREGULATION IN NEUROCOGNITIVE AND NEUROPSYCHIATRIC DISORDERS

13

Michael Geaghan[1,3] and Murray J. Cairns[1,2,3]

*[1]University of Newcastle, Callaghan, NSW, Australia [2]Schizophrenia Research Institute, Sydney, NSW, Australia
[3]Hunter Medical Research Institute, Newcastle, NSW, Australia*

INTRODUCTION—MicroRNAs IN THE BRAIN

The enormous regulatory power of microRNAs (miRNAs) is of particular importance for proper brain development and function. This is highlighted in early work in the zebrafish model, where depletion of functional Dicer protein resulted in severe developmental abnormalities, particularly in the brain (Giraldez et al., 2005). These mutant zebrafish developed asymmetrically, with the brains affected by severely reduced ventricle size and no midbrain−hindbrain boundary. Clearly, miRNAs and their biogenesis are vital for proper development and function of neurons and the brain (Fig. 13.1, and further discussed in Chapter 15: Comparative Functions of miRNAs in Embryonic Neurogenesis/Neuronal Network Formation). In view of this, it is not surprising that numerous studies have implicated them as significant players in the pathophysiology of several neurological conditions, including schizophrenia (SZ), bipolar disorder (BD), major depressive disorder (MDD), and autism spectrum disorders (ASD). There is substantial research suggesting that miRNA genes and their processing machinery are genetically associated with psychiatric syndromes, and this is manifested as changes in miRNA expression. Some studies suggest that they may also play a role in the mechanisms of neuroleptic medications. These lines of evidence are derived from numerous genetic association studies and molecular studies, examining both postmortem brain tissues as well as peripheral tissues such as blood serum, peripheral blood mononuclear cells (PBMCs), and stem cells derived from olfactory epithelium. Together over the last decade, these studies have provided a strong case for miRNA involvement in various neuropsychiatric disorders.

DYSREGULATED MicroRNAs IN NEUROPSYCHIATRIC DISORDERS
SCHIZOPHRENIA

The link between SZ and miRNA dysfunction is well supported by numerous studies over the last decade. One of the earliest studies investigated postmortem brain tissue from the dorsolateral

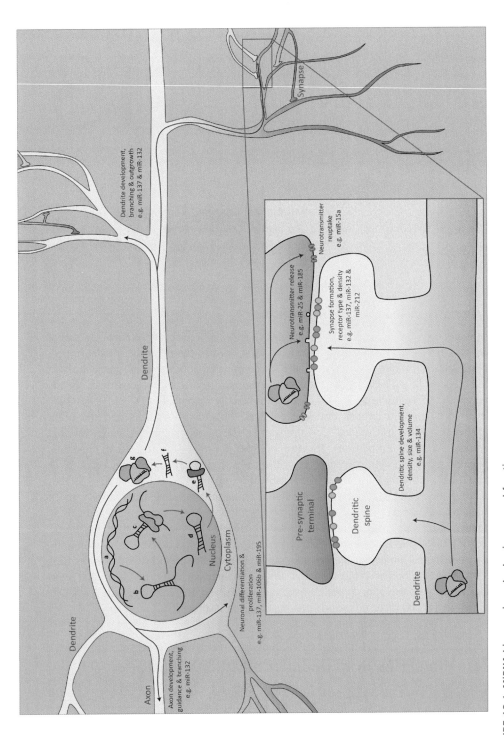

FIGURE 13.1 MiRNA biogenesis and roles in neuronal function

MiRNAs are processed through a number of steps within both the nucleus and the cytoplasm. A miRNA gene (A) is transcribed into a primary transcript (pri-miRNA) (B) containing an inverted repeat, which forms a hairpin. This hairpin is recognized by the microprocessor complex, consisting of DROSHA and DGCR8 (C). The microprocessor cleaves the trailing 3′ and 5′ ends of the transcript, leaving behind the hairpin, known as the precursor miRNA (pre-miRNA) (D). This is exported to the cytoplasm, where DICER removes the loop, leaving behind an imperfectly base paired dsRNA (E). One of these strands is then loaded into the RNA-Induced Silencing Complex (F), which is then able to target mRNA transcripts throughout the neuron and silence translation into protein. In this way, miRNAs are able to influence numerous aspects of neuronal function, including axon development, neuronal proliferation and differentiation, dendrite development, outgrowth and branching, and synaptic formation and function.

prefrontal cortex (DLPFC) Brodmann Area 9 (BA9) (Perkins et al., 2007). The authors used a custom microarray to analyze tissues from SZ subjects and healthy controls, and identified 16 significantly differentially expressed miRNAs, 15 of which were downregulated. Of these, seven were validated by quantitative polymerase chain reaction (qPCR) in a small subcohort of subjects. Using a similar approach, we investigated postmortem miRNA levels in the DLPFC (BA9) as well as the superior temporal gyrus (BA22) (Beveridge et al., 2008, 2010). This study found a large number of elevated miRNA species, including miR-107 and members of the miR-15 family (miR-15a/b, miR-16, and miR-195). These miRNAs are all closely related. For example, miR-107 and miR-195 are thought to be involved in neurodevelopmental processes, with miR-107 being capable of influencing neuronal migration (Moncini et al., 2011), and miR-195 able to target BDNF (brain-derived neurotrophic factor), which is important in neurodevelopment (Mellios et al., 2008). Furthermore, this miRNA family has been identified in other psychiatric disorders (Table 13.1). In addition, our study also found that the miRNA biogenesis gene *DGCR8* was upregulated in both brain regions, and that two other major biogenesis genes, *DICER* and *DROSHA*, were also upregulated within the DLPFC. This suggests that miRNA biogenesis was increased within these brain regions and responsible for the global elevation of miRNA. *DGCR8* is also affected by the hemizygous deletion of 22q11.2 deletion syndrome, which predisposes individuals to DiGeorge syndrome and a high risk of developing SZ (Bassett et al., 2005). More recently, we also observed a general upregulation of miRNAs in the DLPFC (BA46) in postmortem SZ, including miR-107 again, alongside an upregulation of the miRNA biogenesis gene *DICER* (Santarelli et al., 2011). Thus, miRNA biogenesis may play an important role in SZ pathophysiology.

Postmortem studies have also identified SZ-associated dysregulation of miR-132/212 (Kim et al., 2010; Miller et al., 2012). These miRNAs have been shown to have a significant role in neuronal function (discussed further below and in Chapter 12: MiRNA in Neuronal Networks Maturation and Plasticity). Kim et al. (2010) observed several miRNAs, including miR-132-3p, its alternate strand miRNA miR-132-5p, and miR-212 to be upregulated within BA46 of the DLPFC in patients with SZ compared to healthy controls. Using the miRNA target prediction tool miRanda and Ingenuity Pathway Analysis, these miRNAs were predicted to affect pathways involved in nervous system function and disease, including SZ. miR-132-3p and miR-212 were each specifically found negatively correlated with expression of tyrosine hydroxylase and phosphogluconate dehydrogenase, two genes previously observed altered in SZ. Miller et al. (2012) studied the same brain region using a microarray experiment and found miR-132-3p and miR-132-5p to be downregulated in SZ. This downregulation of miR-132-3p in SZ is also supported by another study of PBMCs from both humans with SZ, and rats that showed a SZ-like phenotype (Yu et al., 2015). The human SZ patients also showed increased miR-132-3p levels following 8 weeks' treatment with the antipsychotic risperidone. These studies suggest that this miRNA group is particularly important for SZ, and other studies have also identified its significance in other neuropsychiatric syndromes (Table 13.1).

In another study from our lab, Gardiner et al. (2012) analyzed miRNA expression in PBMCs using both microarrays and qPCR found 83 miRNAs significantly downregulated in SZ. Interestingly, 17 of these miRNAs are located within the imprinted DLK1-DIO3 region at the 14q32 locus and are part of a number of maternally expressed miRNA clusters within the region (Royo and Cavaille, 2008). The majority of these miRNAs were part of the large miR-379/410 cluster, transcription of which is known to be induced by neuronal activity (Fiore et al., 2009). In a similar study, Lai et al. (2011) compared peripheral tissues from SZ and control patients using

Table 13.1 Recurrent miRNA Associated With Neuropsychiatric Disorders

Schizophrenia		Autism Spectrum Disorders		Major Depressive Disorder		Bipolar Disorder	
miRNA	Authors	miRNA	Authors	miRNA	Authors	miRNA	Authors
miR-106b	Perkins et al. (2007), Moreau et al. (2011), Smalheiser et al. (2014)	**miR-106b**	Abu-Elneel et al. (2008), Sarachana et al. (2010)	**miR-107**	Smalheiser et al. (2011), Belzeaux et al. (2012)	let-7b	Miller et al. (2012), Shih et al. (2012)
miR-107	Beveridge et al. (2010), Santarelli et al. (2011)	miR-1286	Vaishnavi et al. (2013), Marrale et al. (2014)	miR-125a	Smalheiser et al. (2011), Cao et al. (2013)	**miR-106b**	Moreau et al. (2011), Shih et al. (2012), Smalheiser et al. (2014)
miR-132-3p	Kim et al. (2010), Miller et al. (2012)	miR-1306	Vaishnavi et al. (2013), Marrale et al. (2014)	**miR-132-3p**	Smalheiser et al. (2011), Li et al. (2013)	**miR-132-3p**	Miller et al. (2012), Walker et al. (2015)
miR-132-5p	Kim et al. (2010), Miller et al. (2012)	**miR-132-3p**	Abu-Elneel et al. (2008), Talebizadeh et al. (2008), Sarachana et al. (2010)	miR-142-3p	Smalheiser et al. (2011), Smalheiser et al. (2012)	miR-133b	Kim et al. (2010), Shih et al. (2012)
miR-134	Santarelli et al. (2011), Gardiner et al. (2012)	miR-146b	Abu-Elneel et al. (2008), Talebizadeh et al. (2008)	**miR-145-5p**	Smalheiser et al. (2011), Belzeaux et al. (2012)	**miR-145-5p**	Kim et al. (2010), Smalheiser et al. (2014)
miR-137	Ripke et al. (2011), Whalley et al. (2012), Green et al. (2013)	**miR-148b**	Abu-Elneel et al. (2008), Sarachana et al. (2010)	miR-182	Smalheiser et al. (2011), Li et al. (2013), Cao et al. (2013)	**miR-148b**	Moreau et al. (2011), Shih et al. (2012)
miR-150	Santarelli et al. (2011), Miller et al. (2012)	miR-149	Vaishnavi et al. (2013), Marrale et al. (2014)	miR-200c	Smalheiser et al. (2011), Belzeaux et al. (2012)	**miR-15a**	Moreau et al. (2011), Shih et al. (2012)

miR-15a	Beveridge et al. (2010), Moreau et al. (2011)	**miR-17**	Kannu et al. (2013), Hemmat et al. (2014)	miR-298	Smalheiser et al. (2011), Cao et al. (2013)	**miR-17**	Shih et al. (2012), Smalheiser et al. (2014)
miR-16	Beveridge et al. (2010), Smalheiser et al. (2014)	miR-185	Sarachana et al. (2010), Vaishnavi et al. (2013), Marrale et al. (2014)	miR-376a-5p	Smalheiser et al. (2011), Smalheiser et al. (2012), Belzeaux et al. (2012)	miR-27b	Miller et al. (2012), Shih et al. (2012)
miR-17	Santarelli et al. (2011), Wong et al. (2013), Smalheiser et al. (2014)	miR-18a	Kannu et al. (2013), Hemmat et al. (2014)	miR-381	Smalheiser et al. (2011), Belzeaux et al. (2012)	miR-29a	Kim et al. (2010), Shih et al. (2012)
miR-181b	Beveridge et al. (2008), Beveridge et al. (2010), Shi et al. (2012)	**miR-195**	Sarachana et al. (2010), Vaishnavi et al. (2013)	miR-494	Smalheiser et al. (2012), Belzeaux et al. (2012)	**miR-29c**	Shih et al. (2012), Banigan et al. (2013), Smalheiser et al. (2014)
miR-195	Perkins et al. (2007), Beveridge et al. (2010), Shi et al. (2012)	miR-199b-5p	Sarachana et al. (2010), Ghahramani Seno et al. (2011)	miR-497	Smalheiser et al. (2011), Smalheiser et al. (2012)	miR-874	Kim et al. (2010), Miller et al. (2012)
miR-212	Perkins et al. (2007), Kim et al. (2010)	miR-19a	Kannu et al. (2013), Hemmat et al. (2014)				

(Continued)

Table 13.1 Recurrent miRNA Associated With Neuropsychiatric Disorders *Continued*

Schizophrenia		Autism Spectrum Disorders		Major Depressive Disorder		Bipolar Disorder	
miRNA	Authors	miRNA	Authors	miRNA	Authors	miRNA	Authors
miR-219-3p	Shi et al. (2012), Smalheiser et al. (2014)	miR-19b-1	Kannu et al. (2013), Hemmat et al. (2014)				
miR-24	Perkins et al. (2007), Xu et al. (2010a–c) Xu et al. (2010a)	miR-200a	Vaishnavi et al. (2013), Marrale et al. (2014)				
miR-26b	Perkins et al. (2007), Beveridge et al. (2010)	miR-200b	Vaishnavi et al. (2013), Marrale et al. (2014)				
miR-29c	Perkins et al. (2007), Beveridge et al. (2010)	miR-20a	Kannu et al. (2013), Hemmat et al. (2014)				
miR-30b	Perkins et al. (2007), Mellios et al. (2012)	miR-211	Miller et al. (2009), Sarachana et al. (2010), Vaishnavi et al. (2013)				
miR-30e	Perkins et al. (2007), Xu et al. (2010b)	miR-23a	Abu-Elneel et al. (2008), Talebizadeh et al. (2008), Sarachana et al. (2010)				
miR-320	Tabares-Seisdedos and Rubenstein (2009), Miller et al. (2012)	miR-320a	Abu-Elneel et al. (2008), Talebizadeh et al. (2008)				

miRNA	References	miRNA	References
miR-34a	Kim et al. (2010), Lai et al. (2011)	miR-429	Vaishnavi et al. (2013), Marrale et al. (2014)
miR-432	Lai et al. (2011), Gardiner et al. (2012)	miR-484	Abu-Elneel et al. (2008), Vaishnavi et al. (2013)
miR-544	Kim et al. (2010), Gardiner et al. (2012)	miR-598	Abu-Elneel et al. (2008), Vaishnavi et al. (2013)
miR-652	Lai et al. (2011), Smalheiser et al. (2011)	miR-649	Vaishnavi et al. (2013), Marrale et al. (2014)
miR-7	Perkins et al. (2007), Beveridge et al. (2010), Kim et al. (2010)	miR-650	Ghahramani Seno et al. (2011), Marrale et al. (2014)
		miR-92a-1	Kannu et al. (2013), Hemmat et al. (2014)
		miR-93	Abu-Elneel et al. (2008), Sarachana et al. (2010)

These miRNAs have been associated with either schizophrenia, bipolar disorder, major depressive disorder, or autism spectrum disorders by multiple studies. The miRNAs in bold have been associated with two or more of these conditions.

both TaqMan Low Density Array and qPCR analyses and observed six miRNAs upregulated and one downregulated. Some of these miRNAs further correlated with various negative symptoms and cognitive dysfunctions, in particular elevated miR-449a correlated with many features of the Wisconsin Card Sorting Test, suggesting an involvement with, or sensitivity to dysfunctions in executive brain function.

Genetic association experiments have also linked several miRNAs to SZ. In 2011, one of the largest genome-wide association studies (GWAS) of SZ was undertaken by the Psychiatric Genome Consortium (PGC), which analyzed 17,836 cases and 33,859 controls (Ripke et al., 2011). Amongst the genome-wide significant single nucleotide polymorphisms (SNPs) was rs1625579, which resides within the intron for the primary transcript of miR-137; this SNP was the strongest new association with SZ identified by the study. Importantly, the same study identified four other loci associated with SZ which were also predicted targets of miR-137. In 2014, the PGC confirmed this association in a much larger study of 36,989 cases and 113,075 controls (Ripke et al., 2014). This strong association of miR-137 with SZ has made it an important focus for further research; such investigations have revealed a number of functional implications relevant to SZ etiology, including a role in proliferation and differentiation of neuronal stem cells in the developing brain (Sun et al., 2011), as well as roles synaptic signaling (Collins et al., 2014). The risk allele rs1625579 has also been observed to correspond with reduced miR-137 expression in the DLPFC and hyperactivation of this region (Guella et al., 2013; van Erp et al., 2014). Although miR-137 has been a major miRNA SNP associated with SZ, it is not the only one. For example, the 2014 PGC study also found a SZ-associated SNP within the *MIR548AJ2* gene (Ripke et al., 2014). Furthermore, there also exist SNPs that affect miRNA function indirectly, such as by interfering with their binding sites. Another study found the SNP rs3822674 which affects a binding site of miR-498 in the complexin 2 (CPLX2) 3′ untranslated region (3′UTR) (Begemann et al., 2010). The authors of this study demonstrated via luciferase assay that the T allele at this SNP induced translational repression in the presence of miR-498, whereas the C allele blocked the miRNA's function, and thus prevented repression. Furthermore, the C allele, in combination with the C and T alleles of SNPs rs1366116 and rs3892909, respectively (also within *CPLX2*), was associated with the poorest cognitive performance within the study.

Environmental factors, including psychosocial stress and maternal immune activation (MIA), also have a substantial impact on miRNA levels and have implications for SZ. For example, in our laboratory we have used the viral mimic polyriboinosinic–polyribocytidilic acid (poly I:C) in rats to explore the miRNA response to the inflammatory cascade related to maternal infections known to be involved in the pathophysiology of SZ. This model induces phenotypic abnormalities in the offspring that mimic phenotypes observed in SZ. We observed 21 differentially expressed miRNA species within the entorhinal cortex after treatment (Hollins et al., 2014). The adolescent rats were also treated with cannabinoid receptor (CB$_1$) agonist HU210—either alone or in combination with poly I:C—as a model for human adolescent cannabis exposure. HU210 exposure alone resulted in 7 dysregulated miRNA species, most of which were also dysregulated in the MIA group. The two-hit poly I:C/HU210 combination model lead to 18 differentially expressed miRNAs. Pathways analysis revealed potential roles within mitogen-activated protein kinase signaling—important for neuronal development and cognition (Samuels et al., 2008)—as well as Wnt signaling—also important in neuronal development and SZ (Miyaoka et al., 1999). Thus, this study highlights the ability for environmental factors to influence miRNA expression and SZ pathophysiology.

Finally, antipsychotic treatments can also affect miRNA expression. One small-scale study assessed the effect of a 6-week treatment with either olanzapine, quetiapine, ziprasidone, or risperidone on serum expression of 9 SZ-linked miRNAs in SZ patients who had either never taken antipsychotics, or had not taken them in the past 3 months (Song et al., 2014). The authors found miR-181b significantly downregulated following antipsychotic treatment, which correlated with improved negative symptoms of SZ. Another study found that miR-326 and miR-9 are capable of targeting the dopamine receptor D2 (*DRD2*), and elevated miR-326 can reduce *DRD2* mRNA and protein levels (Shi et al., 2014). In addition, the minor allele G of the SNP rs1130354, found within the miR-326/miR-9 binding site, can prevent miR-326 targeting the *DRD2* 3'UTR. These findings emphasize the interaction between miRNAs and antipsychotics, since many antipsychotics are dopamine receptor antagonists, and dopamine hyperactivity is thought to be a major part of SZ pathophysiology (Howes et al., 2015).

Overall, there is currently substantial evidence for miRNA involvement in SZ. In particular, some miRNAs, such as miR-132-3p, have been associated with SZ on multiple occasions and are even implicated in other neuropsychiatric disorders (discussed below). This suggests that these miRNAs are particularly important for healthy neuronal function and makes them intriguing targets for further research.

BIPOLAR DISORDER

There is emerging evidence for miRNA involvement in BD. Several studies have now shown significant alterations in miRNA expression levels in cortical brain tissue from affected individuals. Kim et al. (2010) analyzed postmortem DLPFC (BA46) tissue samples of individuals with either BD or SZ, and healthy controls and identified seven elevated and eight downregulated miRNAs in BD cases compared to controls. Interestingly, one of the elevated miRNAs, miR-145-5p, has also been observed by Smalheiser et al. (2014) to be downregulated in the prefrontal cortex (PFC) (BA10) of patients with BD and is thought to regulate neurite outgrowth (Zhang et al., 2011). Moreau et al. (2011) also analyzed miRNA expression in BD using postmortem tissue from BA9 of the PFC. Amongst the dysregulated miRNAs, miR-106b was found downregulated in BD. This miRNA has also been identified in BD and other psychiatric disorders by multiple studies (see Table 13.1). miR-106b may therefore be another particularly important miRNA for psychiatric diseases, and a few studies suggest that miR-106b and related miRNAs are involved in neuronal differentiation and proliferation (Brett et al., 2011; Trompeter et al., 2011).

BD has also been associated with various SNPs affecting miRNA function. In 2013, a large GWAS found 56 SNPs across 5 genomic loci; one of these loci was close to the *MIR2113* gene (Muhleisen et al., 2014). A later study reanalyzed this data, looking at just the known miRNA loci, and identified three promising miRNA loci—miR-499, miR-708, and miR-1908—which all possessed multiple SNPs (Forstner et al., 2015). Pathways analysis predicted a number of neuronal pathways, such as synaptic vesicle transport, brain development, and neuron projection, to be significantly enriched for target genes of these miRNAs. There is also some genetic evidence that miR-137 may be involved in BD as well. One study has reported a significant association of a rare enhancer SNP 1:g. 98515539A > T at the miR-137 locus with both SZ and BD (Duan et al., 2014). Using luciferase reporter and electrophoresis mobility shift assays, the authors demonstrated the risk allele of this rare variant which resulted in a decrease in enhancer activity by interfering with YY1 transcription factor binding and subsequent activation.

Finally, mood stabilizing medications used for the treatment of BD can affect miRNA expression. The synaptic miRNA miR-134, previously observed in SZ (Gardiner et al., 2012; Santarelli et al., 2011) and known to regulate dendritic spine size (Schratt et al., 2006), was found downregulated in peripheral plasma samples of BD patients compared to controls, which was partially reversed after 4 weeks of medication (Rong et al., 2011). In another study, valproic acid was found to induce proteasomal degradation of DICER protein (Zhang et al., 2013). DICER protein levels were decreased significantly after 12 hours of treatment with valproic acid, which could be prevented by inhibition of the proteasome with MG132. In addition, *DICER* mRNA levels were found downregulated after 6 hours of treatment. This loss of DICER protein has enormous implications for miRNA expression; indeed, the same study found 117 dysregulated miRNAs via microarray. Given the newly appreciated role for miRNAs in BD, it is possible that valproic acid's effect on miRNA biogenesis may in fact be part of its therapeutic mechanism.

MAJOR DEPRESSIVE DISORDER

Dysregulation of miRNAs within the brain has also been linked to MDD in recent years. Early evidence for this was a genetic association between depression and a SNP—rs1653625—in the 3′UTR of P2RX7 (purinergic receptor P2x, ligand-gated ion channel 7) (Rahman et al., 2010), a gene thought to be involved in modulating synaptic neurotransmission (Sperlagh et al., 2006). This SNP occurs within the putative miRNA target site for miR-1302 and miR-625, and the C allele was found to be significantly overrepresented in MDD. Another interesting MDD-associated SNP is rs112439044, located within the pre-miR-30e gene, the T allele of which was slightly more common in MDD patients than controls (Xu et al., 2010c). Furthermore, the heterozygous C/T genotype was correlated with a longer P300 waveform latency, which is a correlate of slower cognitive functioning.

Two particularly significant variants have been identified in the miRNA biogenesis pathway genes *AGO1* (Argonaute 1) (rs636832) and *DGCR8* (rs3757) (He et al., 2012), further emphasize the importance of miRNAs for MDD. Furthermore, a postmortem study by Smalheiser et al. (2012) has found a global downregulation of miRNAs within BA9 of the PFC in depressed suicide subjects. A SNP in the *DICER1* gene—rs10144436—has also been significantly associated with posttraumatic stress disorder comorbid with depression and downregulation of *DICER1* expression (Wingo et al., 2015). Another study has also identified *DICER1* downregulation in MDD (Aston et al., 2005). Moreover, *DICER1* has been shown to be regulated through the WNT pathway by β-catenin (Dias et al., 2014). β-Catenin overexpression in mouse nucleus accumbens during accelerated social defeat stress resulted in reduced development of social avoidance and antidepressant- and anxiolytic-like effects in other tests. Using chromatin immunoprecipitation, the authors found that β-catenin targeted and elevated the expression of *DICER1* following chronic social defeat stress. When *DICER1* was knocked down, the animals displayed social avoidance. This suggests that this β-catenin-*DICER1* interaction is important for stress resilience.

Another study by Smalheiser et al. (2011) established a mouse model to assess the relationship between frontal cortical miRNA expression and learned helplessness—an analogue for depressive symptoms that can be measured in animals. The mice were subjected to repeat, inescapable shock, and miRNA expression was compared between those that displayed learned helplessness and those that did not. Those mice that did not show learned helplessness displayed a significant global downregulation of miRNA expression as an adaptive response, whereas the mice that did show learned helplessness did not show this miRNA dysregulation. Importantly, three of the affected molecules, miR-96, miR-182, and miR-183, are parts of a polycistronic miRNA cluster thought to be involved in regulating genes in step with the circadian clock (Xu et al., 2007). Disruption of the circadian clock is thought to be an important factor in a number of disorders, including depression (Mendlewicz, 2009). Thus, this result suggests that miRNA perturbation of circadian clock machinery may play a role in the pathophysiology of depression.

miR-132-3p is also associated with depression. Elevated serum levels of miR-132-3p, as well as miR-182, have been observed in MDD by qPCR analysis (Li et al., 2013). Both of these miRNAs target BDNF, which was found to be downregulated in serum in depression cases. The interaction between miR-132-3p and BDNF, as well as with CREB (cAMP response element-binding protein) and glucocorticoids, has been proposed to be significant in development of some cases of depression, as well as the comorbidity of cardiovascular diseases with depression (Zheng et al., 2013). Furthermore, miR-132-3p is another miRNA implicated in regulating the circadian clock (Cheng et al., 2007).

Antidepressants also interact with miRNAs. One study found miR-1202 significantly downregulated in the ventrolateral PFC in postmortem tissue from depressed individuals, whereas in individuals with antidepressant treatment history, this downregulation was smaller (Lopez et al., 2014). Furthermore, miR-1202 expression was negatively correlated with that of metabotropic glutamate receptor 4 (*GRM4*)—a gene involved in neurotransmission and thought to be important for anxiety-related disorders (Davis et al., 2013). The authors further demonstrated that chronic (15 days) treatment of human neural progenitor cells (NPCs) in vitro with either imipramine or citalopram elevated miR-1202 expression and subsequently decreased *GRM4* mRNA and protein expression. miR-335 also interacts with *GRM4* in antidepressant treatment (Li et al., 2015). In this study, miR-335 was found downregulated in blood samples from MDD patients. Using in vitro luciferase reporter assays, the authors showed that miR-335 targets *GRM4* 3′UTR. The authors also used human NPCs to show that chronic (7 days) treatment with citalopram elevated miR-335 and simultaneously downregulated *GRM4* mRNA. Thus, miRNA-*GRM4* interaction appears to be important in antidepressant treatment. These studies demonstrate the importance of miRNAs not only in the pathophysiology of depression, but also their role in the treatment of this severe psychiatric disorder.

AUTISM SPECTRUM DISORDERS

At present, the evidence for miRNA involvement in ASD is not as extensive as that for SZ, BD, and MDD. In an early, small-scale study, Talebizadeh et al. (2008) analyzed lymphoblastoid cell lines derived from six subjects with autism and six controls, and the majority of the identified were predicted to target autism-related genes. In addition, miR-132-3p was found dysregulated, once again emphasizing its importance in psychiatric disorders. Another study also looked at lymphoblastoid cell lines derived from sets of twins and siblings, and amongst the findings was the differential expression of miRNAs including miR-106b, miR-107, and miR-195, all of which are also implicated in SZ (Sarachana et al., 2010). Importantly, the differentially expressed miRNAs from this study were predicted to target a wide range of genes that were enriched for neurological functions. Many of the predicted target genes were also found differentially expressed in a parallel microarray experiment, and a significant proportion of these were inversely correlated with the miRNAs that targeted them. A few postmortem tissue studies have also directly investigated the differential miRNA expression in ASD. For example, Ander et al. (2015) investigated miRNA expression profiles in the superior temporal sulcus (STS) and primary auditory cortex (PAC) of ASD cases compared to controls and found miR-4753-5p and miR-1 altered in the STS, and miR-664-3p, miR-4709-3p, miR-297, and miR-4742-3p altered in the PAC. These miRNAs were further predicted to target nervous system pathways, as well as cell cycle and cell—cell signaling pathways.

Several studies have also identified autism-related SNPs and copy number variants (CNVs) that affect miRNA genes and their function. One analysis of 378 CNVs that had been associated with ASD on multiple occasions found 41 (around 11%) to contain a total of 71 miRNAs (Vaishnavi et al., 2013). These miRNAs were predicted to form a complex network with many autism-associated genes involved in processes including synaptic transmission, nervous system development, and behavior. Amongst the top 10 most connected miRNAs were miR-195, miR-497, and miR-34a, which are also associated with SZ. Two other studies have identified a microduplication of the miR-17-92 cluster on chromosome 13q31.3 in patients with autistic traits (Hemmat et al., 2014, Kannu et al., 2013).

MiRNAs OF INTEREST AND THEIR IMPLICATIONS FOR NEURONAL FUNCTION AND PSYCHIATRIC PATHOPHYSIOLOGY

Within each of these conditions, a number of miRNAs stand out as being particularly significant, as they have been identified by multiple, independent studies (Table 13.1). Furthermore, a number of miRNAs have been associated with multiple conditions, suggesting their particular importance in regulating neuronal functions. In addition, various components of the miRNA biogenesis machinery are implicated in the pathophysiology of disorders like SZ and major depression. As such, these miRNAs and miRNA biogenesis genes are of particular interest with regards to how they regulate cellular functions within the brain, and how this contributes to psychiatric disease.

miR-137

Since miR-137 has such a strong genetic association with SZ, it has become a significant focus of miRNA research within the context of this disorder. In mice, miR-137 overexpression inhibits neuronal maturation and development of dendrites, partially by targeting Mind Bomb-1 (*Mib1*) (Smrt et al., 2010). Other neurogenesis genes including *c-KIT*, *YBX1*, *AKT2*, *CDC42*, *CDK6*, and *TGFβ2* are also known targets of miR-137 (Lim et al., 2005; Tamim et al., 2014). This miRNA has also been shown to target and repress transcription factors *Klf4* and *Tbx3* in embryonic stem cells; these genes are important for self-renewal and pluripotency, and by targeting these genes, miR-137 can promote differentiation (Jiang et al., 2013). SZ is considered a disorder of neurodevelopment (Marenco and Weinberger, 2000), and as such, miR-137's capacity to regulate neuronal maturation and differentiation further highlights its importance in the context of this condition.

miR-137 is also associated with deficits in brain function and neuronal signaling. Green et al. (2013) investigated the SZ risk SNP rs1625579 in 617 cases and 764 controls and found an association with a cognitive deficit subtype of SZ; specifically, the G allele, in combination with a high score for negative symptoms, significantly predicted membership within the cognitive deficit subtype. More recently, the risk allele T/T genotype at this SNP was associated with miR-137 downregulation in the DLPFC, and hyperactivation of this brain region, a correlate of SZ (Guella et al., 2013; van Erp et al., 2014). Finally, miR-137 is predicted to target genes such as GABA and glutamate receptor subunits *GABRA1* and *GRIN2A*, which are vital components of synaptic signaling and related pathways such as axonal guidance and long-term potentiation (LTP) (Wright et al., 2013). This is especially important for SZ, since disruptions to glutamatergic and GABAergic synapses and perturbations of LTP are highly significant SZ pathophysiology (Frantseva et al., 2008; Seshadri et al., 2013). Altogether, the current evidence suggests that miR-137 is extremely important for normal regulation of neurodevelopment and neuronal activity, as well as for healthy brain function overall, and that disruption of this miRNA may affect neurodevelopmental processes and synaptic activity, which may in turn contribute SZ pathophysiology.

miR-132/212

As discussed above, the miR-132/212 cluster has been associated with multiple neuropsychiatric disorders, including SZ, BD, MDD, and ASD (Table 13.1). As such, their regulatory roles in the brain have been a focal point of many studies and are explored in depth in Chapter 12, MiRNA in Neuronal Networks Maturation and Plasticity. This miRNA group is heavily involved in neurite and synapse regulation (Magill et al., 2010; Pathania et al., 2012), which is important in the pathophysiology of psychiatric diseases such as SZ and depression (Marsden, 2013; Seshadri et al., 2013). It is also known to regulate SZ-associated genes, including *DNMT3A*, *GATA2*, and *DPYSL3*, in response to synaptic activity (Miller et al., 2012). BDNF expression is also affected by miR-132-3p and has implications for depression and BD (Cunha et al., 2006; Li et al., 2013). In both disorders, serum levels of BDNF have been observed downregulated during depressive episodes, and manic episodes in BD are also accompanied by miR-132-3p downregulation. miR-212 may also indirectly regulate BDNF via regulation of methyl-CpG binding protein 2 (MeCP2) (Im et al., 2010). Furthermore, BDNF in turn regulates the miR-132/212 cluster (Remenyi et al., 2010). BDNF is an important neurotrophin, and thus, its interaction with the miR-132/212 cluster is particularly attractive for neuropsychiatric research.

miR-15/107 FAMILY

The miR-15/107 family consists of a number of miRNAs with extremely similar seed regions, sharing the AGCAGCA sequence. As seen in Table 13.1, miR-15a, miR-195, and miR-107 are all associated with multiple neuropsychiatric disorders. Current research suggests that these miRNAs are involved in regulating some aspects of neurodevelopment and synaptic signaling. For example, miR-15a has been shown to regulate the expression of the serotonin transporter gene *SLC6A4*, which could be important in disorders such as depression (Moya et al., 2013). The miR-15/107 family also regulates neurodevelopmental processes. For example, miR-195 and methyl-CpG binding protein 1 (MDB1) form a regulatory loop which regulates the balance between proliferation and differentiation of NPCs (Liu et al., 2013). When miR-195 levels are high, MDB1 is lowly expressed, preventing differentiation, while low miR-195 levels allow MDB1 expression to increase, leading to enhanced differentiation, and furthermore feeds back and downregulates miR-195. miR-107 and related miR-103 also regulate neurodevelopmental processes by targeting cyclin-dependent kinase regulatory subunit 1 (CDK5R1) (Moncini et al., 2011). These miRNAs were found to downregulate CDK5R1 expression, leading to reduced neuronal migration in cell culture. One study has also found that miR-195 targets the 3′UTR of BDNF (Mellios et al., 2008), suggesting that interactions between the miR-15 family and BDNF may be involved in psychiatric disorders.

miRNA BIOGENESIS GENES

Since various miRNAs have important roles in neurodevelopment and brain function, it follows that the cellular machinery responsible for their production would also be critical for healthy brain development. Indeed, several miRNA biogenesis genes are important for brain structure and function. *DGCR8*, which codes for one of the microprocessor components, is located within the 22q11.2 region. As discussed above, a microdeletion at this region, known as 22q11.2 deletion syndrome, causes DiGeorge syndrome and predisposes individuals to a high risk of developing SZ. Homologous models of this microdeletion in mice result in *Dgcr8* haploinsufficiency, altered miRNA biogenesis, and impaired behavior and cognitive function, including hyperactivity, reduced fear conditioning, and impaired spatial working memory (Stark et al., 2008). This mouse model also displayed reduced numbers and sizes of hippocampal dendritic spines and lower dendritic complexity. Similar models also display altered synaptic function, including an age-dependent increase in hippocampal LTP as a result of the *Dgcr8* deletion (Earls et al., 2012). This change in LTP is in part due to the loss of miR-25 and miR-185, which both regulate the sarco/endoplasmic reticulum Ca^{2+}-ATPase 2 (SERCA2) calcium pump. In their absence, SERCA2 is overexpressed, leading to elevated Ca^{2+} concentration in the endoplasmic reticulum, slower Ca^{2+} uptake from the presynaptic region, and increased neurotransmitter release, resulting in elevated LTP. Other studies have found *Dgcr8* loss in mice results in reduced frequency of excitatory postsynaptic currents (EPSCs) and inhibitory postsynaptic currents) within the PFC (Hsu et al., 2012; Schofield et al., 2011). 22q11.2 deletion mouse models also display an elevation of dopamine D2 receptors (Drd2) in auditory thalamocortical projection neurons due to *Dgcr8* loss (Chun et al., 2014), resulting in presynaptic deficits in signaling between ventral medial geniculate nucleus neurons and layer 3/4 neurons of the auditory cortex, leading to reduced EPSCs in the cortical neurons. These projection neurons are more sensitive to Drd2-inhibiting antipsychotics, application of which reverses these synaptic deficits, with EPSCs returning to wild-type levels. This suggests reductions in miRNA biogenesis may result in DRD2 elevation, and in turn contribute to the development of psychosis in 22q11.2 deletion syndrome.

Collectively, these studies emphasize how critical normal miRNA expression is for neurodevelopment, and show that global miRNA depletion can lead to abnormal brain development.

PERSPECTIVE

Our current understanding of the roles that miRNAs play in neuropsychiatric disorders is still slowly developing. However, from the current research, a few trends are emerging. First of all, several miRNAs have been repeatedly associated with one or more psychiatric disorders and appear to share similar roles in regulating neurodevelopmental processes and synaptic activity. This suggests that these miRNAs are particularly important for more general brain function and development, potentially acting as molecular "hubs" for psychiatric disorders like SZ which are neurodevelopmental in origin. This also makes miRNA an interesting focus in a clinical context, with potential both as a diagnostic tool and as a target for novel treatments. Numerous studies have already associated altered peripheral miRNA expression with various psychiatric disorders. These miRNAs may therefore be useful as biomarkers that can be used clinically to aid diagnosis of these conditions, as well as potentially identifying individuals at risk for developing psychiatric disease. However, the use of miRNAs as biomarkers is still a difficult problem. Firstly, there have been numerous miRNAs associated with various psychiatric disorders to date. Determining which miRNAs are useful biomarkers and which are not is a difficult task. Furthermore, there is significant heterogeneity between different studies regarding the degree and direction of miRNA dysregulation. For example, miR-106b-5p was found to be upregulated in postmortem PFC tissue (BA9) from individuals with SZ by Perkins et al. (2007), whereas Moreau et al. (2011) found that the same miRNA in the same brain region was slightly downregulated. This disagreement between numerous studies further confounds the issue of using these miRNAs in a clinical setting, and highlights the need for more studies that can shed light on the miRNA expression profiles across numerous brain regions. This would help progress our understanding of how miRNAs are involved in the pathophysiology of these disorders and aid the identification of suitable psychiatric disease biomarkers.

Further, understanding the miRNA expression profile throughout the brain in both healthy and affected individuals may also help identify novel treatments that make use of miRNAs or miRNA antagonists to target perturbed cellular pathways. Such treatments are an appealing concept, although currently the efficacy with which these molecules can be delivered is a severely limiting factor. However, there has been progress made in clinical trials where cancer patients were intravenously administered with small interfering RNA (siRNA) nanoparticles—small interfering RNAs, which are chemically identical to synthetic miRNAs. The siRNAs successfully targeted and downregulated both the mRNA and protein of the gene *RRM2*, a known anticancer target (Davis et al., 2010; Xu et al., 2010b). There has also been success with inhibiting miRNAs in vivo; expression of miR-15b within the hearts of mice was reduced following injection with anti-miR-15b oligonucleotides (Hullinger et al., 2012). However, these preliminary successes with miRNA delivery may not easily translate to the brain. Krutzfeldt et al. (2007) observed that intravenous administration of anti-miR-16 oligonucleotides was unsuccessful in affecting brain miRNA levels; however, direct injection into the cerebral cortex was effective. Infusion of anti-miRNAs into the ventricles of the brain has also been found to be effective (Ouyang et al., 2012; Yin et al., 2010). As such, there is potential for miRNAs and their antagonists to be used therapeutically for treatment of psychiatric disorders; however, the current feasibility of this is still largely unknown.

Over the last decade, a substantial body of research has emerged suggesting that miRNAs have a very significant role in brain development and function, and as such their dysregulation may be involved in the development of numerous neuropsychiatric conditions, including SZ, mood disorders, and ASD. Several studies support a genetic association between psychiatric disorders and miRNA and their targets, and many others that have demonstrated significant perturbations of miRNA expression within both brain and peripheral tissue that correlate with these conditions. In addition, miRNA biogenesis genes have been associated with some of these conditions. Dysregulation of the major miRNA biogenesis genes *DGCR8*, *DROSHA*, and *DICER* in SZ, and the association of SNPs in *DGCR8* and *AGO1* in MDD strongly supports the hypothesis that changes to miRNA expression and function are an important aspect of psychiatric pathophysiology. This area of study is still very much a growing field, and there is still a great deal that is unknown and confounded by conflicting reports. However, it is clear that these fascinating molecules are a vital component of posttranscriptional regulatory networks within the brain and have a huge potential for aiding our understanding of the development of neuropsychiatric disorders, and may even be useful as diagnostic tools and novel treatments for these disorders.

REFERENCES

Abu-Elneel, K., Liu, T., Gazzaniga, F.S., Nishimura, Y., Wall, D.P., Geschwind, D.H., et al., 2008. Heterogeneous dysregulation of microRNAs across the autism spectrum. Neurogenetics 9, 153–161.

Ander, B.P., Barger, N., Stamova, B., Sharp, F.R., Schumann, C.M., 2015. Atypical miRNA expression in temporal cortex associated with dysregulation of immune, cell cycle, and other pathways in autism spectrum disorders. Mol. Autism 6, 37.

Aston, C., Jiang, L., Sokolov, B.P., 2005. Transcriptional profiling reveals evidence for signaling and oligodendroglial abnormalities in the temporal cortex from patients with major depressive disorder. Mol. Psychiatry 10, 309–322.

Banigan, M.G., Kao, P.F., Kozubek, J.A., Winslow, A.R., Medina, J., Costa, J., et al., 2013. Differential expression of exosomal microRNAs in prefrontal cortices of schizophrenia and bipolar disorder patients. PLoS One 8, e48814.

Bassett, A.S., Chow, E.W., Husted, J., Weksberg, R., Caluseriu, O., Webb, G.D., et al., 2005. Clinical features of 78 adults with 22q11 deletion syndrome. Am. J. Med. Genet. A 138, 307–313.

Begemann, M., Grube, S., Papiol, S., Malzahn, D., Krampe, H., Ribbe, K., et al., 2010. Modification of cognitive performance in schizophrenia by complexin 2 gene polymorphisms. Arch. Gen. Psychiatry 67, 879–888.

Belzeaux, R., Bergon, A., Jeanjean, V., Loriod, B., Formisano-Treziny, C., Verrier, L., et al., 2012. Responder and nonresponder patients exhibit different peripheral transcriptional signatures during major depressive episode. Transl. Psychiatry 2, e185.

Beveridge, N.J., Gardiner, E., Carroll, A.P., Tooney, P.A., Cairns, M.J., 2010. Schizophrenia is associated with an increase in cortical microRNA biogenesis. Mol. Psychiatry 15, 1176–1189.

Beveridge, N.J., Tooney, P.A., Carroll, A.P., Gardiner, E., Bowden, N., Scott, R.J., et al., 2008. Dysregulation of miRNA 181b in the temporal cortex in schizophrenia. Hum. Mol. Genet. 17, 1156–1168.

Brett, J.O., Renault, V.M., Rafalski, V.A., Webb, A.E., Brunet, A., 2011. The microRNA cluster miR-106b~25 regulates adult neural stem/progenitor cell proliferation and neuronal differentiation. Aging 3, 108–124.

Cao, M.Q., Chen, D.H., Zhang, C.H., Wu, Z.Z., 2013. [Screening of specific microRNA in hippocampus of depression model rats and intervention effect of Chaihu Shugan San]. Zhongguo Zhong yao za zhi 38, 1585–1589.

Cheng, H.Y., Papp, J.W., Varlamova, O., Dziema, H., Russell, B., Curfman, J.P., et al., 2007. microRNA modulation of circadian-clock period and entrainment. Neuron 54, 813−829.

Chun, S., Westmoreland, J.J., Bayazitov, I.T., Eddins, D., Pani, A.K., Smeyne, R.J., et al., 2014. Specific disruption of thalamic inputs to the auditory cortex in schizophrenia models. Science 344, 1178−1182.

Collins, A.L., Kim, Y., Bloom, R.J., Kelada, S.N., Sethupathy, P., Sullivan, P.F., et al., 2014. Transcriptional targets of the schizophrenia risk gene MIR137. Translational psychiatry 4, e404.

Cunha, A.B., Frey, B.N., Andreazza, A.C., Goi, J.D., Rosa, A.R., Goncalves, C.A., et al., 2006. Serum brain-derived neurotrophic factor is decreased in bipolar disorder during depressive and manic episodes. Neurosci. Lett. 398, 215−219.

Davis, M.E., Zuckerman, J.E., Choi, C.H., Seligson, D., Tolcher, A., Alabi, C.A., et al., 2010. Evidence of RNAi in humans from systemically administered siRNA via targeted nanoparticles. Nature 464, 1067−1070.

Davis, M.J., Iancu, O.D., Acher, F.C., Stewart, B.M., Eiwaz, M.A., Duvoisin, R.M., et al., 2013. Role of mGluR4 in acquisition of fear learning and memory. Neuropharmacology 66, 365−372.

Dias, C., Feng, J., Sun, H., Shao, N.Y., Mazei-Robison, M.S., Damez-Werno, D., et al., 2014. beta-catenin mediates stress resilience through Dicer1/microRNA regulation. Nature 516, 51−55.

Duan, J., Shi, J., Fiorentino, A., Leites, C., Chen, X., Moy, W., et al., 2014. A rare functional noncoding variant at the GWAS-implicated MIR137/MIR2682 locus might confer risk to schizophrenia and bipolar disorder. Am. J. Hum. Genet. 95, 744−753.

Earls, L.R., Fricke, R.G., Yu, J., Berry, R.B., Baldwin, L.T., Zakharenko, S.S., 2012. Age-dependent microRNA control of synaptic plasticity in 22q11 deletion syndrome and schizophrenia. J. Neurosci. 32, 14132−14144.

Fiore, R., Khudayberdiev, S., Christensen, M., Siegel, G., Flavell, S.W., Kim, T.K., et al., 2009. Mef2-mediated transcription of the miR379-410 cluster regulates activity-dependent dendritogenesis by fine-tuning Pumilio2 protein levels. EMBO J. 28, 697−710.

Forstner, A.J., Hofmann, A., Maaser, A., Sumer, S., Khudayberdiev, S., Muhleisen, T.W., et al., 2015. Genome-wide analysis implicates microRNAs and their target genes in the development of bipolar disorder. Transl. Psychiatry 5, e678.

Frantseva, M.V., Fitzgerald, P.B., Chen, R., Moller, B., Daigle, M., Daskalakis, Z.J., 2008. Evidence for impaired long-term potentiation in schizophrenia and its relationship to motor skill learning. Cereb. Cortex 18, 990−996.

Gardiner, E., Beveridge, N.J., Wu, J.Q., Carr, V., Scott, R.J., Tooney, P.A., et al., 2012. Imprinted DLK1-DIO3 region of 14q32 defines a schizophrenia-associated miRNA signature in peripheral blood mononuclear cells. Mol. Psychiatry 17, 827−840.

Ghahramani Seno, M.M., Hu, P., Gwadry, F.G., Pinto, D., Marshall, C.R., Casallo, G., et al., 2011. Gene and miRNA expression profiles in autism spectrum disorders. Brain Res. 1380, 85−97.

Giraldez, A.J., Cinalli, R.M., Glasner, M.E., Enright, A.J., Thomson, J.M., Baskerville, S., et al., 2005. MicroRNAs regulate brain morphogenesis in zebrafish. Science 308, 833−838.

Green, M.J., Cairns, M.J., Wu, J., Dragovic, M., Jablensky, A., Tooney, P.A., et al., 2013. Genome-wide supported variant MIR137 and severe negative symptoms predict membership of an impaired cognitive subtype of schizophrenia. Mol. Psychiatry 18, 774−780.

Guella, I., Sequeira, A., Rollins, B., Morgan, L., Torri, F., van Erp, T.G., et al., 2013. Analysis of miR-137 expression and rs1625579 in dorsolateral prefrontal cortex. J. Psychiatr Res. 47, 1215−1221.

He, Y., Zhou, Y., Xi, Q., Cui, H., Luo, T., Song, H., et al., 2012. Genetic variations in microRNA processing genes are associated with susceptibility in depression. DNA Cell Biol. 31, 1499−1506.

Hemmat, M., Rumple, M.J., Mahon, L.W., Strom, C.M., Anguiano, A., Talai, M., et al., 2014. Short stature, digit anomalies and dysmorphic facial features are associated with the duplication of miR-17∼92 cluster. Mol. Cytogenet. 7, 27.

Hollins, S.L., Goldie, B.J., Carroll, A.P., Mason, E.A., Walker, F.R., Eyles, D.W., et al., 2014. Ontogeny of small RNA in the regulation of mammalian brain development. BMC Genomics 15, 777.

Howes, O., McCutcheon, R., Stone, J., 2015. Glutamate and dopamine in schizophrenia: an update for the 21st century. J. Psychopharmacol. 29, 97−115.

Hsu, R., Schofield, C.M., Dela Cruz, C.G., Jones-Davis, D.M., Blelloch, R., Ullian, E.M., 2012. Loss of microRNAs in pyramidal neurons leads to specific changes in inhibitory synaptic transmission in the prefrontal cortex. Mol. Cell. Neurosci. 50, 283−292.

Hullinger, T.G., Montgomery, R.L., Seto, A.G., Dickinson, B.A., Semus, H.M., Lynch, J.M., et al., 2012. Inhibition of miR-15 protects against cardiac ischemic injury. Circ. Res. 110, 71−81.

Im, H.I., Hollander, J.A., Bali, P., Kenny, P.J., 2010. MeCP2 controls BDNF expression and cocaine intake through homeostatic interactions with microRNA-212. Nat. Neurosci. 13, 1120−1127.

Jiang, K., Ren, C., Nair, V.D., 2013. MicroRNA-137 represses Klf4 and Tbx3 during differentiation of mouse embryonic stem cells. Stem Cell Res. 11, 1299−1313.

Kannu, P., Campos-Xavier, A.B., Hull, D., Martinet, D., Ballhausen, D., Bonafe, L., 2013. Post-axial polydactyly type A2, overgrowth and autistic traits associated with a chromosome 13q31.3 microduplication encompassing miR-17−92 and GPC5. Eur. J. Med. Genet. 56, 452−457.

Kim, A.H., Reimers, M., Maher, B., Williamson, V., McMichael, O., McClay, J.L., et al., 2010. MicroRNA expression profiling in the prefrontal cortex of individuals affected with schizophrenia and bipolar disorders. Schizophr. Res. 124, 183−191.

Krutzfeldt, J., Kuwajima, S., Braich, R., Rajeev, K.G., Pena, J., Tuschl, T., et al., 2007. Specificity, duplex degradation and subcellular localization of antagomirs. Nucleic Acids Res. 35, 2885−2892.

Lai, C.Y., Yu, S.L., Hsieh, M.H., Chen, C.H., Chen, H.Y., Wen, C.C., et al., 2011. MicroRNA expression aberration as potential peripheral blood biomarkers for schizophrenia. PLoS ONE 6, e21635.

Li, J., Meng, H., Cao, W., Qiu, T., 2015. MiR-335 is involved in major depression disorder and antidepressant treatment through targeting GRM4. Neurosci. Lett. 606, 167−172.

Li, Y.J., Xu, M., Gao, Z.H., Wang, Y.Q., Yue, Z., Zhang, Y.X., et al., 2013. Alterations of serum levels of BDNF-related miRNAs in patients with depression. PLoS ONE 8, e63648.

Lim, L.P., Lau, N.C., Garrett-Engele, P., Grimson, A., Schelter, J.M., Castle, J., et al., 2005. Microarray analysis shows that some microRNAs downregulate large numbers of target mRNAs. Nature 433, 769−773.

Liu, C., Teng, Z.Q., McQuate, A.L., Jobe, E.M., Christ, C.C., von Hoyningen-Huene, S.J., et al., 2013. An epigenetic feedback regulatory loop involving microRNA-195 and MBD1 governs neural stem cell differentiation. PLoS ONE 8, e51436.

Lopez, J.P., Lim, R., Cruceanu, C., Crapper, L., Fasano, C., Labonte, B., et al., 2014. miR-1202 is a primate-specific and brain-enriched microRNA involved in major depression and antidepressant treatment. Nat. Med. 20, 764−768.

Magill, S.T., Cambronne, X.A., Luikart, B.W., Lioy, D.T., Leighton, B.H., Westbrook, G.L., et al., 2010. microRNA-132 regulates dendritic growth and arborization of newborn neurons in the adult hippocampus. Proc. Natl. Acad. Sci. U. S. A. 107, 20382−20387.

Marenco, S., Weinberger, D.R., 2000. The neurodevelopmental hypothesis of schizophrenia: following a trail of evidence from cradle to grave. Dev. Psychopathol. 12, 501−527.

Marrale, M., Albanese, N.N., Cali, F., Romano, V., 2014. Assessing the impact of copy number variants on miRNA genes in autism by Monte Carlo simulation. PLoS ONE 9, e90947.

Marsden, W.N., 2013. Synaptic plasticity in depression: molecular, cellular and functional correlates. Prog. Neuropsychopharmacol. Biol. Psychiatry 43, 168−184.

Mellios, N., Galdzicka, M., Ginns, E., Baker, S.P., Rogaev, E., Xu, J., et al., 2012. Gender-specific reduction of estrogen-sensitive small RNA, miR-30b, in subjects with schizophrenia. Schizophr. Bull. 38, 433−443.

Mellios, N., Huang, H.S., Grigorenko, A., Rogaev, E., Akbarian, S., 2008. A set of differentially expressed miRNAs, including miR-30a-5p, act as post-transcriptional inhibitors of BDNF in prefrontal cortex. Hum. Mol. Genet. 17, 3030−3042.

Mendlewicz, J., 2009. Disruption of the circadian timing systems: molecular mechanisms in mood disorders. CNS Drugs 23 (Suppl 2), 15−26.

Miller, B.H., Zeier, Z., Xi, L., Lanz, T.A., Deng, S., Strathmann, J., et al., 2012. MicroRNA-132 dysregulation in schizophrenia has implications for both neurodevelopment and adult brain function. Proc. Natl. Acad. Sci. U. S. A. 109, 3125−3130.

Miller, D.T., Shen, Y., Weiss, L.A., Korn, J., Anselm, I., Bridgemohan, C., et al., 2009. Microdeletion/duplication at 15q13.2q13.3 among individuals with features of autism and other neuropsychiatric disorders. J. Med. Genet. 46, 242−248.

Miyaoka, T., Seno, H., Ishino, H., 1999. Increased expression of Wnt-1 in schizophrenic brains. Schizophr. Res. 38, 1−6.

Moncini, S., Salvi, A., Zuccotti, P., Viero, G., Quattrone, A., Barlati, S., et al., 2011. The role of miR-103 and miR-107 in regulation of CDK5R1 expression and in cellular migration. PLoS ONE 6, e20038.

Moreau, M.P., Bruse, S.E., David-Rus, R., Buyske, S., Brzustowicz, L.M., 2011. Altered microRNA expression profiles in postmortem brain samples from individuals with schizophrenia and bipolar disorder. Biol. Psychiatry 69, 188−193.

Moya, P.R., Wendland, J.R., Salemme, J., Fried, R.L., Murphy, D.L., 2013. miR-15a and miR-16 regulate serotonin transporter expression in human placental and rat brain raphe cells. Int. J. Neuropsychopharmacol. 16, 621−629.

Muhleisen, T.W., Leber, M., Schulze, T.G., Strohmaier, J., Degenhardt, F., Treutlein, J., et al., 2014. Genome-wide association study reveals two new risk loci for bipolar disorder. Nat. Commun. 5, 3339.

Ouyang, Y.B., Lu, Y., Yue, S., Xu, L.J., Xiong, X.X., White, R.E., et al., 2012. miR-181 regulates GRP78 and influences outcome from cerebral ischemia in vitro and in vivo. Neurobiol. Dis. 45, 555−563.

Pathania, M., Torres-Reveron, J., Yan, L., Kimura, T., Lin, T.V., Gordon, V., et al., 2012. miR-132 enhances dendritic morphogenesis, spine density, synaptic integration, and survival of newborn olfactory bulb neurons. PLoS ONE 7, e38174.

Perkins, D.O., Jeffries, C.D., Jarskog, L.F., Thomson, J.M., Woods, K., Newman, M.A., et al., 2007. microRNA expression in the prefrontal cortex of individuals with schizophrenia and schizoaffective disorder. Genome Biol. 8, R27.

Rahman, O.A., Sasvari-Szekely, M., Szekely, A., Faludi, G., Guttman, A., Nemoda, Z., 2010. Analysis of a polymorphic microRNA target site in the purinergic receptor P2RX7 gene. Electrophoresis 31, 1790−1795.

Remenyi, J., Hunter, C.J., Cole, C., Ando, H., Impey, S., Monk, C.E., et al., 2010. Regulation of the miR-212/132 locus by MSK1 and CREB in response to neurotrophins. Biochem. J. 428, 281−291.

Ripke, S., Neale, B.M., Corvin, A., Walters, J.T.R., Farh, K.-H., Holmans, P.A., et al., 2014. Biological insights from 108 schizophrenia-associated genetic loci. Nature 511, 421−427.

Ripke, S., Sanders, A., Kendler, K., Levinson, D., Sklar, P., Holmans, P., et al., 2011. Genome-wide association study identifies five new schizophrenia loci. Nat. Genet. 43, 969−976.

Rong, H., Liu, T.B., Yang, K.J., Yang, H.C., Wu, D.H., Liao, C.P., et al., 2011. MicroRNA-134 plasma levels before and after treatment for bipolar mania. J. Psychiatr. Res. 45, 92−95.

Royo, H., Cavaille, J., 2008. Non-coding RNAs in imprinted gene clusters. Biol. Cell 100, 149−166.

Samuels, I.S., Karlo, J.C., Faruzzi, A.N., Pickering, K., Herrup, K., Sweatt, J.D., et al., 2008. Deletion of ERK2 mitogen-activated protein kinase identifies its key roles in cortical neurogenesis and cognitive function. J. Neurosci. 28, 6983−6995.

Santarelli, D.M., Beveridge, N.J., Tooney, P.A., Cairns, M.J., 2011. Upregulation of dicer and microRNA expression in the dorsolateral prefrontal cortex Brodmann area 46 in schizophrenia. Biol. Psychiatry 69, 180−187.

Sarachana, T., Zhou, R., Chen, G., Manji, H.K., Hu, V.W., 2010. Investigation of post-transcriptional gene regulatory networks associated with autism spectrum disorders by microRNA expression profiling of lymphoblastoid cell lines. Genome Med. 2, 23.

Schofield, C.M., Hsu, R., Barker, A.J., Gertz, C.C., Blelloch, R., Ullian, E.M., 2011. Monoallelic deletion of the microRNA biogenesis gene Dgcr8 produces deficits in the development of excitatory synaptic transmission in the prefrontal cortex. Neural Dev. 6, 11.

Schratt, G.M., Tuebing, F., Nigh, E.A., Kane, C.G., Sabatini, M.E., Kiebler, M., et al., 2006. A brain-specific microRNA regulates dendritic spine development. Nature 439, 283–289.

Seshadri, S., Zeledon, M., Sawa, A., 2013. Synapse-specific contributions in the cortical pathology of schizophrenia. Neurobiol. Dis. 53, 26–35.

Shi, S., Leites, C., He, D., Schwartz, D., Moy, W., Shi, J., et al., 2014. MicroRNA-9 and microRNA-326 regulate human dopamine D2 receptor expression, and the microRNA-mediated expression regulation is altered by a genetic variant. J. Biol. Chem. 289, 13434–13444.

Shi, W., Du, J., Qi, Y., Liang, G., Wang, T., Li, S., et al., 2012. Aberrant expression of serum miRNAs in schizophrenia. J. Psychiatr Res. 46, 198–204.

Shih, W.L., Kao, C.F., Chuang, L.C., Kuo, P.H., 2012. Incorporating information of microRNAs into pathway analysis in a genome-wide association study of bipolar disorder. Front. Genet. 3, 293.

Smalheiser, N.R., Lugli, G., Rizavi, H.S., Torvik, V.I., Turecki, G., Dwivedi, Y., 2012. MicroRNA expression is down-regulated and reorganized in prefrontal cortex of depressed suicide subjects. PLoS ONE 7, e33201.

Smalheiser, N.R., Lugli, G., Rizavi, H.S., Zhang, H., Torvik, V.I., Pandey, G.N., et al., 2011. MicroRNA expression in rat brain exposed to repeated inescapable shock: differential alterations in learned helplessness vs. non-learned helplessness. Int. J. Neuropsychopharmacol. 14, 1315–1325.

Smalheiser, N.R., Lugli, G., Zhang, H., Rizavi, H., Cook, E.H., Dwivedi, Y., 2014. Expression of microRNAs and other small RNAs in prefrontal cortex in schizophrenia, bipolar disorder and depressed subjects. PLoS ONE 9, e86469.

Smrt, R.D., Szulwach, K.E., Pfeiffer, R.L., Li, X., Guo, W., Pathania, M., et al., 2010. MicroRNA miR-137 regulates neuronal maturation by targeting ubiquitin ligase mind bomb-1. Stem Cells 28, 1060–1070.

Song, H.T., Sun, X.Y., Zhang, L., Zhao, L., Guo, Z.M., Fan, H.M., et al., 2014. A preliminary analysis of association between the down-regulation of microRNA-181b expression and symptomatology improvement in schizophrenia patients before and after antipsychotic treatment. J. Psychiatr. Res. 54, 134–140.

Sperlagh, B., Vizi, E.S., Wirkner, K., Illes, P., 2006. P2X7 receptors in the nervous system. Prog. Neurobiol. 78, 327–346.

Stark, K.L., Xu, B., Bagchi, A., Lai, W.S., Liu, H., Hsu, R., et al., 2008. Altered brain microRNA biogenesis contributes to phenotypic deficits in a 22q11-deletion mouse model. Nat. Genet. 40, 751–760.

Sun, G., Ye, P., Murai, K., Lang, M.F., Li, S., Zhang, H., et al., 2011. miR-137 forms a regulatory loop with nuclear receptor TLX and LSD1 in neural stem cells. Nature communications 2, 529.

Tabares-Seisdedos, R., Rubenstein, J.L., 2009. Chromosome 8p as a potential hub for developmental neuropsychiatric disorders: implications for schizophrenia, autism and cancer. Mol. Psychiatry 14, 563–589.

Talebizadeh, Z., Butler, M.G., Theodoro, M.F., 2008. Feasibility and relevance of examining lymphoblastoid cell lines to study role of microRNAs in autism. Autism Res. 1, 240–250.

Tamim, S., Vo, D.T., Uren, P.J., Qiao, M., Bindewald, E., Kasprzak, W.K., et al., 2014. Genomic analyses reveal broad impact of miR-137 on genes associated with malignant transformation and neuronal differentiation in glioblastoma cells. PLoS ONE 9, e85591.

Trompeter, H.I., Abbad, H., Iwaniuk, K.M., Hafner, M., Renwick, N., Tuschl, T., et al., 2011. MicroRNAs MiR-17, MiR-20a, and MiR-106b act in concert to modulate E2F activity on cell cycle arrest during neuronal lineage differentiation of USSC. PLoS ONE 6, e16138.

Vaishnavi, V., Manikandan, M., Tiwary, B.K., Munirajan, A.K., 2013. Insights on the functional impact of microRNAs present in autism-associated copy number variants. PLoS ONE 8, e56781.

van Erp, T.G., Guella, I., Vawter, M.P., Turner, J., Brown, G.G., McCarthy, G., et al., 2014. Schizophrenia miR-137 locus risk genotype is associated with dorsolateral prefrontal cortex hyperactivation. Biol. Psychiatry 75, 398−405.

Walker, R.M., Rybka, J., Anderson, S.M., Torrance, H.S., Boxall, R., Sussmann, J.E., et al., 2015. Preliminary investigation of miRNA expression in individuals at high familial risk of bipolar disorder. J. Psychiatr. Res. 62, 48−55.

Whalley, H.C., Papmeyer, M., Romaniuk, L., Sprooten, E., Johnstone, E.C., Hall, J., et al., 2012. Impact of a microRNA MIR137 susceptibility variant on brain function in people at high genetic risk of schizophrenia or bipolar disorder. Neuropsychopharmacology 37, 2720−2729.

Wingo, A.P., Almli, L.M., Stevens, J.J., Klengel, T., Uddin, M., Li, Y., et al., 2015. DICER1 and microRNA regulation in post-traumatic stress disorder with comorbid depression. Nat. Commun. 6, 10106.

Wong, J., Duncan, C.E., Beveridge, N.J., Webster, M.J., Cairns, M.J., Weickert, C.S., 2013. Expression of NPAS3 in the human cortex and evidence of its posttranscriptional regulation by miR-17 during development, with implications for schizophrenia. Schizophr. Bull. 39, 396−406.

Wright, C., Turner, J.A., Calhoun, V.D., Perrone-Bizzozero, N., 2013. Potential impact of miR-137 and its targets in schizophrenia. Front. Genet. 4, 58.

Xu, B., Karayiorgou, M., Gogos, J.A., 2010a. MicroRNAs in psychiatric and neurodevelopmental disorders. Brain Res. 1338, 78−88.

Xu, S., Witmer, P.D., Lumayag, S., Kovacs, B., Valle, D., 2007. MicroRNA (miRNA) transcriptome of mouse retina and identification of a sensory organ-specific miRNA cluster. J. Biol. Chem. 282, 25053−25066.

Xu, Y., Li, F., Zhang, B., Zhang, K., Zhang, F., Huang, X., et al., 2010b. MicroRNAs and target site screening reveals a pre-microRNA-30e variant associated with schizophrenia. Schizophr. Res. 119, 219−227.

Xu, Y., Liu, H., Li, F., Sun, N., Ren, Y., Liu, Z., et al., 2010c. A polymorphism in the microRNA-30e precursor associated with major depressive disorder risk and P300 waveform. J. Affect. Disord. 127, 332−336.

Yin, K.J., Deng, Z., Huang, H., Hamblin, M., Xie, C., Zhang, J., et al., 2010. miR-497 regulates neuronal death in mouse brain after transient focal cerebral ischemia. Neurobiol. Dis. 38, 17−26.

Yu, H.C., Wu, J., Zhang, H.X., Zhang, G.L., Sui, J., Tong, W.W., et al., 2015. Alterations of miR-132 are novel diagnostic biomarkers in peripheral blood of schizophrenia patients. Prog. Neuropsychopharmacol. Biol. Psychiatry 63, 23−29.

Zhang, H.Y., Zheng, S.J., Zhao, J.H., Zhao, W., Zheng, L.F., Zhao, D., et al., 2011. MicroRNAs 144, 145, and 214 are down-regulated in primary neurons responding to sciatic nerve transection. Brain Res. 1383, 62−70.

Zhang, Z., Convertini, P., Shen, M., Xu, X., Lemoine, F., de la Grange, P., et al., 2013. Valproic acid causes proteasomal degradation of DICER and influences miRNA expression. PLoS ONE 8, e82895.

Zheng, Z., Zeng, Y., Huang, H., Xu, F., 2013. MicroRNA-132 may play a role in coexistence of depression and cardiovascular disease: a hypothesis. Med. Sci. Monit. 19, 438−443.

CIRCULAR RNAs EXPRESSION, FUNCTION, AND REGULATION IN NEURAL SYSTEMS

14

Ivano Legnini and Irene Bozzoni
Sapienza University of Rome, Rome, Italy

A BRIEF HISTORY OF CIRCULAR RNA DISCOVERY

RNA is the core biochemical species involved in the genetic information flow as defined by the central dogma of molecular biology. Conversion of a particular genotype into its related phenotype relies on RNA in a triple fashion: (1) the *messengers* able to translate a DNA sequence into a protein are made of RNA, (2) the translation process itself is completed by the ribosome/tRNA system, an RNA-based molecular machine, and (3) RNA itself can be an *effector* or a *regulator* of a given biological process, as proteins do. A key feature of messenger RNAs as well as of regulatory RNAs is their limited stability. Although the intrinsic property of genomes is their ability to preserve themselves through chemical stability and ability to replicate, the transcriptome has to be dynamic at both the biosynthetic and turnover stages in order to allow a living organism to respond to environmental changes. This is particularly true for complex multicellular organisms, where cell and tissue diversity are primarily established thanks to imperfect segmentation of cytoplasmic components, such as proteins and RNAs, and further achieved by means of what we call *regulation of gene expression*. In Eukaryotes, 5′ and 3′ termini of RNAs are the substrate for most RNA-degrading enzymes (exoribonucleases). Although those tails are protected in a variety of ways (e.g., 7-methylguanylate cap and polyadenosine tail in messenger RNAs), they can be exposed to exoribonucleases for allowing fast clearance of virtually any given RNA, therefore providing a robust tool of gene expression regulation.

Very recently, a whole new structural class of RNA molecules, called *circular RNAs*, has caught the attention of life scientists because of their exceptional stability given by the lack of 5′ and 3′ termini. Those covalently closed polyribonucleotides originate through a variety of mechanisms, can be detected in all domains of life, and can play important biological roles. The first one conforms to their stability, as they constitute the self-replicating genome of viroids. Viroids are probably the smallest replicating objects observed in nature (with the controversial exception of prions), as they are just made of a relatively short nude circular RNA ($\sim 200-400$ nucleotides), able to infect their plant host. They use endogenous RNA polymerase II for triggering rolling-circle amplification of their genome which is then again transcribed into the original strand and subsequently chopped and self-ligated into multiple new viroids (reviewed in Flores et al., 2009). Although viroid replication requires a host, they represent a beautiful subject for abiogenesis studies and

Essentials of Noncoding RNA in Neuroscience. DOI: http://dx.doi.org/10.1016/B978-0-12-804402-5.00014-5

speculation, as they are genomes, enzymes, and replicating objects at the same time. Viroids were discovered 40 years ago thanks to different techniques that relied on their infectivity and chemical structure. Treatment of infective particles of the *Potato Spindle Tuber Viroid* with an exoribonuclease did not affect their infectivity Diener and Smith (1971); later on, viroid particles were directly observed with electron microscopy (Sanger et al., 1976) and sequenced (Gross et al., 1978). Another source of a typical RNA circle was discovered in the ciliated protozoan *Tetrahymena thermophila*, again thanks to direct visualization through electron microscopy, in addition to an important feature of circular RNAs, that is their peculiar electrophoretic mobility (Grabowski et al., 1981). An intervening sequence in *T. thermophila* ribosomal RNA precursor is excised as a linear RNA, that is able to circularize by autocatalysis (Zaug et al., 1983). The ability to circularize was found to be general among nuclear group I introns, as they can use the hydrolyzed 3′ splice site for a nucleophilic attack to the 5′ splice site, resulting in full length intronic circular RNAs (Nielsen, 2003). Those peculiar RNAs are widely expressed in Archea, mostly coming from noncoding RNA genes, such as tRNAs and ribosomal RNAs (Danan et al., 2011), and they can be translated into proteins (Kjems and Garrett, 1988; Burggraf et al., 1993; Dalgaard and Garrett, 1992). Coding circular RNAs represent an intriguing alternative to common prokaryotic and eukaryotic linear messengers for several reasons. Although the chance of fast turnover can be fundamental for complex responses to environmental stresses or global rearrangements in the cellular transcriptome, the expression of housekeeping genes is often linked to highly stable mRNAs. Moreover, stability can be required for translation of mRNAs whose renewal is somehow impaired (e.g., in cells with inactive nuclei or in specific subcellular compartments, such as neuronal synapses). The chance of circular RNA translation in higher Eukaryotes and some speculative consequences will be more deeply discussed later in this chapter. Another interesting function of RNA circularization was firstly described in *Cyanidioschyzon merolae*, a unicellular red alga having peculiar tRNA genes where the sequence encoding 3′ arm lies upstream to that of 5′ arm in the genome (*permuted tRNA genes*). Therefore, circularization is necessary for restoring the required mature sequence, which is additionally processed by endonucleases for relieving the mature tRNA ends (Soma et al., 2007).

In addition to such variety of circular RNA forms, several studies in the former few years reported the unexpected occurrence of thousands of circular RNAs containing exonic sequences in Eukaryotes. Those molecules appear to be expressed among disparate phyla, and in virtually all the classic model organisms used in genetics and molecular biology, from flies to mammals. Actually, the presence of circularized exons in mammals was discovered many years ago for a few, well-characterized genes (Burd et al., 2010; Capel et al., 1993; Cocquerelle et al., 1993; Hansen et al., 2011; Nigro et al., 1991; Surono 1999). But only with the advent of next generation sequencing technologies, a few recent studies were able to discover their genome-wide preponderance, thus bringing those new molecules to the fore (Salzman et al., 2012; Jeck et al., 2012; Memczak et al., 2013). High-throughput sequencing of cDNA fragments, indeed, can detect exon circularization thanks to the mapping of spliced reads on the genome. When a sequencing read (or a fragment, in the case of *Paired End* sequencing) is split and its 5′ portion maps downstream to its 3′ mate, this event can be explained as the result of RNA circularization. By using different approaches of cDNA library preparation and data analysis, the aforementioned studies reported convergent results: thousands of genes can produce circular RNAs as the result of exon circularization, those circular RNAs can be very highly expressed, representing the main product of a gene in several cases, and their production can be conserved among species, indicating the chance of a biologically relevant role. Adopting the notation of the majority of authors, from now on, we will call these exon-derived, abundant, and conserved molecules *circRNAs*.

CircRNA METABOLISM
EUKARYOTIC CircRNAs ARE PRODUCED BY THE SPLICEOSOME

As stated in the previous paragraph, circRNAs are made of exonic sequences. The first identifications were fortuitous consequences of studies on aberrant pre-mRNA splicing events. Therefore, the link between splicing and circRNAs has always been clear, although the reasons why certain exons are circularized and other are not remained elusive for long time. CircRNA production occurs via a particular splicing reaction called *back-splicing*, catalyzed by the canonical spliceosome (Fig. 14.1) (Starke et al., 2015). For all we know, back-splicing is a very simple process, basically identical to linear splicing, where the 5′ and 3′ splice sites involved do not reside within the same intron but in two distinct introns. Particularly, a 5′ splice site is used by the spliceosome in conjunction with a 3′ splice site that lies in one of the upstream introns (circRNAs can be made of many exons). Given the chance of such process, one may actually ask why exon circularization does occur for specific exons and not for all of them. Indeed, two close splice sites are present at the borders of any given exon; therefore, their usage by the spliceosome might seem easier than that of two splice sites far away one from each other in introns, which are typically longer than exons. However, the spliceosome is recruited co-transcriptionally, and RNA Pol II elongation rate is tightly coupled with the kinetics of the splicing reaction (Ardehali and Lis, 2009; Görnemann et al., 2005; Lacadie and Rosbash, 2005). Such argument is further strengthened by the observation that a slower transcription rate (which typically increases splicing efficiency) allows linear splicing to overcome back-splicing. Indeed, slow-polymerase mutant flies produce significantly less circular RNAs, suggesting that linear splicing competes with circRNA biogenesis (Ashwal-Fluss et al., 2014). However, exon definition and splicing regulation are much more complicated, given the intervention of specific sequences, such as exonic and intronic splicing enhancers and silencers, RNA-binding proteins (RBPs), splicing factors and chromatin states.

SPECIFIC SEQUENCES AND POSITIONAL BIASES ASSOCIATED WITH EXON CIRCULARIZATION

An important observation about circularization trends in eukaryotic genomes concerns the size of circRNA-flanking introns, which are typically longer than average (Ashwal-Fluss et al., 2014; Venø et al., 2015; Westholm et al., 2014). This constitutes an apparent contradiction with the correlation between transcription rate and circRNA synthesis, as increased intron length should mimic slow polymerase activity from the kinetic point of view. Anyway, large introns are typically spliced less efficiently than small ones, pointing to the fact that efficient assembly of the spliceosome has to deal with the physical distance of 5′ and 3′ splice sites. An interesting confirmation comes from *Schizosaccharomyces pombe*, where introns are typically shorter than in higher Eukaryotes, and

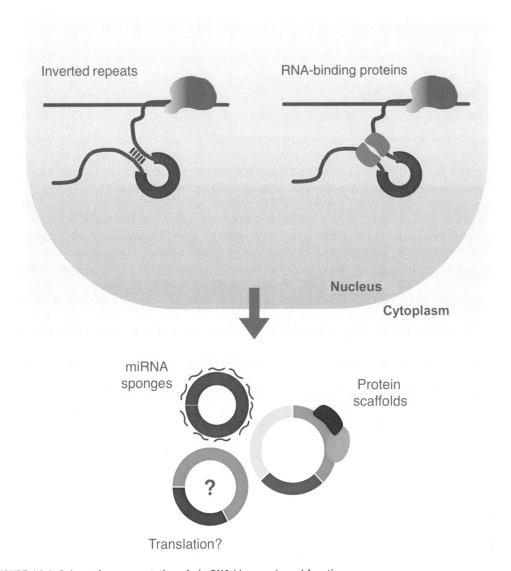

FIGURE 14.1 Schematic representation of circRNA biogenesis and function

During transcription in the nucleus (primary transcript in red, RNA Pol II in green), intronic sequences can fold thanks to inverted repeats (left) or the action of RNA-binding proteins (right) and bring the 5′ and 3′ splice site of two introns in close proximity (thin red line), promoting exon circularization (thick red line). The mature circRNA is then exported to the cytoplasm where it can exert its role as a microRNA sponge, protein scaffold, and perhaps other unknown biological functions.

back-splicing should be unlikely to occur. Indeed, circular RNAs in yeast are much less abundant than in mammals, and it seems that they are produced through an exon-containing lariat intermediate, therefore allowing their post-transcriptional maturation (Barrett et al., 2015).

A second important bias of introns that flank circularized exons is their increased content in repetitive elements. One of the first circular RNAs characterized in mammals comes from the *Sex Determining Region Y* gene (SRY), and it was demonstrated that inverted repeats surrounding its mature sequence are sufficient to direct circularization (Capel et al., 1993; Hacker et al., 1995; Jeske et al., 1995). Such observation was later generalized, with the finding that Alu repeats are enriched in circRNA-flanking introns, and that the frequency of inverted Alu repeats correlates with circRNA expression (Jeck et al., 2012; Liang and Wilusz, 2014). Moreover, the construction of plasmids expressing the circRNA *CDR1as* (which will be further described later in this chapter) required the insertion of long inverted repeats in order to obtain efficient back-splicing (Hansen et al., 2013). Alu repeats are not the only repetitive elements correlated with circRNA production. The presence of any complementary sequences able to base-pair after transcription is indeed sufficient to trigger exon circularization (Zhang et al., 2014), in a way that makes the propensity of an exon to circularize even predictable by measuring the probability of the flanking introns to base-pair via Reverse Complementary Matches or RCMs (Ivanov et al., 2015). It is thought that those inverted repeats are necessary to bring the two splice sites involved in circRNA production in close proximity, thereby suggesting a model of circRNA biogenesis for which inefficient linear splicing (due to transcription kinetics and/or intron length) favors circularization when two splice sites are brought together by RCMs or other factors.

RNA-BINDING PROTEINS AND RNA EDITING REGULATE CircRNA SYNTHESIS

Back-splicing is not favored solely by Alu repeats and other intronic sequences. As observed for alternative splicing, it seems that RBPs can play a fundamental role promoting or blocking exon circularization (Fig. 14.1). The first circRNA biogenesis factor to be described was *Muscleblind*, a splicing factor derived from the Mbl gene in flies and MBNL1-3 in mammals. Interestingly, the second exon of Mbl/MBNL1 forms a very abundant circRNA, perhaps the most abundant one in fly heads. Moreover, its circularization is directly induced by the Mbl protein itself, which is able to bind both the flanking introns and the mature circRNA (Ashwal-Fluss et al., 2014). Another circRNA-promoting factor is the RBP *Quaking* (QKI), a splicing factor, which promotes myelinization and oligodendrocyte differentiation (Haroutunian 2006; Chen et al. 2007) and found to be involved in diseases, such as ataxia, schizophrenia, and glioblastoma (Chénard and Richard, 2008). Its discovery as a circRNA regulator followed the observation that epithelial-to-mesenchimal transition (EMT) of immortalized human mammary cells is accompanied by a consistent increase of circRNA production (Conn et al., 2015). An RNAi-based screening of RBPs followed by measurement of circRNA synthesis by means of reporter vectors indeed revealed that QKI, which is upregulated in EMT, is able to bind introns and promote exon circularization. Finally, the role of many hnRNPs as well as SR proteins was found to be involved in circRNA production in flies (Kramer et al., 2015).

A second important hint to the definition of transacting factors for exon circularization came from the study of adenosine-to-inosine (A-to-I) editing by *Adenosine deaminase acting on RNA* (*ADAR*) enzymes. The ADAR family includes a group of enzymes responsible for adenosine-to-inosine conversion in double-stranded RNA. This modification is one of the most common editing events found in animals and has a fundamental impact on RNA function due to inosine inability to form Watson–Crick pairing with thymine, being its hydrogen bonds propensity similar to that of guanine. Although most of A-to-I sites are found in nontranslated regions of mRNAs as well as in repetitive elements, a well-characterized function of ADAR editing is in codon conversion, such as in the subunits of α-amino-3-hydroxy-5-methyl-4-isoxazolepropionic acid (AMPA) and kainate glutamate receptors, GluA2 and GluK1-2 (Sommer et al., 1991). Indeed, ADAR activity on glutamate receptors is observed during neuron maturation, converting Ca^{2+} permeable complexes to impermeable, an event that is required for postnatal survival of mice (Higuchi et al., 2000). In general, ADAR expression is enriched in the nervous system, particularly in neurons, and editing activity is strongly regulated in brain development (Jacobs et al., 2009). A-to-I editing is particularly frequent in Alu elements (Levanon et al., 2004; Kim, 2004), and there are several interpretations about its functional meaning (Levanon and Eisenberg, 2014). A direct consequence of inosine substitution in Alu elements is their reduced intermolecular base-pairing. One could therefore predict that exon circularization, which is favored by pairing of inverted Alu sequences in introns, should be inhibited by ADAR activity. The first confirmation of such hypothesis came with the finding that A-to-I conversions are enriched in sequences surrounding circRNA exons (Ivanov et al., 2015). Knock-down of ADAR1 and 2 in HEK293 cells, indeed, produced a consistent increase in circRNA expression, suggesting that A-to-I editing, probably by unfolding Alu-mediated secondary structures, prevents exon circularization. This is also true in in vitro models of neurogenesis, such as human SH-SY5Y and murine P19 cells, where ADAR1 is naturally downregulated upon induction of differentiation, and circRNA production is increased in parallel (Rybak-Wolf et al., 2015). Also in *Drosophila*, dADAR activity, which is regulated by temperature, is correlated with circRNA synthesis (Rybak-Wolf et al., 2015).

ElciRNAs AND ciRNAs

The failure of certain steps in the splicing process can lead to the synthesis of peculiar circular RNA products. CircRNAs are typically made by more than one exon (on average, 3, with a high variance, Rybak-Wolf et al., 2015). It follows that there can be introns within the predicted mature circRNA sequence. Those introns are in general efficiently removed; therefore, circRNAs harbor only exonic sequences (Westholm et al., 2014). However, RNA Polymerase II CLIP-seq revealed the existence of circular RNAs carrying unspliced introns which remain bound to the locus of transcription. These molecules, called *Exon-Intron-circular RNAs* (ElciRNAs), seem to be able to promote their host-gene transcription in a spliceosome-dependent manner (Li et al., 2015). Another class of unusual circular RNAs in animals is represented by *circular intronic RNAs* or ciRNAs. Normally, when a splicing reaction is completed, the excised intron has the form of a lariat, given by the linking of the 5′ splice site to an adenosine at the branch point site. Intron lariats accumulation is usually rapidly cleared by a debranching activity, able to hydrolyze the 2′−5′ phosphodiester bond at the branch site. CiRNAs are simply nondebranched lariats, where the free single strand RNA terminus is degraded by exonucleases. As ElciRNAs, those molecules are thought to be responsible for regulating transcription of their parental gene (Zhang et al., 2013).

SUBCELLULAR LOCALIZATION OF CircRNAs

Although EIciRNAs and ciRNAs remain bound to their site of transcription, the majority of exon-derived circRNAs are exported to the cytoplasm (Salzman et al., 2012; Jeck et al., 2012). Cytoplasmic accumulation can be achieved in two alternative ways: (1) an active export system (as for most RNAs) or (2) slow nonspecific molecular-diffusion (e.g., an escape following nuclear membrane break-up during mitosis). Transcripts produced by RNA Pol II are co-transcriptionally assembled into export-competent Ribonucleoparticles (RNPs), thanks to the concerted activity of several steps of maturation and quality control. Most studies about RNA export have been conducted in yeast, where mutations in elongation, termination, and export factors produce overlapping phenotypes, indicating the need of multiple steps of maturation in order to acquire the proper set of proteins for cytoplasmic translocation (Aguilera, 2005). Multiple adaptor proteins are recruited to nascent mRNA to form the transport/export complex (TREX), which is able to direct the RNP to the nucleopore permeability barrier through the export adaptor Mex67-Mtr2 (Delaleau and Borden, 2015). Although TREX is recruited during the capping and elongation checkpoint in yeast, its interaction with RNPs in metazoans occurs primarily through the Exon Junction Complex (EJC). The EJC is deposited after splicing on nascent RNAs, therefore presumably also on circRNAs, providing a possible explanation for their export. On the other hand, given the very high stability of circRNAs and their expression level usually lower than their linear counterpart, it can be inferred that, on average, they have a very slow biosynthetic rate. Therefore, the cytoplasmic accumulation observed at the steady state might also follow slow kinetics, pointing to the diffusion model. However, whether those molecules are actively exported through the nucleopore or not remains to be tested with appropriate model systems and experimental approaches.

One intriguing fact about circRNA localization is that they are not simply dissolved in the cytosol, but can be specifically located at subcellular compartments. You et al. (2015) conducted RNAseq analysis on murine synaptosomes and microdissected neuropili from the hippocampus and compared circRNA expression levels to that measured in whole hippocampal homogenate and microdissected somata. Strikingly, they observed that most circRNAs are enriched in the synaptic fractions with respect to whole cells or somata, with substantial overlap between the two different preparations. A similar pattern was observed also in preparations from rat tissues. Moreover, direct visualization by in situ hybridization confirmed cytoplasmic and dendritic localization of a subset of abundant circRNAs. Rybak-Wolf et al. (2015) confirmed those results by sequencing total RNA from synaptosome fractions obtained from whole brain lysates, finding that also in this case circRNAs are substantially enriched in synapses with respect to cell body and cytoplasm. A striking example of specific synaptic localization is given by a circRNA coming from the Stau2 gene, whose linear product, instead, is primarily localized in the cytoplasm. This suggests that it might exist a molecular transport system for circRNAs, as it occurs for protein and mRNA axonal and dendritic transport.

THE ENIGMA OF CircRNA TURNOVER

A curious exception to the cytoplasmic localization of circRNAs is given by the circular product coming from a transcript antisense to the *Cerebellar Degeneration-Related Protein 1* locus, CDR1as. Although this circRNA is predominantly expressed in the cytoplasm of neurons, it was shown that its turnover is delegated to a nuclear microRNA, miR-671 (Hansen et al., 2011). A single binding site with almost perfect complementarity to miR-671 allows Ago2 slicer-dependent cleavage of CDR1as, allowing fast and efficient turnover of such circular RNA. MiR-671 nuclear localization suggests that CDR1as could be imported into the nucleus in tissues or conditions where its function as a microRNA sponge (discussed later) becomes unneeded, or it may be targeted in the nucleus prior to cytoplasmic translocation.

In general, circular RNA turnover is an unsolved problem, as it has been shown in multiple works that they have an unusually long half-life (Jeck et al., 2012; Memczak et al., 2013), exceeding the typical length of 24—48 hours of experiments requiring the block of transcription, and currently there are not any known active degradation pathways. Circular RNA high stability has also been indirectly observed in some physiological phenomena, such as the unusual enrichment of circular transcripts in platelets, which are anucleated and thus unable to renew degraded transcripts (Alhasan et al., 2015), or the fact that circRNAs tend to accumulate over time during *Drosophila* lifespan (Westholm et al., 2014). It is currently unknown whether any circRNA-specific RNase exists in cells, although there are dozens of endonucleases annotated in Eukaryotes. Those enzymes, however, typically evolved as biosynthetic factors (e.g., *Nob1* in rRNA maturation) or gene expression regulators (e.g., *Argonaute* proteins), whereas the designed general turnover and RNA quality surveillance systems are 5′ and 3′ exoribonucleases (XRNs and the Exosome). Actually, the core subunit of the Exosome, *Dis3*, is able to hydrolyze also internal ribonucleotides in vitro (Lebreton et al., 2008), therefore constituting a good candidate for future studies.

A fascinating alternative to nucleolytic targeting for circRNA turnover has been recently suggested by Lasda and Parker (2016). They observed that circRNAs are enriched in extracellular vesicles produced by cultured cells and proposed that clearance of excess molecules might be achieved through expulsion to the extracellular space. CircRNA enrichment in vesicles, on the other hand, could indicate their involvement in a cell-to-cell communication system, as previously observed for other transcripts, like microRNAs (Zhang et al., 2015).

CircRNA EXPRESSION PATTERNS
CircRNAs ARE REGULATED IN TIME AND SPACE

Since the very first high-throughput studies about circRNAs, their strong tissue/state-specific expression appeared clear. Salzman et al. (2012, 2013) reported that circular RNAs can represent the predominant splice isoform of hundreds of genes and that this feature is cell-type specific. Later on, Memczak et al. (2013) conducted a similar analysis not only on cell lines, but also on *C. elegans,* with a focus on its early developmental stages, where they found circular RNAs being expressed in the oocyte but silenced in one- or two-cell embryo and the opposite. In a detailed analysis of neuronal circRNAs, Rybak-Wolf et al. compared circRNA expression in different tissues from human and mouse samples and confirmed the strong tissue specificity. Although expression of circular RNAs follows the tissue-specific transcription of their parental genes, it has been observed that also post-transcriptional regulation can occur, determining whether a given locus produces a circular or linear product in response to the cellular environment. For example, circularization of exon 2 of the Mbl gene in *Drosophila* can be as efficient in fly heads as almost absent in S2 cells (Ashwal-Fluss et al., 2014). Similarly, circular RNAs coming for example from the Elf2 gene or from the Stau2 gene in mouse can be expressed less or more than their linear counterparts in testis or heart with respect to brain.

As expected, circRNA expression is also modulated in response to cell-state transitions. For example, EMT of human mammary cells is accompanied by a global increase in circRNA expression, which is thought to be related to QKI overexpression in EMT as previously described (Conn et al., 2015). Also hypoxia, a key angiogenetic stimulus, triggers a global rearrangement in circRNA expression level in HUVEC (*Human Umbilical Vein Endothelial Cells*), which seems uncoupled from the accompanying variation in the transcriptional landscape of these cells (Boeckel et al., 2015).

A Key Point for understanding the relevance of circRNA function and regulation relates to their specific and strong enrichment in neuronal tissues.

NEURAL ENRICHMENT OF CircRNAs

Alternative splicing events are particularly abundant in the Central Nervous System (CNS), with thousands of specific splice sites and exons being detected only in neuronal tissues. Those events are often conserved and play fundamental roles in neurogenesis and physiology. They are usually triggered by the expression of neuronal-enriched splicing factors (e.g., nSR100, NOVA etc.) and are often able to reshape the binding properties of the encoded proteins (Raj and Blencowe, 2015). Such wide repertoire of transcript isoforms is reflected in an equivalent great diversity of circRNAs. Transcriptome analysis by total RNA sequencing of five tissues derived from adult mice revealed a strong prevalence of circRNAs in brain, with more than 6000 different species, with respect to heart, lung, liver, and testis, that express in average 1000–3000 circular RNAs (You et al., 2015). To understand the amplitude of this phenomenon, it is sufficient to note that, in brain, circa 20% of protein-coding genes harbor at least one circRNA. Such disparity is explained both by the number of genes that produce circRNAs, which are more in brain, but also by the propensity of the same genes to make circular products, which is higher in brain than in other tissues. Interestingly, circRNA-producing genes are enriched in Gene Onthology terms related to synaptic functions and structural components. In parallel, as previously reported, circular RNAs themselves are recovered in dendrites, thus indicating their possible role in neuron connection and communication. Indeed, the response of hippocampal neurons to bicuculline suggests a role of circRNAs in synaptic plasticity. Bicuculline is an antagonist of $GABA_A$ receptor and triggers a rearrangement in circRNA expression involving more than 40 different transcripts.

An even deeper study of circRNAs in neuronal tissues involved the analysis of 29 different CNS-related samples, from mouse and human (Rybak-Wolf et al., 2015). Here, the numbers are even more striking—more than 15000 circRNAs recovered in mouse and more than 65000 in human samples, which is much more than what observed in other tissues like thyroid, liver, and muscle. Among those, a few hundreds were expressed at impressive levels, with more than 100 reads spanning the back-spliced junction, indicating a concentration similar to that of the highest expressed protein-coding genes. In human samples, there's an average of 3 circRNAs per gene, with approximately 2000 genes producing more than 10 circRNAs. Moreover, there are consistent differences between different regions in the brain. Analysis of data from olfactory bulb, prefrontal cortex, hippocampus, and cerebellum in mouse revealed strong differential expression of circular transcripts, with an overall enrichment in cerebellum, which has a higher density of neuronal cells than other cell types, again indicating a global neuronal enrichment of circRNAs.

Similar analyses were applied to embryonic development of the CNS and in vitro neuronal differentiation systems. In the mouse hippocampus, a clear shift is observed in circRNA production at early postnatal stages with respect to embryonic samples, with a global induction of RNA circularization at P10, concomitantly with the beginning of synaptogenesis (You et al., 2015). Similarly, in vitro differentiation systems like mouse P19 cells, a cell line derived from an embryo carcinoma, and human SH-SY5Y, a cell line derived from neuroblastoma, display a global induction of circRNA expression when stimulated with retinoic acid.

Those dramatic changes of circRNA expression both in time and space during neurogenesis were also observed in pig brains, where circa 9000 circRNAs were annotated, with a peak of expression in midembryogenesis (E60) and with a consistent enrichment in cortex and cerebellum (Venø et al., 2015).

BIOLOGICAL FUNCTION OF CircRNAs
TECHNICAL CHALLENGES IN CircRNA FUNCTIONAL STUDIES

Perhaps the most relevant question that has still to be addressed about circular RNAs relates to their function. The active circularization mechanisms discussed previously in this chapter, as well as circRNA evolutionary conservation and their regulated expression patterns, suggest that these molecules are not stochastic errors of the splicing apparatus and could have specific molecular functions. However, with the exception of a few cases, this has still to be proven. The main issue about the study of circRNA function is the scarcity of molecular tools able to perturb their expression levels. Overexpression of circRNAs in cultured cells and living animals has been achieved through many different approaches that can be virtually applied to any desired candidate (Ashwal-Fluss et al., 2014; Conn et al., 2015; Hansen et al., 2013; Liang and Wilusz, 2014). Those methods involve the use of plasmids for transient expression, carrying the mature circRNA sequence flanked either by the genomic context of well-expressed circular RNAs or by artificially modified introns with long inverted repeats. Although these approaches can produce abundant circular molecules, they usually have low specificity, producing also large quantities of linear precursors, thus frustrating the chance to observe circRNA-specific effects. On the other hand, also circRNA knock-down by RNA interference has been successfully exploited (Boeckel et al., 2015; Du et al., 2016). Again, the effectiveness of the approach is theoretically vitiated by the chance of off-target effects, due to the limited space for designing circular-specific siRNAs. Most circular RNAs are indeed produced from genes that encode also a linear product, which contains the entire circRNA sequence, and the only genuine circRNA-specific sequence is the back-spliced junction. Most of the siRNAs targeting such sequence, anyway, could also bind to the linear RNA in a microRNA-like fashion. Therefore, detection of circRNA-specific effects must involve the use of additional precautions, like complementation of the phenotype or a detailed investigation of off-target effects. A third approach might represent a winning strategy, reconciling high efficiency with specific targeting. *Clustered regularly-interspaced short palindromic repeats* (CRISPR)/*Cas9* have been successfully adapted for fast and efficient genome editing of cultured cells and animals (reviewed in Wright et al.,

2016). The CRISPR/Cas9 system relies on the activity of a nuclease (Cas9) whose cut site is programmed by a short RNA guide that can be targeted to any DNA sequence flanked by an NGG trinucleotide. Since the introduction of this technology, both deletions by nonhomologous end joining and insertions through homology-directed repair can be achieved with unprecedented efficiency, at the cost of a couple of plasmids. We suggest that CRISPR/Cas9 can be used for circular RNA-specific deletion targeted on intronic sequences, either deleting the introns surrounding a given candidate, or mutating the inverted repeats responsible for exon circularization.

CircRNAs AS microRNA INHIBITORS

The first established function of a circRNA is that of CDR1as, a sponge for miR-7 in neuronal tissues (Fig. 14.1). As previously described, CDR1as was identified as an antisense circRNA to the CDR1 transcript, whose turnover is triggered by miR-671. Two different studies (Hansen et al., 2013; Memczak et al., 2013) found that CDR1as RNA is densely bound by Ago2 as assayed by CLIP techniques, and this binding is mostly explained by the presence of 74 distinct sites complementary to the miR-7 seed. The frequency of miR-7 binding sites in this region is unmatched with any other eukaryotic gene, and the strong conservation of the majority of those sites indicates a functional meaning of such binding. Ectopic expression of CDR1as (named also *ciRS-7* for *circular RNA sponge for miR-7*) blocks miR-7-mediated repression of known targets such as SNCA, EGFR, and IRS2 in HeLa cells. In vivo, CDR1as and miR-7 are colocalized in the brain of both developing and adult mice, with the highest circRNA expression in hippocampus and neocortex. Due to the lack of specific approaches for circRNA perturbation in vivo, the role of CDR1as in brain physiology and development is unclear. However, the use of *Danio rerio* as an in vivo model system allowed at least confirming its ability to inhibit miR-7 activity. Although zebrafish has lost the CDR1 locus, miR-7 is expressed at high level in the brain and its block by antisense morpholinos induced a phenotype in neurogenesis, strongly reducing the midbrain size. As expected, the overexpression of CDR1as in zebrafish embryos phenocopied the miR-7 loss-of-function phenotype (Memczak et al., 2013). Interestingly, miR-7 is involved in Parkinson's disease (Junn et al., 2009), but it's also well-known for being deregulated in cancer, thus pointing to a possible oncogenic role of CDR1as, which is indeed found expressed at high levels in neuroblastoma, astrocytoma, renal cell, and lung carcinomas (Peng et al., 2015). Another famous circRNA, the one coming from the *SRY* gene, has multiple binding sites for a microRNA, miR-138 (Hansen et al., 2013). As for CDR1as, circRNA SRY is recovered by immunoprecipitating Ago2 upon miR-138 expression, and its sequence responds to miR-138 expression in the context of a reporter vector (Hansen et al., 2013). Although CDR1as and SRY represent important paradigms, it seems that sequestering microRNAs is not a general phenomenon for circular RNAs, at least in mammals (Guo et al., 2014; Rybak-Wolf et al., 2015; Westholm et al., 2014; You et al., 2015).

Additional functional circular RNAs include circ-ZNF292, exhibiting pro-angiogenic activity in cultured endothelial cells, and circ-FOXO3, which is able to block cell-cycle progression by binding CDK2 and p21 in mouse fibroblasts. It is interesting to note that a general mechanism for circRNA function has not been identified up to now. It seems more likely that thanks to their peculiar biochemical properties, including stability and possibility of binding proteins and nucleic acids, those molecules initially arising from aberrant splicing events have been evolutionary selected for disparate functions that still have to be discovered. Therefore, we can conclude that, even if a fraction has been converted into potent microRNA sponges by convergent evolution, circRNAs do not seem to be a coherent functional class of RNAs.

TRANSLATION OF CircRNAs
ARTIFICIAL CIRCULAR RNAs CAN BE TRANSLATED IN EUKARYOTES

As suggested in the first paragraph of this chapter, circRNAs might represent a fascinating alternative to linear messenger RNAs. Exon circularization could add new reading frames to a coding sequence, expanding the repertoire of protein isoforms, especially truncated proteins or single domains. Moreover, their strong stability could be important for cell types or compartments (e.g., distant synapses) where mRNA renewal is particularly difficult. Their localization in extracellular vesicles could constitute an intriguing mechanism of cell-to-cell communication through the exchange of genetic information, as happens for microRNAs. However, standard eukaryotic translation requires the formation of an initiation complex that involves both the 5′ cap and the polyA tail. Internal initiation, on the other hand, is a fundamental alternative mechanism of translation, which works with the direct recruitment of the ribosome on specific RNA structures called *Internal Ribosome Entry Sites* (IRES). IRES usage is typical of viruses, which are often able to block cap-dependent translation and to take advantage of the cellular protein-synthesis apparatus. But also many eukaryotic mRNAs are able to recruit the ribosome through this alternative route in particular physiological conditions (e.g., angiogenesis, heat shock response and other stress responses, Komar and Hatzoglou, 2011).

Mechanistically, internal translation of circular RNAs should be equivalent to that of linear RNAs, with the 40 S subunit of the ribosome directly recruited at an IRES sequence. The first evidence that this mechanism actually works comes from an old study reporting that synthetic circular templates carrying the IRES from the Encephalomyocarditis virus can be engaged by the ribosome for multiple rounds of translation in vitro (Chen and Sarnow, 1995). More recently, similar approaches were applied to cell cultures with the transfection of circRNA-producing plasmids carrying a GPF-coding sequence preceded by a viral IRES. The back-spliced circular RNA was able to be translated, confirming that free ends are not required, as expected, for efficient ribosome recruitment (Conn et al., 2015; Wang and Wang, 2014). Abe et al., (2015) used a different approach and tested synthetic RNA circles carrying an infinite open-reading frame, with multiple Flag-tag encoding sequences organized in tandem in order to preserve the frame after circularization. Such templates were translated both in reticulocyte extracts and in living HeLa cells, suggesting that circular RNA translation might occur in vivo even with the absence of an IRES, through a spurious and probably inefficient recruitment of the ribosome.

ARE NATURAL CircRNAs TRANSLATED?

The lack of a general sequence or structural motif for cellular IRESes hampers the chance to study circRNA translation genome-wide. However, Ribosome footprinting followed by RNA sequencing (*Riboseq* or *Ribosome Profiling*, Ingolia et al., 2009) has proven to be a powerful technique for monitoring translation of a given model system with a relatively simple procedure. RNase I digestion of cytoplasmic extracts, which leaves intact RNA fragments protected by the ribosome

(*Ribosome Protected Fragments* or *RPFs*), allows the production of large libraries of cDNA fragments constituting a representative sample of the cellular translatome. Riboseq has been adopted in a few works trying to address the question of whether endogenous circular RNAs are translated. Guo et al. (2014) searched for RPFs mapping on the back-spliced junction of 717 circRNAs annotated in human U2OS cells, a cell line originated from an osteosarcoma, and found zero matches. Similarly, after failing to map mass spectrometry-derived peptides to circRNA-specific sequences, You et al. (2015) searched for RPFs mapping to back-spliced junctions in rat brain and, again, found no matches. Although these observations could discourage further studies on natural circRNA coding potential, there are few arguments pointing to the chance that internal translation of circular molecules might occur. First, the studies on synthetic circular templates that were mentioned in the previous paragraph state that in theory circRNAs can work as efficient messenger RNAs. Second, internal translation is usually triggered in particular conditions, especially as a response to various forms of stress. For example, two coeval studies reported that m6a methylation in the 5′ UTR of mRNAs, a particular RNA modification catalyzed posttranscriptionally by RNA methyl-transferases, responds to heat shock stress and is able to trigger cap-independent translation of certain mRNAs, such as HSP70 (Meyer et al., 2015; Zhou et al., 2015). Riboseq approaches should therefore be directed to particular cell types and experimental conditions that allow mapping the translatome after the switch toward particular translation initiation pathways. Given the diversity and abundance of circRNA sequences, if just a subset of them were found to function as messenger RNAs, an entire new portion of the eukaryotic proteome would be disclosed, with a myriad of new possible functions. This chance demands to make an effort towards a better understanding of circRNA-coding ability, keeping in mind, at the same time, that many other biological roles could be attributed to these enigmatic molecules.

ACKNOWLEDGMENTS

This work was partially supported by grants from: ERC-2013 (AdG 340172−MUNCODD), Human Frontiers Science Program Award RGP0009/2014, Epigen-Epigenomics Flagship Project, AFM-Telethon (17835), AriSLA full grant 2014 "ARCI," PRIN and Parent Project Italia.

REFERENCES

Abe, N., Matsumoto, K., Nishihara, M., Nakano, Y., Shibata, A., Maruyama, H., et al., 2015. Rolling circle translation of circular RNA in living human cells. Sci. Rep 5, 16435. Available from: http://dx.doi.org/10.1038/srep16435.

Aguilera, A., 2005. Cotranscriptional mRNP assembly: from the DNA to the nuclear pore. Curr. Opin. Cell Biol. 17, 242−250. Available from: http://dx.doi.org/10.1016/j.ceb.2005.03.001.

Alhasan, A., Izuogu, O., Al-Balool, H., Steyn, J., Evans, A., Colzani, M., et al., 2015. Circular RNA enrichment in platelets is a signature of transcriptome degradation. Blood 127, e1−e11. Available from: http://dx.doi.org/10.1182/blood-2015-06-649434.

Ardehali, M., Lis, J., 2009. Tracking rates of transcription and splicing in vivo. Nat. Struct. Mol. Biol. 16, 1123−1124. Available from: http://dx.doi.org/10.1038/nsmb1109-1123.

Ashwal-Fluss, R., Meyer, M., Pamudurti, N., Ivanov, A., Bartok, O., Hanan, M., et al., 2014. circRNA biogenesis competes with Pre-mRNA splicing. Mol. Cell 56, 55−66. Available from: http://dx.doi.org/10.1016/j.molcel.2014.08.019.

Barrett, S., Wang, P., Salzman, J., 2015. Circular RNA biogenesis can proceed through an exon-containing lariat precursor. eLife 4, e07540. Available from: http://dx.doi.org/10.7554/elife.07540.

Boeckel, J., Jaé, N., Heumüller, A., Chen, W., Boon, R., Stellos, K., et al., 2015. Identification and characterization of hypoxia-regulated endothelial circular RNANovelty and significance. Circ. Res. 117, 884−890. Available from: http://dx.doi.org/10.1161/circresaha.115.306319.

Burd, C., Jeck, W., Liu, Y., Sanoff, H., Wang, Z., Sharpless, N., 2010. Expression of linear and novel circular forms of an INK4/ARF-associated non-coding RNA correlates with atherosclerosis risk. PLoS Genet. 6, e1001233. Available from: http://dx.doi.org/10.1371/journal.pgen.1001233.

Burggraf, S., Larsen, N., Woese, C., Stetter, K., 1993. An intron within the 16 S ribosomal RNA gene of the archaeon pyrobaculum aerophilum. Proc. Natl. Acad. Sci. U. S. A. 90, 2547−2550. Available from: http://dx.doi.org/10.1073/pnas.90.6.2547.

Capel, B., Swain, A., Nicolis, S., Hacker, A., Walter, M., Koopman, P., et al., 1993. Circular transcripts of the testis-determining gene SRY in adult mouse testis. Cell 73, 1019−1030. Available from: http://dx.doi.org/10.1016/0092-8674(93)90279-y.

Chen, C., Sarnow, P., 1995. Initiation of protein synthesis by the eukaryotic translational apparatus on circular RNAs. Science 268, 415−417. Available from: http://dx.doi.org/10.1126/science.7536344.

Chen, Y., Tian, D., Ku, L., Osterhout, D., Feng, Y., 2007. The selective RNA-binding protein quaking I (QKI) is necessary and sufficient for promoting oligodendroglia differentiation. J. Biol. Chem. 282, 23553−23560. Available from: http://dx.doi.org/10.1074/jbc.m702045200.

Chénard, C., Richard, S., 2008. New implications for the QUAKING RNA binding protein in human disease. J. Neurosci. Res. 86, 233−242. Available from: http://dx.doi.org/10.1002/jnr.21485.

Cocquerelle, C., Mascrez, B., Hétuin, D., Bailleul, B., 1993. Mis-Splicing yealds circular RNA molecules. FASEB J. 7, 155−160.

Conn, S., Pillman, K., Toubia, J., Conn, V., Salmanidis, M., Phillips, C., et al., 2015. The RNA binding protein quaking regulates formation of circRNAs. Cell 160, 1125−1134. Available from: http://dx.doi.org/10.1016/j.cell.2015.02.014.

Dalgaard, J.Z., Garrett, R.A., 1992. Protein-coding introns from the 23S rRNA-encoding gene form stable circles in the hyperthermophilic archaeon *Pyrobaculum organotrophum*. Gene 121, 103−110.

Danan, M., Schwartz, S., Edelheit, S., Sorek, R., 2011. Transcriptome-wide discovery of circular RNAs in *Archaea*. Nucleic Acids Res. 40, 3131−3142. Available from: http://dx.doi.org/10.1093/nar/gkr1009.

Delaleau, M., Borden, K., 2015. Multiple export mechanisms for mRNAs. Cells 4, 452−473. Available from: http://dx.doi.org/10.3390/cells4030452.

Diener, T., Smith, D., 1971. Potato spindle tuber viroid. Virology 46, 498−499. Available from: http://dx.doi.org/10.1016/0042-6822(71)90052-3.

Du, W., Yang, W., Liu, E., Yang, Z., Dhaliwal, P., Yang, B., 2016. Foxo3 circular RNA retards cell cycle progression via forming ternary complexes with p21 and CDK2. Nucleic Acids Res. 44, 2846−2858. Available from: http://dx.doi.org/10.1093/nar/gkw027.

Flores, R., Gas, M., Molina-Serrano, D., Nohales, M., Carbonell, A., Gago, S., et al., 2009. Viroid replication: rolling-circles, enzymes and ribozymes. Viruses 1, 317−334. Available from: http://dx.doi.org/10.3390/v1020317.

Görnemann, J., Kotovic, K., Hujer, K., Neugebauer, K., 2005. Cotranscriptional spliceosome assembly occurs in a stepwise fashion and requires the cap binding complex. Mol. Cell 19, 53−63. Available from: http://dx.doi.org/10.1016/j.molcel.2005.05.007.

Grabowski, P., Zaug, A., Cech, T., 1981. The intervening sequence of the ribosomal RNA precursor is converted to a circular RNA in isolated nuclei of tetrahymena. Cell 23, 467−476. Available from: http://dx.doi.org/10.1016/0092-8674(81)90142-2.

Gross, H., Domdey, H., Lossow, C., Jank, P., Raba, M., Alberty, H., et al., 1978. Nucleotide sequence and secondary structure of potato spindle tuber viroid. Nature 273, 203−208. Available from: http://dx.doi.org/10.1038/273203a0.

Guo, J.U., Agarwal, V., Guo, H., Bartel, D.P., 2014. Expanded identification and characterization of mammalian circular RNAs. Genome Biol 15, 409.

Hacker, A., Capel, B., Goodfellow, P., Lovell-Badge, R., 1995. Expression of SRY, the mouse sex determining gene. Development 121, 1603−1614.

Hansen, T., Jensen, T., Clausen, B., Bramsen, J., Finsen, B., Damgaard, C., et al., 2013. Natural RNA circles function as efficient microRNA sponges. Nature 495, 384−388. Available from: http://dx.doi.org/10.1038/nature11993.

Hansen, T., Wiklund, E., Bramsen, J., Villadsen, S., Statham, A., Clark, S., et al., 2011. miRNA-dependent gene silencing involving Ago2-mediated cleavage of a circular antisense RNA. EMBO J. 30, 4414−4422. Available from: http://dx.doi.org/10.1038/emboj.2011.359.

Haroutunian, V., 2006. The human homolog of the QKI gene affected in the severe dysmyelination "quaking" mouse phenotype: downregulated in multiple brain regions in schizophrenia. Am. J. Psychiatry 163, 1834. Available from: http://dx.doi.org/10.1176/appi.ajp.163.10.1834.

Higuchi, M., Maas, S., Single, F., Hartner, J., Rozov, A., Burnashev, N., et al., 2000. Point mutation in an AMPA receptor gene rescues lethality in mice deficient in the RNA-editing enzyme ADAR2. Nature 6, 78−81.

Ingolia, N., Ghaemmaghami, S., Newman, J., Weissman, J., 2009. Genome-wide analysis in vivo of translation with nucleotide resolution using ribosome profiling. Science 324, 218−223. Available from: http://dx.doi.org/10.1126/science.1168978.

Ivanov, A., Memczak, S., Wyler, E., Torti, F., Porath, H., Orejuela, M., et al., 2015. Analysis of intron sequences reveals hallmarks of circular RNA biogenesis in animals. Cell Rep. 10, 170−177. Available from: http://dx.doi.org/10.1016/j.celrep.2014.12.019.

Jacobs, M., Fogg, R., Emeson, R., Stanwood, G., 2009. ADAR1 and ADAR2 expression and editing activity during forebrain development. Dev. Neurosci. 31, 223−237. Available from: http://dx.doi.org/10.1159/000210185.

Jeck, W., Sorrentino, J., Wang, K., Slevin, M., Burd, C., Liu, J., et al., 2012. Circular RNAs are abundant, conserved, and associated with Alu repeats. RNA 19, 141−157. Available from: http://dx.doi.org/10.1261/rna.035667.112.

Jeske, Y., Bowles, J., Greenfield, A., Koopman, P., 1995. Expression of a linear SRY transcript in the mouse genital ridge. Nat. Genet. 10, 480−482. Available from: http://dx.doi.org/10.1038/ng0895-480.

Junn, E., Lee, K., Jeong, B., Chan, T., Im, J., Mouradian, M., 2009. Repression of synuclein expression and toxicity by microRNA-7. Proc. Natl. Acad. Sci. U. S. A. 106, 13052−13057. Available from: http://dx.doi.org/10.1073/pnas.0906277106.

Kim, D., 2004. Widespread RNA editing of embedded Alu elements in the human transcriptome. Genome Res. 14, 1719−1725. Available from: http://dx.doi.org/10.1101/gr.2855504.

Kjems, J., Garrett, R., 1988. Novel splicing mechanism for the ribosomal RNA intron in the archaebacterium desulfurococcus mobilis. Cell 54, 693−703. Available from: http://dx.doi.org/10.1016/0092-8674(81)90142-2.

Komar, A., Hatzoglou, M., 2011. Cellular IRES-mediated translation. Cell Cycle 10, 229−240. Available from: http://dx.doi.org/10.4161/cc.10.2.14472.

Kramer, M., Liang, D., Tatomer, D., Gold, B., March, Z., Cherry, S., et al., 2015. Combinatorial control of *Drosophila* circular RNA expression by intronic repeats, hnRNPs, and SR proteins. Genes Dev. 29, 2168−2182. Available from: http://dx.doi.org/10.1101/gad.270421.115.

Lacadie, S., Rosbash, M., 2005. Cotranscriptional spliceosome assembly dynamics and the role of U1 snRNA:5′ss base pairing in yeast. Mol. Cell 19, 65−75. Available from: http://dx.doi.org/10.1016/j.molcel.2005.05.006.

Lasda, E., Parker, R., 2016. Circular RNAs co-precipitate with extracellular vesicles: a possible mechanism for circRNA clearance. PLoS ONE 11, e0148407. Available from: http://dx.doi.org/10.1371/journal.pone.0148407.

Lebreton, A., Tomecki, R., Dziembowski, A., Séraphin, B., 2008. Endonucleolytic RNA cleavage by a eukaryotic exosome. Nature 456, 993−996. Available from: http://dx.doi.org/10.1038/nature07480.

Levanon, E., Eisenberg, E., 2014. Does RNA editing compensate for Alu invasion of the primate genome? BioEssays 37, 175−181. Available from: http://dx.doi.org/10.1002/bies.201400163.

Levanon, E., Eisenberg, E., Yelin, R., Nemzer, S., Hallegger, M., Shemesh, R., et al., 2004. Systematic identification of abundant A-to-I editing sites in the human transcriptome. Nat. Biotechnol. 22, 1001−1005. Available from: http://dx.doi.org/10.1038/nbt996.

Li, Z., Huang, C., Bao, C., Chen, L., Lin, M., Wang, X., et al., 2015. Exon-intron circular RNAs regulate transcription in the nucleus. Nat. Struct. Mol. Biol. 22, 256−264. Available from: http://dx.doi.org/10.1038/nsmb.2959.

Liang, D., Wilusz, J., 2014. Short intronic repeat sequences facilitate circular RNA production. Genes Dev. 28, 2233−2247. Available from: http://dx.doi.org/10.1101/gad.251926.114.

Memczak, S., Jens, M., Elefsinioti, A., Torti, F., Krueger, J., Rybak, A., et al., 2013. Circular RNAs are a large class of animal RNAs with regulatory potency. Nature 495, 333−338. Available from: http://dx.doi.org/10.1038/nature11928.

Meyer, K., Patil, D., Zhou, J., Zinoviev, A., Skabkin, M., Elemento, O., et al., 2015. 5′ UTR m6A promotes cap-independent translation. Cell 163, 999−1010. Available from: http://dx.doi.org/10.1016/j.cell.2015.10.012.

Nielsen, H., 2003. The ability to form full-length intron RNA circles is a general property of nuclear group I introns. RNA 9, 1464−1475. Available from: http://dx.doi.org/10.1261/rna.5290903.

Nigro, J., Cho, K., Fearon, E., Kern, S., Ruppert, J., Oliner, J., et al., 1991. Scrambled exons. Cell 64, 607−613. Available from: http://dx.doi.org/10.1016/0092-8674(91)90244-s.

Peng, L., Yuan, X., Li, G., 2015. The emerging landscape of circular RNA ciRS-7 in cancer (Review). Oncol. Rep. Available from: http://dx.doi.org/10.3892/or.2015.3904.

Raj, B., Blencowe, B., 2015. Alternative splicing in the mammalian nervous system: recent insights into mechanisms and functional roles. Neuron 87, 14−27. Available from: http://dx.doi.org/10.1016/j.neuron.2015.05.004.

Rybak-Wolf, A., Stottmeister, C., Glažar, P., Jens, M., Pino, N., Giusti, S., et al., 2015. Circular RNAs in the mammalian brain are highly abundant, conserved, and dynamically expressed. Mol. Cell 58, 870−885. Available from: http://dx.doi.org/10.1016/j.molcel.2015.03.027.

Salzman, J., Chen, R., Olsen, M., Wang, P., Brown, P., 2013. Cell-type specific features of circular RNA expression. PLoS Genet. 9, e1003777. Available from: http://dx.doi.org/10.1371/journal.pgen.1003777.

Salzman, J., Gawad, C., Wang, P., Lacayo, N., Brown, P., 2012. Circular RNAs are the predominant transcript isoform from hundreds of human genes in diverse cell types. PLoS ONE 7, e30733. Available from: http://dx.doi.org/10.1371/journal.pone.0030733.

Sanger, H., Klotz, G., Riesner, D., Gross, H., Kleinschmidt, A., 1976. Viroids are single-stranded covalently closed circular RNA molecules existing as highly base-paired rod-like structures. Proc. Natl. Acad. Sci. U. S. A. 73, 3852−3856. Available from: http://dx.doi.org/10.1073/pnas.73.11.3852.

Soma, A., Onodera, A., Sugahara, J., Kanai, A., Yachie, N., Tomita, M., et al., 2007. Permuted tRNA genes expressed via a circular RNA intermediate in *Cyanidioschyzon merolae*. Science 318, 450−453. Available from: http://dx.doi.org/10.1126/science.1145718.

Sommer, B., Köhler, M., Sprengel, R., Seeburg, P., 1991. RNA editing in brain controls a determinant of ion flow in glutamate-gated channels. Cell 67, 11−19. Available from: http://dx.doi.org/10.1016/0092-8674 (91)90568-j.

Starke, S., Jost, I., Rossbach, O., Schneider, T., Schreiner, S., Hung, L., et al., 2015. Exon circularization requires canonical splice signals. Cell Rep. 10, 103−111. Available from: http://dx.doi.org/10.1016/j. celrep.2014.12.002.

Surono, A., 1999. Circular dystrophin RNAs consisting of exons that were skipped by alternative splicing. Hum. Mol. Genet. 8, 493−500. Available from: http://dx.doi.org/10.1093/hmg/8.3.493.

Venø, M., Hansen, T., Venø, S., Clausen, B., Grebing, M., Finsen, B., et al., 2015. Spatio-temporal regulation of circular RNA expression during porcine embryonic brain development. Genome Biol. 16. Available from: http://dx.doi.org/10.1186/s13059-015-0801-3.

Wang, Y., Wang, Z., 2014. Efficient backsplicing produces translatable circular mRNAs. RNA 21, 172−179. Available from: http://dx.doi.org/10.1261/rna.048272.114.

Westholm, J., Miura, P., Olson, S., Shenker, S., Joseph, B., Sanfilippo, P., et al., 2014. Genome-wide analysis of *Drosophila* circular RNAs reveals their structural and sequence properties and age-dependent neural accumulation. Cell Rep. 9, 1966−1980. Available from: http://dx.doi.org/10.1016/j.celrep.2014.10.062.

Wright, A., Nuñez, J., Doudna, J., 2016. Biology and applications of CRISPR systems: harnessing nature's toolbox for genome engineering. Cell 164, 29−44. Available from: http://dx.doi.org/10.1016/j. cell.2015.12.035.

You, X., Vlatkovic, I., Babic, A., Will, T., Epstein, I., Tushev, G., et al., 2015. Neural circular RNAs are derived from synaptic genes and regulated by development and plasticity. Nat. Neurosci. 18, 603−610. Available from: http://dx.doi.org/10.1038/nn.3975.

Zaug, A., Grabowski, P., Cech, T., 1983. Autocatalytic cyclization of an excised intervening sequence RNA is a cleavage−ligation reaction. Nature 301, 578−583. Available from: http://dx.doi.org/10.1038/301578a0.

Zhang, J., Li, S., Li, L., Li, M., Guo, C., Yao, J., et al., 2015. Exosome and exosomal microRNA: trafficking, sorting, and function. Genomics Proteomics Bioinformatics 13, 17−24. Available from: http://dx.doi.org/ 10.1016/j.gpb.2015.02.001.

Zhang, X., Wang, H., Zhang, Y., Lu, X., Chen, L., Yang, L., 2014. Complementary sequence-mediated exon circularization. Cell 159, 134−147. Available from: http://dx.doi.org/10.1016/j.cell.2014.09.001.

Zhang, Y., Zhang, X., Chen, T., Xiang, J., Yin, Q., Xing, Y., et al., 2013. Circular intronic long noncoding RNAs. Mol. Cell 51, 792−806. Available from: http://dx.doi.org/10.1016/j.molcel.2013.08.017.

Zhou, J., Wan, J., Gao, X., Zhang, X., Jaffrey, S., Qian, S., 2015. Dynamic m6A mRNA methylation directs translational control of heat shock response. Nature 526, 591−594. Available from: http://dx.doi.org/ 10.1038/nature15377.

COMPARATIVE FUNCTIONS OF miRNAs IN EMBRYONIC NEUROGENESIS AND NEURONAL NETWORK FORMATION

15

Emma Ristori and Stefania Nicoli

Yale School of Medicine, New Haven, CT, United States

INTRODUCTION

EVOLUTION OF THE CENTRAL NERVOUS SYSTEM IN METAZOA

The main role of the central nervous system (CNS) is to integrate and process sensory information coming from the periphery and initiate an appropriate response. Anatomically, the CNS is comprised of nervous tissue organized in interconnected agglomerations of functionally specialized neurons (nuclei and ganglia) that connect to the periphery via bundle of nerve fibers. From an evolutionary point of view, the nervous system makes its first appearance at the origin of the animal kingdom (also called Metazoa), and shows different shapes and different degrees of complexity through evolution of invertebrate and vertebrate organisms (Hill et al., 2012), reflecting their adaptation to a wide range of environments. Animals with a nervous system include Cnidaria (e.g. hydra, jellyfish, anemones, and corals) and all Bilateria, whereas Porifera (sponges) do not possess a nervous system (Fig. 15.1).

The simplest form of nervous system in multicellular organisms is found in Cnidaria. For example, the hydra shows a loosely organized system of grouped sensory neurons called the *nerve net* that controls the contraction and the expansion of the gastrovascular cavity (Kelava et al., 2015) (Fig. 15.1).

The complexity of the nervous system increases in Bilateria animals where we observe two main evolutionary trends: the centralization and the cephalization of the nervous system (Arendt et al., 2008; Hill et al., 2012) (Fig. 15.1). The term *centralization* refers to the structural organization of the nervous system in which neurons are collected into central integrating areas rather than being randomly dispersed. *Cephalization* refers to the concentration of nervous structures and functions at one end of the body, in particular the head.

The centralization of the nervous system is anatomically evident in all animals with bilateral symmetry by the presence of a nerve cord running through the length of the entire body and formed by aggregation of neurons in longitudinally arranged clusters organized into central ganglia and peripheral sensory organs. In invertebrates (that include the clade Protostomia and some phyla of the superphylum Deuterostomia), the nerve cord is localized ventrally. In vertebrates (phylum

Essentials of Noncoding RNA in Neuroscience. DOI: http://dx.doi.org/10.1016/B978-0-12-804402-5.00015-7

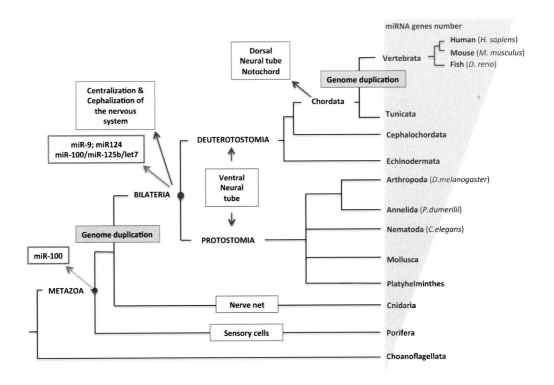

FIGURE 15.1

MiRNA genes and CNS evolution.

Chordata) (Fig. 15.1) the nerve cord is dorsal and surrounded by a notochord, a rod-like skeletal structure along the length of the body that guides and organizes the formation and differentiation of the nervous system in the brain and spinal cord during embryogenesis.

Cephalization, the other general evolutionary trend in nervous system organization, involves varying degrees of anterior concentration of nervous system organization. In invertebrate deuterostomes and protostomes, cephalization increases with evolution and the enlargement of the anterior ganglia that receives sensory inputs and controls feeding, gives rise to the first brain. For example, flarworms (Platyhelminthes) (Fig. 15.1) have a simple ladder-like nervous system with a small ganglionic brain. The complexity of mollusks' nervous system (e.g. octopuses, squids, snails, slugs, clams and mussels, etc.) correlates with habitat as well as phylogeny: small moving mollusks (e.g. clams) have little or no cephalization and simple sensory organs and the nervous system is organized in a network of paired ganglia. On the contrary, cephalopods (e.g. octopus) present the most sophisticated invertebrate nervous system characterized by a large brain, large eyes, and high conduction giant axons. Other invertebrates, such as roundworms (Nematoda) (Fig. 15.1) have both a dorsal and a ventral nerve cord connected to a circular brain (nerve ring) that allows the animal to detect the environment and to react to it, while Annelida (segmented worms) and Arthropoda (insects and crustaceans) (Fig. 15.1) present an anterior ganglionic brain connected to a ventral nerve cord that consists of paired segmental ganglia. Insects brain is formed by three pairs of fused ganglia: *protocerebrum*, *deuterocerebrum*, and *tritocerebrum*. The ganglia are located dorsally and each pair controls a limited spectrum of activities. Giant fiber systems (axons) connect the brain with legs and wings and allow rapid conduction of nerve impulses. They also possess a greater variety of sensory receptors than any other group of organisms, including vertebrates. Diversely, Echinoderms (e.g. sea urchin and starfish) (Fig. 15.1) have secondarily evolved radial symmetry and like Cnidaria, they show an uncentralized nervous system characterized by a set of net-like radial nerves connecting to a central nerve ring surrounding the mouth. In vertebrates, both centralization and cephalization acquire the highest complexity. The brain and the spinal cord, protected by vertebrae, form the CNS, whereas the nerves and the ganglia form the peripheral nervous system. Anatomically, the vertebrate brain is subdivided in three portions, highly conserved from fish to mammals: *prosencephalon* (forebrain), *mesencephalon* (midbrain), and *rhombencephalon* (hindbrain). More complex brains are further subdivided for integration of complex tasks. For example, mammals have two new structures, the *neocerebellum*, which plays an important role in skilled motor functions, and the *neocortex*, which is involved in higher functions such as sensory perception, spatial reasoning, conscious thought, and language.

This wide variety of anatomical structures and functions is controlled by processes that take place during the development of the mature CNS, such as neurogenesis, a multistep process that promotes the generation of differentiated neurons starting from a pool of neural stem cells (NSCs). Neurogenesis is a fundamental process for both embryonic neurodevelopment and adult brain plasticity and requires the precise coordination of multilayered gene regulatory networks. The process of neurogenesis begins when ectodermal cells (in Cnidaria also endodermal cells) acquire neurogenic potential that can be either spread out over all the ectoderm resulting in the formation of a nerve plexus, as in the case of Cnidaria or restricted to a particular area called neuroectoderm in Chordata. In *Drosophila* and other invertebrates, the nervous system arises from a subset of ventral neuroectodermal cells (neuroblasts) that divide asymmetrically in a stem-like fashion. In vertebrates, the entire neuroectoderm (neuroepithelium) folds into the embryo to form the dorsal neural tube. This is followed by the separation of neural cells from the ectoderm and by their migration, proliferation, and differentiation (Hartenstein and Wodarz, 2013; Hartenstein and Stollewerk, 2015; Richards and Rentzsch, 2014).

Homologies in the regulation of self-renewal, multipotency and cell fate determination of NSCs have been shown in *Drosophila* and vertebrates (Brand and Livesey, 2011). The cellular mechanisms that control neurogenesis are highly conserved in the animal kingdom, starting from Cnidaria, before the rise of bilaterian animals (Hartenstein and Stollewerk, 2015). For example, the SoxB (Sry-related box genes) family of genes in both fly and vertebrates are expressed in the ectoderm and are important for providing neurogenic potential, while maintaining the proliferative state of the neuroectoderm (Sasai, 2001). Proneural genes encoding for bHLH (basic helix−loop−helix) transcription factors, such as Achaete/Scute family and atonal family, trigger the differentiation of neural progenitors (NPs) to postmitotic neurons in both vertebrates and invertebrates (Quan and Hassan, 2005). Moreover, proneural genes, such as bHLH, promote in both vertebrates and invertebrates a mechanism known as lateral inhibition. In cells differentiating in neurons, proneural genes induce the expression of the ligand delta that activates the Notch receptor located on membrane of the neighbor cells. The activation of Notch represses the expression of proneural genes in the neighbor cells, preventing those cells to become neurons and thus controlling the spatial patterning of NPs (Hartenstein and Wodarz, 2013).

Together with transcription factors, small noncoding RNAs, in particular microRNAs (miRNAs), have been shown to be tightly involved in controlling multiple aspects of neural development. Moreover the miRNA−mRNA interactions described during neurogenesis are highly evolutionary conserved in vertebrates and invertebrates. In this chapter, we will discuss the importance of evolutionary conserved miRNAs in controlling neural development, focusing specifically on their regulation of embryonic neurogenesis.

MiRNAs AND NEURAL DEVELOPMENT
MiRNA BIOGENESIS AND FUNCTION

MiRNAs are 20−22 nucleotide (nt) small noncoding RNA that finely tune gene expression at the post-transcriptional level (Bartel, 2004). miRNA biogenesis starts in the nucleus, where RNA polymerase II transcribes the majority of miRNAs coding genes to generate a primary miRNA (pri-miRNA). In the canonical processing pathway, the pri-miRNA is cleaved by the RNAse III enzyme Drosha and its cofactor DGCR8 (DiGeorge syndrome critical region gene 8) into a 70−100-nt-long precursor miRNA (pre-miRNA) hairpin. The pre-miRNA is exported to the cytoplasm by the nuclear transporter protein Exportin V, and is then cleaved by RNAse III enzyme Dicer into a double stranded mature miRNA. In the noncanonical processing pathway, intron-encoded miRNAs (mirtrons) are excised by the spliceosome and directly processed by Dicer. One strand from the mature duplex is then loaded into the RNA-induced silencing complex (RISC), which mediates the binding of the miRNA with its target mRNA, typically at the 3′ untranslated region (3′UTR) and induces the target mRNA cleavage by Argonaute (Ago), the catalytic component of RISC. This interaction leads to mRNA degradation or translational inhibition (Fig. 15.2). The imperfect base-paring nature of target-miRNA binding, occurring only in the seed region (positioned at nt 2−7), allows a single miRNA to potentially target and regulate the expression of hundreds of genes. In addition, many mRNAs possess multiple binding sites in their 3′UTRs that can potentially be targeted by multiple miRNAs (Bartel, 2009).

CONSERVATION AND EVOLUTION OF miRNA GENES IN METAZOA

MiRNA genes originated early during metazoan evolution (Grimson et al., 2008) (Fig. 15.1). The expression of specific miRNA families in the brain is conserved in both primitive protostomes and primitive deuterostomes (Christodoulou et al., 2010). The number of miRNA genes largely increases during the evolution of the metazoan kingdom and positively correlates with the increased morphological complexity of multicellular organisms. For example, subsets of miRNAs, such as the neuronal specific miR-9 and the cardiovascular specific miR-126, are found only in Bilateria and not in simpler organisms such as Porifera and Cnidaria that don't present heart and brain (Fig. 15.1), suggesting a role of miRNAs in the origin of complex organs (Sempere et al., 2007). Once a miRNA is integrated into a gene regulatory pathway,

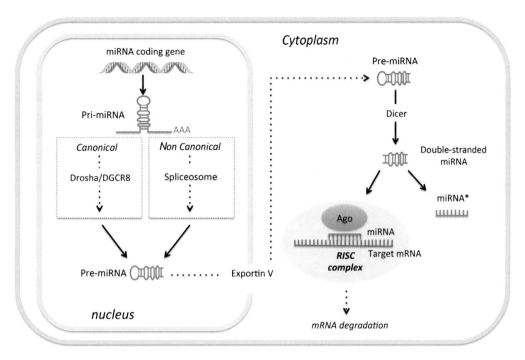

FIGURE 15.2

MiRNA biogenesis.

it shows rare secondary loss as well as rare mutations in the mature sequences (Sempere et al., 2006). Despite this high level of conservation, miRNAs are subject to variation over time and members of miRNA families have undergone modifications to the mature sequence such as seed shifts, nucleotide substitutions, and insertions, giving rise to different seed sequences and thus the potential to target a diverse set of transcripts (Kosik, 2009; Wheeler et al., 2009). Gene duplication is one of the mechanisms used by evolution to expand the repertoire of miRNA genes in Metazoa. Families of paralogous genes with significant sequence homology and identical seed regions on the mature sequence arise through gene duplication that can happen either on the same chromosome location (tandem duplication) or on a different chromosome (nonlocal duplication) (Hertel et al., 2006). At least two gene duplication events occurred through miRNA evolution, one at the base of the bilaterian lineage and a second expansion has been shown among vertebrates (Fig. 15.1). Another mechanism that increases miRNA diversity is the alternative processing by Drosha and Dicer of the hairpin structure that leads to the emergence of novel miRNAs, often arising as additional members of existing clusters (Hertel and Stadler, 2015). Therefore, the miRNA repertoire evolves to exert new specific functions by acquisition of new miRNAs over time in addition to modification of preexisting miRNAs. A study on the evolution of human miRNAs revealed that mutations in both precursor and mature sequences are more frequent in young miRNA genes than in old miRNA groups, suggesting different evolutionary pressures on miRNA genes originated in different ages. For example, severe phenotypes are more often observed in knockouts for older rather than new miRNAs (Zhu et al., 2012). Recently emerged miRNAs show high tissue specificity compared to the broad expression of old miRNAs suggesting that new miRNAs might play a role in new specie-specific gene regulatory pathways resulting from increased morphological complexity. Moreover, recently evolved miRNAs are more subjected to negative selection because their random target interactions often induce detrimental regulatory effects. However, when a new miRNA induces a rare beneficial target interaction, this successful miRNA results integrate into gene regulatory networks acquiring new targets and more functional importance over time (Bartel and Chen, 2004; Mohammed et al., 2014; Zhu et al., 2012).

MiRNA-CONTROLLED NEURAL REGULATORY GENE NETWORKS

The ability of miRNAs to regulate gene expression confers robustness to biological processes (Ebert and Sharp, 2012). Biological robustness has been defined as the ability of a biological system to maintain its function despite of endogenous and exogenous perturbations (Kitano, 2004). Recent studies suggested that the involvement of miRNAs into feedback and feed-forward loops of gene regulatory networks provide developmental robustness by buffering fluctuations in protein levels (Ebert and Sharp, 2012; Kitano, 2004; Li et al., 2009; Siciliano et al., 2013; Tsang et al., 2007) (Fig. 15.3A−C). miRNAs bind to their target genes forming simple feedback network motifs and working as genetic switches by repressing their targets. Feedback loops can be either positive or negative, when the target itself negatively regulates the miRNA's expression (Fig. 15.3A). For example, during murine neurogenesis, miR-9 negatively regulates the nuclear receptor Tlx (also

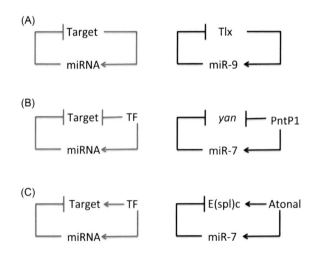

FIGURE 15.3

Neural miRNAs are wired in feedback and feed-forward gene regulatory loops (A) example of negative feedback loop; (B) example of coherent feed-forward loop; (C) example of incoherent feed-forward loop.

known as NR2E1, Nuclear receptor subfamily 2 group E member 1) to inhibit NSCs proliferation and promote neural differentiation. Tlx is an essential regulator of NSC self-renewal and can repress miR-9 expression forming a negative feedback loop that controls the balance between NSC proliferation and differentiation (Zhao et al., 2009; Fig. 15.3A).

In addition, miRNAs can take part in more complex feed-forward regulatory loops that involve the interaction with transcription factors or epigenetic regulators and allow gene expression tuning or buffering. Depending on the effect on transcriptional regulation, these loops can be divided into coherent and incoherent modes. In the coherent feed-forward loop, the miRNA reinforces the transcriptional repression on the target gene. Thus, the miRNA and its targets result mutually exclusive (Fig. 15.3B). This type of miRNA-target gene interaction functions to tune protein output and/or to precisely remove proteins that should not be expressed in that cell type. For example, miR-7 forms a coherent feed-forward loop with its target gene *yan* and the transcription factor Pnt-P1 (PointedP1) to control *Drosophila* photoreceptor determination. The transcription factor Pnt-P1 directly represses *yan* and enhances miR-7 transcription, which in turn represses *yan* in photoreceptor cells (Li et al., 2009; Fig. 15.3B). In the incoherent feed-forward loop, one miRNA and transcription factor have opposite effects on the target gene, resulting in coexpression of the miRNA and its target (Fig. 15.3C). In this case, miRNAs maintain uniform protein output over time, buffering the gene expression noise due to the stochastic nature of transcription, translation, and degradation steps of the target genes. (Li et al., 2009; Osella et al., 2011; Tsang et al., 2007). For example, during *Drosophila* sensory organ development, precursor cells express the *atonal* gene that will promote sensory organ determination. *Atonal* directly activates both miR-7 and the *E(spl)* (enhancer of split) gene, that feeds back to repress *atonal*. miR-7 targets the *E(spl)* gene and in doing so modulates the *E(spl)* repression effect on *atonal*, whereas indirectly promoting *atonal* transcription (Li et al., 2009; Fig. 15.3C). The involvement of miRNAs in simple and complex neural gene regulatory networks is critical for neurogenesis as it provides a fine control of the spatiotemporal synthesis of proteins, thus allowing the precise regulation of NSC proliferation and self-renewal as well as cell lineage commitment and neural differentiation.

THE CONTRIBUTION OF miRNAs TO NEUROGENESIS

In vertebrates, the neurogenesis process initiates in the neuroepithelium, area enriched of multipotent NSCs, characterized by self-renewal potential. In response to specific stimuli, NSCs can commit to a specific cellular fate that will become NPs and subsequently differentiate into neurons. During differentiation, NPs divide asymmetrically producing one daughter cell that maintains its self-renewal potential and another daughter cell that differentiates, giving rise to all cell types that constitute the CNS such as neurons, astrocytes, and oligodendrocytes. In *Drosophila*, neurogenesis starts from the neuroectoderm, where a small group of ectodermal epithelial cells (neuroblasts) form the proneural cluster by acquiring neural competence upon induction of proneural genes. Neuroblasts undergo repeated asymmetric divisions to generate neurons and/or glial cells.

In these developmental contexts, proliferation and gene expression regulation must be tightly controlled in a spatial and temporal manner, and miRNAs have been proposed as key regulators of neurogenesis. miRNAs are in fact highly expressed in the nervous system, such as in the human brain where miRNAs regulate two-third of all the genes (Krol et al., 2010). Many studies analyzed the role of miRNAs at various stages of neural development and maturation and pointed out the importance of a temporal and spatial miRNAs mediated control of neural differentiation and specification. In fact, the expression of neural-enriched miRNAs is often restricted to specific areas at different developmental stages, as described later in this chapter (Bak et al., 2008; Kosik, 2006; Kosik and Krichevsky, 2005; Lang and Shi, 2012; Olsen et al., 2009). The overall impact of miRNAs as essential regulators of neural development was first demonstrated by global miRNA loss-of-function experiments obtained by gene deletion of key components of the miRNA processing machinery that leads to the global loss of miRNA expression. For example, loss of maternal-zygotic expression of Dicer, leads to brain defects in zebrafish, characterized by reduced ventricle size and loss of the midbrain–hindbrain boundary (MHB) (Giraldez et al., 2005). In mouse, the constitutive loss of Dicer leads to early embryonic lethality, before the onset of neurogenesis, suggesting a role for miRNAs in maintaining stem cell populations (Bernstein et al., 2003), whereas the conditional knockout mice show a smaller cortex (De Pietri Tonelli et al., 2008) and an altered expansion and differentiation of the NSC population (Saurat et al., 2013). In *Drosophila*, mutations of Dicer1 and Pasha (also known as DGCR8) but not Ago1 and Ago2 lead to the disruption of olfactory projection neurons, suggesting a function for miRNAs in establishing wiring specificity in the nervous system (Berdnik et al., 2008).

Even though the loss of global miRNA population, achieved by the disruption of miRNA biogenesis, proves their broad importance in neural development, it is critical to distinguish the role of individual miRNAs in controlling neural regulatory networks. In the next part of the chapter, we will discuss the functions of individual miRNAs that are enriched in the CNS during neurogenesis and evolutionary conserved from invertebrates to vertebrates.

EVOLUTIONARY CONSERVED miRNAs IN VERTEBRATE AND INVERTEBRATE NEURAL STEM CELLS

Among the entire miRNA repertoire, miR-100 is believed to be the most evolutionary ancient miRNA in Metazoa (Grimson et al., 2008; Sempere et al., 2006; Wheeler et al., 2009). The miR-100 gene is located in the Hox gene cluster, which is essential for shaping the body plan of the embryo, and thus suggests the importance of this ancient miRNA during neural development. miR-100 evolved from a single copy found in *Cnidaria* to form the miR-10/100 cluster in Bilateria (Wheeler et al., 2009). It has been shown that miR-10 targets different sets of Hox genes in *Caenorhabditis elegans*, *Drosophila*, zebrafish, and mouse to control the anterior–posterior patterning (Tehler et al., 2011). In humans, miR-10 is specifically expressed in the hindbrain and spinal cord where it influences caudalization of human NPs by targeting wnt1 (Wnt family member 1) and MSX1 (Msh Homeobox 1) genes, which have a pivotal role during early development (Jonsson et al., 2015).

A comparative study analyzing the expression of ancient animal miRNAs in Cnidaria, deuterostomes and in the marine anellid *Platynereis dumerilii*, a protostome retaining ancestral bilaterian features, showed that miR-100 expression was initially limited to small groups of neurosecretory cells located around the mouth, together with other two ancient miRNAs miR-125 and let-7

(Christodoulou et al., 2010). miRNAs that make their appearance in Bilateria evolved often in a conserved tissue-specific context. For example, the brain-specific miR-124 was found in differentiating neurons and in the ventral nerve cord in the marine annelid *P. dumerilii* and maintains similar specificity in vertebrates, where it is expressed in the developing CNS to maintain neuronal identity. However, even if during the evolution of the phylum Chordata some of these miRNAs maintain the ancestral bilaterian function, others likely evolve new functions and new tissue-specificity in correlation with increased morphological complexity. For example, miRNA miR-9/miR-9* is specifically located in chemosensory sense organs in invertebrate protostomes, and maintains the expression in sensory organs in the fly and in sensory cells of the brain olfactory centers in mouse. In vertebrates, however, miR-9 shows a novel expression pattern and functions as it becomes strictly CNS-specific and broadly expressed in neuronal precursors where it plays a key role in regulating neurogenesis (Christodoulou et al., 2010; Lagos-Quintana et al., 2002; Li et al., 2006). During evolution, the spatial and temporal specific context of miRNA expression and function usually changes. For example, miR-100 and the related miR-125 and let-7 show a conserved role in developmental timing from nematodes to vertebrates, deciding whether an organism will develop past the neonatal stage during late embryonic developmental transitions. However, in insects, the function of these miRNAs seems to affect more metamorphosis progression (Pasquinelli et al., 2000; Reinhart et al., 2000; Rubio and Belles, 2013). The differences in spatiotemporal expression of these conserved miRNAs during evolution result in significant differences in functions exerted in different organisms and are often linked to the increasing morphological complexity of the nervous system, from the simple nerve net found in Cnidaria to the appearance of a centralized nervous system in Bilateria.

In the following part of the chapter, we will illustrate the conserved and novel roles of the highly evolutionary conserved miRNAs let-7, miR-125, miR-9, and miR-124 in controlling neurogenesis of different organism models.

let-7

The let-7 (*lethal-7*) family is highly conserved through the bilaterian animal phylum and it is one of the first miRNAs discovered in *C. elegans* (Reinhart et al., 2000). Multiple events of duplication led to the formation of the large let-7 family in vertebrates with 18 family members identified in zebrafish. All the let-7 family members share a high level of sequence similarity and 100% identity in the seed region, illustrating the strong selective pressure on the mature sequence and suggesting a conserved function among isoforms. The let-7 family has been found to play a role in both early and late developmental timing decisions. While in invertebrates let-7 exerts a broad role in shaping development, understanding the functional role of the let-7 family in vertebrate development is still limited because of the redundancy of the let-7 mature isoforms. However, various studies showed that let-7 induces cell-cycle exit and promotes differentiation during neurogenesis in vertebrates. During CNS development, let-7 expression begins in committed NPs and becomes highly expressed in postmitotic neurons. In vertebrate NSC niches, let-7 promotes differentiation and neuronal maturation by silencing pluripotency factors and suppressors of neural differentiation. Several let-7 family members such as let-7-a, let-7-c, and let-7-e have been found to be upregulated during mouse brain development (Wulczyn et al., 2007). let-7-b has been found to regulate NSCs proliferation and differentiation, targeting distinct molecules in NSCs across a spectrum of developmental stages.

For example, in the embryonic mouse brain, let-7-b regulates NSC fate decision and promotes neurogenesis by targeting the nuclear receptor Tlx (NR2E1, Nuclear receptor subfamily 2 group E member 1), which is known to regulate cell-cycle progression in NSCs and its downstream effector *cyclin-D1* (Li et al., 2008; Zhao et al., 2010). In addition, let-7-b has been shown to target the transcription factor Hmga2 (high-mobility group AT-hook 2) to reduce the self-renewal of NSCs in the aging brain (Nishino et al., 2008). In invertebrates, let-7 expression starts during the larva-to-adult transition. In *C. elegans*, let-7 is a key regulator of developmental timing by negatively regulating the heterochronic gene Lin41. Mutant worms for let-7 die prematurely failing to progress from the larval stage to adult stage, indicating the importance of let-7 regulation during developmental progression (Reinhart et al., 2000). Moreover, let-7 regulates neuronal temporal patterning in worms by targeting *hbl-1* (hunch-back-like 1) in ventral nerve cord neurons (Lin et al., 2003). In *Drosophila*, let-7 mutants exhibit multiple defects arising during metamorphosis, including defects in the maturation of the neuromuscular junction. It has been shown that in the postembryonic brain, let-7 regulates cell fate of neuroblasts of the mushroom bodies, cells similar to mammalian neuronal progenitors that can generate different neuron types in response to developmentally regulated steroid expression (Kucherenko et al., 2012).

One of the most evolutionary conserved pathways involving let-7 family is the let-7/Lin28 axis, which is critical to the neural cell fate decision and developmental progression as it serves as a switch in deciding whether NSCs will differentiate in mature neuron. In both *C. elegans* and mouse, the conserved RNA-binding protein Lin28 controls let-7 biogenesis and therefore let-7 mature levels. Lin28 selectively represses the maturation of let-7 family by binding to a region of the pre-let-7 called precursor element, and subsequently promoting its degradation. In addition, Lin28 contains let-7 binding sites, making its expression sensitive to let-7 levels in the cell (Rehfeld et al., 2015). During mouse neurogenesis, Lin28 is present through the neural tube until embryonic day 11.5 and its expression drops in committed radial glial progenitors. The transcription factor Sox2 (SRY box 2) has been placed upstream the Lin28/let-7 interaction, as it positively regulates Lin28 expression in NPs cells and promotes their proliferation (Cimadamore et al., 2013). Therefore, Lin28 and let-7 form a double-negative feedback loop that controls two mutually exclusive states, a stem-like state characterized by high levels of Lin28 and low levels of let-7 and a commitment-like state, characterized by low levels of Lin28 and high levels of let-7 (Rehfeld et al., 2015).

miR-125

miR-125 is another ancient miRNA that clusters together with miR-100 and let-7 in the same primary transcript (Rubio and Belles, 2013). miR-125 was found expressed together with miR-100 and let-7 in neurosecretory cells of the foregut of the bilaterian ancestor (Christodoulou et al., 2010). miR-125 is the *C. elegans* lin-4 homologue and is present in a single copy in *Drosophila*. In vertebrates, gene duplication events led to the formation of subfamilies with different chromosome locations and seed regions, miR-125-a and miR-125-b in humans and mouse and to a third subfamily, miR-125-c in zebrafish.

Similar to let-7, miR-125 has been shown to play a key role in developmental timing. In fact, mutant flies lacking miR-125 show widespread defects during metamorphosis, such as a delay in the neuromuscular junction formation. Moreover lin-4 has been shown to be involved in axon guidance during the formation of neuronal circuits in *C. elegans* (Zou et al., 2012). In vertebrates, the subfamily miR-125-b is the most highly enriched in the CNS in both zebrafish and mouse and its expression gradually accumulates during development in parallel with neuronal maturation, suggesting its key role in neuronal differentiation (Le et al., 2009; Smirnova et al., 2005). In fact, miR-125-b downregulates a series of proteins that have negative effects on differentiation and upregulates genes that promote neuronal differentiation. Thus, miR-125b plays a key role in the molecular cascade that contributes to the irreversible commitment of pluripotent human stem cells to the neural lineage (Boissart et al., 2012; Stappert et al., 2013). In addition, it has been reported that miR-125-b is expressed in neural stem/progenitor cells where it plays a key role in coordinating differentiation by repressing *Nestin*, an essential stem cells regulator. In doing so, miR-125-b suppresses the self-renewal of NSCs and promotes their differentiation and migration (Cui et al., 2012).

miR-9

miR-9 is one of the most highly expressed miRNAs in the vertebrate brain. Since its first appearance in the bilaterian line (Wheeler et al., 2009), miR-9 activity shows high versatility, exerting different and often opposite functions in a species- and cellular-specific context, while maintaining high sequence conservation. Duplication events led to the presence of multiple copies of miR-9 in the animal genome. For example, the *Drosophila* genome encodes five different copies (miR-9a-b-c, miR-79, and miR-4), whereas only 3 copies are present in the human genome (miR-9-1, miR-9-2, and miR-9-3). The presence of multiple independent copies can be linked to function-specificity between copies. Different model organisms show a similar spatiotemporal miR-9 expression pattern, predominantly associated with ventricular NPs areas in which its expression is dependent on the Notch signaling activity (Coolen et al., 2013). Several studies identified feedback loops between miR-9 and its target genes, some of which promote proliferation, others differentiation. The miR-9 mediated repression of transcription factors promoting proliferation explains its prodifferentiation effect on NPs. Within target, genes promoting progenitor proliferation is Zic5 (zinc finger transcription factor), Hes 1 (Notch signaling effector hairy and enhancer of split-1), the homeobox factor Gsx2, FoxG1 (Forkhead transcription factor), and the nuclear receptor Nr2e1/TLX (see Table 15.1). In addition, miR-9 has been shown to target the prodifferentiation gene elavl3 (ELAV-like neuron-specific RNA binging protein 3) (Coolen et al., 2012). Consequently, miR-9 regulates the expression of factors promoting antagonist fates, while maintaining neuronal progenitors in an ambivalent state.

In mammals, it has been largely demonstrated that miR-9 is implicated in neurogenesis and especially in regulation of NPs proliferation. In mouse, miR-9 expression starts in the brain around embryonic day 10 and increases during embryonic brain development, but decreases after birth (De Pietri Tonelli et al., 2008). Studies in miR-9-2 and miR-9-3 double knockouts show that miR-9 regulates both NPs proliferation and differentiation by regulating the expression of multiple transcription factors in a context-specific manner (Shibata et al., 2011). In this double mutant mouse, the increase of Foxg1 (Forkhead box protein G1) and Gsx2 (GS homeobox 2) expression levels leads to an increase in NPs proliferation in the pallial and subpallial region suggesting that miR-9 promotes early newborn neuron production by repressing Foxg1 and Gsx2 at early neurogenic stages (Shibata et al., 2011). In a previous study, Zhao and colleagues show that miR-9 targets the nuclear receptor Tlx (NR2E1) in mouse NSC isolated from adult mouse forebrains. Antisense knockdown of miR-9 leads to an increase of NSC proliferation, whereas the overexpression of miR-9 results in

Table 15.1 Conserved miRNAs and Their Functions During Neural Development

Conserved miRNA	Target Genes	Organism	Function During Neural Development	References
miR-100	Hox genes	*C. elegans*	Anterior-posterior patterning	Tehler et al. (2011)
		D. melanogaster	Anterior-posterior patterning	Tehler et al. (2011)
		D. rerio	Anterior-posterior patterning	Tehler et al. (2011)
		M. musculus	Anterior-posterior patterning	Tehler et al. (2011)
miR-10	wnt1/MSX1	*H. sapiens*	Anterior-posterior patterning	Jonsson et al. (2015)
lin-4	LIN-14	*C. elegans*	Control of axon guidance	Zou et al. (2012)
miR-125-b	Nestin	*R. rattus*	Promotes neuronal differentiation	Cui et al. (2012)
let-7	lin-41	*C. elegans*	Controls developmental timing	Hreinhart et al. (2000)
let-7-b	TLX	*M. musculus*	Promotes neuronal differentiation	Li et al. (2008)
	cyclinD1	*M. musculus*	Promotes neuronal differentiation	Zhao et al. (2010)
	Hmga2	*M. musculus*	Promotes neuronal differentiation	Nishino et al. (2008)
miR-9	senseless	*D. melanogaster*	Inhibits neurogenesis in nonneuronal tissues	Li et al. (2006)
	GSX2/FoxG1	*M. musculus*	Promotes progenitor proliferation	Coolen et al. (2012)
	TLX	*M. musculus*	Promotes progenitor proliferation	Shibata et al. (2011)
	Zinc5/Hes1	*D. rerio*	Promotes progenitor proliferation	Zhao et al. (2009)
	elavl3	*D. rerio*	Promotes progenitor proliferation	Coolen et al. (2012)
	hairy1	*X. tropicalis*	Promotes neuronal differentiation	Bonev et al. (2011)
miR-124	unc 13	*C. elegans*	No contribution to neurogenesis	Clark et al. (2010)
	PTBP1	*M. musculus*	Promotes neuronal differentiation	Makeyev et al. (2007)
	Sox9	*M. musculus*	Promotes neuronal differentiation	Cheng et al. (2009)
	REST/SCP1	*G. gallus*	Promotes neuronal differentiation	Visvanathan et al. (2007)
	NeuroD1	*X. tropicalis*	Promotes progenitor proliferation	Liu et al. (2011)

proliferation deficiency and precocious differentiation that can be rescued by introducing a Tlx expression vector that is not regulated by miR-9. In addition, Tlx can transcriptionally inhibit miR-9 genes, suggesting that miR-9 and Tlx form a negative feedback regulatory loop to control the balance of proliferation and differentiation of NSCs (Zhao et al., 2009).

In zebrafish, miR-9 expression is found in the telencephalon at 24 hours post fertilization and later spread throughout the CNS. In the developing brain, miR-9 is expressed in the neurogenic ventricular zone, while it is specifically excluded from boundary regions such as the MHB and the rhombomere boundaries containing pools of long-lasting progenitors. Later in neurogenesis, miR-9 controls the organizing activity and the progenitors' state of the MHB by targeting several components of the Fgf (Fibroblast growth factor) signaling pathway (Leucht et al., 2008). miR-9 is also involved in the regulation of embryonic NPs stages in the neurogenic areas of the zebrafish embryonic brain where it plays a key role in neural cell fate decision, by promoting the transition from nonneurogenic to neurogenic progenitors. Nonneurogenic progenitors express the notch effector

Hes1 that represses proneural genes such as *ascl1* (a vertebrate homolog of the *Drosophila Achaete/Scute* genes). When these cells start expressing miR-9, it dampens Her1 expression and promotes these progenitors to an ambivalent state in which they can either proliferate to maintain the progenitor pool or differentiate into newborn neurons (Coolen et al., 2012). Specific miR-9 levels are therefore important to promote neurogenesis. In the developing zebrafish hindbrain, miR-9 biogenesis levels are indirectly controlled by another miRNA, miR-107. miR-107 expression gradually increases following the differentiation of neuronal progenitors into neurons. By targeting *Dicer*, miR-107 gradually dampens miR-9 expression and modulates its proproliferation and prodifferentiation activity in these cell types, thus allowing the proper progression of neurogenesis (Ristori et al., 2015). Moreover, miR-9 targets such as Tlx (Nr2e1), REST (RE1-Silencing Transcription factor), and Hes1 can repress miR-9 transcription through negative feedback loops modulating the effect of miR-9 on the network (Bonev et al., 2012; Laneve et al., 2010; Zhao et al., 2009). In *Xenopus*, miR-9 shows a regional-specific effect on NPs. It promotes neurogenesis in the forebrain−hindbrain boundary by targeting *hairy1* (member of the *hes* family genes). *hairy1* mediates the effect of miR-9 on progenitor proliferation through Fgf8 signaling in the forebrain and the *Wnt* signaling in the hindbrain (Bonev et al., 2011).

In *Drosophila*, miR-9-a controls the development of peripheral nervous sensory organ (Li et al., 2006). However, its role seems to be different from vertebrates. miR-9-a expression in *Drosophila* is not as prominent as in vertebrates, and it is not expressed in the sensory organ precursors that will give rise to neurons, but in nonneuronal epidermal cells instead. Studies on miR-9a mutants showed that the main function of this miRNA is to inhibit neurogenesis in nonneuronal tissues by suppressing the proneural gene *senseless* during neuronal precursors specification (Li et al., 2006). *Drosophila* sensory organ precursors express miR-7, which complements miR-9a function by promoting the expression of the proneural genes *Atonal* and *Senseless* and therefore neurogenesis (Li et al., 2009).

miR-124

miR-124 is specifically expressed in the CNS and is one of the most abundant miRNAs in the adult brain (Lagos-Quintana et al., 2002). During neurogenesis, this miRNA is expressed at low levels in progenitor cells and its expression gradually increases and accumulates in differentiating and mature neurons (Deo et al., 2006). The neuronal pattern expression of miR-124 is conserved from nematodes to humans; however, it does not seem to be massively involved in *C. elegans* neurogenesis. It is expressed only by a subset of sensory neurons where it contributes to regulating gene expression in this cell type, but its contribution on neurogenesis has not been shown. This suggests that this miRNA might be evolved to have different targets and functions (Clark et al., 2010). In general, miR-124 has been shown to promote cell differentiation and to inhibit cell proliferation. Different mechanisms for miR-124 neuronal differentiation induction have been identified (Table 9.1).

In mouse embryonic development, miR-124 promotes neuronal differentiation by targeting PTBP1 (Polypirimidine Tract Binding Protein 1), a repressor of alternative pre-mRNA splicing (Makeyev et al., 2007). During adult neurogenesis, miR-124 promotes neuronal fate by targeting Sox9 (SRY box 9), an important transcription factor implicated in glial cell specification (Cheng et al., 2009). During chick embryonic neurogenesis, miR-124 suppresses the function of REST and the phosphatase SCP1 (Small-C-terminal Phosphatase 1) complex known to silence proneuronal genes in nonneuronal cells. Interestingly miR-124 as well as miR-9 expression levels are suppressed by the REST complex in nonneuronal cells preventing neuronal differentiation. During neurogenesis, miR-124 feeds back to the REST complex by targeting SCP1, promoting neurogenesis. This suggests that the miR-124/REST-SCP1 axis has a crucial role in determining neuronal fate (Visvanathan et al., 2007).

In contrast, miR-124 exerts a different role during *Xenopus* eye development by promoting cell proliferation and repressing neurogenesis. At early stages of neuronal development, miR-124 plays an anti-differentiation role by targeting the transcription factor NeuroD1 (Neurogenic Differentiation 1) (Liu et al., 2011). This finding supports against the importance of the temporal and spatial context of the miRNA expression pattern, suggesting that miR-124 could have specific functions at different developmental stages by enhancing proliferation and inhibiting neurogenesis at the onset of neurogenesis while playing an opposite role in later stages.

CONCLUSIONS

Comparative embryology studies have shown how the CNS develops from simple structures such as *nerve nets* found in Cnidaria to more complex structures such as the mammalian convoluted brains. Even if different taxa show a wide anatomical diversification, molecular genetic studies reveal a striking conservation in the expression patterns and functions of genes involved in neural development. Moreover, posttranscriptional regulation mechanisms, such as miRNA regulation, that have evolved to confer robustness to developmental programs, are highly conserved through the animal kingdom. miRNAs have emerged as key regulators of neural development. In particular, "ancient" miRNAs are wired in evolutionary conserved neuronal regulatory networks that control different aspects of the neurogenic process in both invertebrates and vertebrates. During evolution, new miRNAs appeared and acquired specific functions that correlate with an increase in morphological complexity of multicellular organisms.

In this chapter, we showed the importance of some of the most highly conserved miRNAs in maintaining the balance between stem cell self-renewal, proliferation, and differentiation during embryonic development and adult life. miRNAs are expressed in a very precise regional, cellular, subcellular, and temporal patterns in both developing and adult brains, reflecting their influence in controlling different aspects of neurogenesis in a context-dependent manner in both vertebrates and invertebrates. To date, there has been difficulty in attributing phenotypes observed in miRNA knockouts in vivo, as most of the mutant models are viable with no obvious developmental and physiological defects. This could be related to the fact that miRNAs are incorporated into complex transcriptional circuits regulated by different factors that can compensate for miRNA loss. Moreover, the presence of miRNA families makes it hard to create efficient knockouts.

Therefore, novel technologies are greatly needed to address the function of miRNAs in different aspects of neural development. Among these innovative approaches, CRISPR/Cas9 genome editing technology is one of the most promising. This genome editing strategy allows the simultaneous targeting of genes on different chromosomes, thus allowing the generation of miRNA families knockout organisms. Therefore, complex regulatory pathways can be dissected to have a better understanding of novel and conserved miRNA-mediated mechanisms crucial for embryonic and adult neurogenesis. miRNAs have been found associated with several neurodegenerative diseases and neuropsychiatric conditions. In fact, some brain-enriched miRNAs, such as miR-9, miR-107, and miR-124, have been consistently identified as dysregulated in Alzheimer's disease (Femminella et al., 2015). Moreover, a subset of miRNAs, such as miR-107, miR-219, and miR-132, has been shown to be implicated in the dysfunction of glutamatergic neurotransmission in schizophrenia (Zhang et al., 2015). Recent studies investigate the use of miRNAs as a biomarker for CNS malignancies, neurological, and psychiatric diseases. Even with some limitations this approach could allow the early detection of the disease and therefore increasing the efficacy and outcome of treatments (Cairns, 2015; Rao et al., 2013; Van Giau and An, 2016; Wei et al., 2015).

ACKNOWLEDGMENTS

We thank Dionna Kasper and Saliha Yilmaz for helpful comments.

REFERENCES

Arendt, D., Denes, A.S., Jekely, G., Tessmar-Raible, K., 2008. The evolution of nervous system centralization. Philos. Trans. R. Soc. London, Ser. B: Biol. Sci. 363 (1496), 1523−1528.

Bak, M., Silahtaroglu, A., Moller, M., Christensen, M., Rath, M.F., Skryabin, B., et al., 2008. MicroRNA expression in the adult mouse central nervous system. RNA 14 (3), 432−444.

Bartel, D.P., 2004. MicroRNAs: genomics, biogenesis, mechanism, and function. Cell 116 (2), 281−297.

Bartel, D.P., 2009. MicroRNAs: target recognition and regulatory functions. Cell 136 (2), 215−233.

Bartel, D.P., Chen, C.Z., 2004. Micromanagers of gene expression: the potentially widespread influence of metazoan microRNAs. Nat. Rev. Genet. 5 (5), 396−400.

Berdnik, D., Fan, A.P., Potter, C.J., Luo, L., 2008. MicroRNA processing pathway regulates olfactory neuron morphogenesis. Curr. Biol. 18 (22), 1754−1759.

Bernstein, E., Kim, S.Y., Carmell, M.A., Murchison, E.P., Alcorn, H., Li, M.Z., et al., 2003. Dicer is essential for mouse development. Nat. Genet. 35 (3), 215−217.

Boissart, C., Nissan, X., Giraud-Triboult, K., Peschanski, M., Benchoua, A., 2012. miR-125 potentiates early neural specification of human embryonic stem cells. Development 139 (7), 1247−1257.

Bonev, B., Pisco, A., Papalopulu, N., 2011. MicroRNA-9 reveals regional diversity of neural progenitors along the anterior posterior axis. Dev. Cell 20 (1), 19−32.

Bonev, B., Stanley, P., Papalopulu, N., 2012. MicroRNA-9 Modulates Hes1 ultradian oscillations by forming a double-negative feedback loop. Cell Rep. 2 (1), 10−18.

Brand, A.H., Livesey, F.J., 2011. Neural stem cell biology in vertebrates and invertebrates: more alike than different? Neuron 70 (4), 719−729.

Cairns, M.J., 2015. Circulating miRNA biomarkers for schizophrenia? Am. J. Psychiatry 172 (11), 1059−1061.

Cheng, L.C., Pastrana, E., Tavazoie, M., Doetsch, F., 2009. miR-124 regulates adult neurogenesis in the subventricular zone stem cell niche. Nat. Neurosci. 12 (4), 399−408.

Christodoulou, F., Raible, F., Tomer, R., Simakov, O., Trachana, K., Klaus, S., et al., 2010. Ancient animal microRNAs and the evolution of tissue identity. Nature 463 (7284), 1084−1088.

Cimadamore, F., Amador-Arjona, A., Chen, C., Huang, C.T., Terskikh, A.V., 2013. SOX2-LIN28/let-7 pathway regulates proliferation and neurogenesis in neural precursors. Proc. Natl. Acad. Sci. U. S. A. 110 (32), E3017−E3026.

Clark, A.M., Goldstein, L.D., Tevlin, M., Tavare, S., Shaham, S., Miska, E.A., 2010. The microRNA miR-124 controls gene expression in the sensory nervous system of *Caenorhabditis elegans*. Nucleic Acids Res. 38 (11), 3780−3793.

Coolen, M., Katz, S., Bally-Cuif, L., 2013. miR-9: a versatile regulator of neurogenesis. Front. Cell. Neurosci. 7, 220.

Coolen, M., Thieffry, D., Drivenes, O., Becker, T.S., Bally-Cuif, L., 2012. miR-9 controls the timing of neurogenesis through the direct inhibition of antagonistic factors. Dev. Cell 22 (5), 1052−1064.

Cui, Y., Xiao, Z., Han, J., Sun, J., Ding, W., Zhao, Y., et al., 2012. MiR-125b orchestrates cell proliferation, differentiation and migration in neural stem/progenitor cells by targeting Nestin. BMC Neurosci. 13, 116.

De Pietri Tonelli, D., Pulvers, J.N., Haffner, C., Murchison, E.P., Hannon, G.J., Huttner, W.B., 2008. miRNAs are essential for survival and differentiation of newborn neurons but not for expansion of neural progenitors during early neurogenesis in the mouse embryonic neocortex. Development 135 (23), 3911−3921.

Deo, M., Yu, J.Y., Chung, K.H., Tippens, M., Turner, D.L., 2006. Detection of mammalian microRNA expression by in situ hybridization with RNA oligonucleotides. Dev. Dyn. 235 (9), 2538−2548.

Ebert, M.S., Sharp, P.A., 2012. Roles for microRNAs in conferring robustness to biological processes. Cell 149 (3), 515−524.

Femminella, G.D., Ferrara, N., Rengo, G., 2015. The emerging role of microRNAs in Alzheimer's disease. Front. Physiol. 6, 40.

Giraldez, A.J., Cinalli, R.M., Glasner, M.E., Enright, A.J., Thomson, J.M., Baskerville, S., et al., 2005. MicroRNAs regulate brain morphogenesis in zebrafish. Science 308 (5723), 833−838.

Grimson, A., Srivastava, M., Fahey, B., Woodcroft, B.J., Chiang, H.R., King, N., et al., 2008. Early origins and evolution of microRNAs and Piwi-interacting RNAs in animals. Nature 455 (7217), 1193−1197.

Hartenstein, V., Stollewerk, A., 2015. The evolution of early neurogenesis. Dev. Cell 32 (4), 390−407.

Hartenstein, V., Wodarz, A., 2013. Initial neurogenesis in *Drosophila*. Wiley Interdiscip. Rev. Dev. Biol. 2 (5), 701−721.

Hertel, J., Lindemeyer, M., Missal, K., Fried, C., Tanzer, A., Flamm, C., et al., 2006. Students of Bioinformatics Computer Labs 2004 and 2005, The expansion of the metazoan microRNA repertoire. BMC Genomics 7, 25.

Hertel, J., Stadler, P.F., 2015. The expansion of animal microRNA families revisited. Life (Basel) 5 (1), 905−920.

Hill, R.W., Wyse, G.A., Anderson, M., 2012. Animal Physiology. Third ed. Sinauer Associates, Inc., pp. 1−800.

Jonsson, M.E., Nelander Wahlestedt, J., Akerblom, M., Kirkeby, A., Malmevik, J., Brattaas, P.L., et al., 2015. Comprehensive analysis of microRNA expression in regionalized human neural progenitor cells reveals microRNA-10 as a caudalizing factor. Development 142 (18), 3166−3177.

Kelava, I., Rentzsch, F., Technau, U., 2015. Evolution of eumetazoan nervous systems: insights from cnidarians. Philos. Trans. R. Soc. London, Ser. B: Biol. Sci. 370, 1684.

Kitano, H., 2004. Biological robustness. Nat. Rev. Genet. 5 (11), 826−837.

Kosik, K.S., 2006. The neuronal microRNA system. Nat. Rev. Neurosci. 7 (12), 911−920.

Kosik, K.S., 2009. MicroRNAs tell an evo-devo story. Nat. Rev. Neurosci. 10 (10), 754−759.

Kosik, K.S., Krichevsky, A.M., 2005. The elegance of the microRNAs: a neuronal perspective. Neuron 47 (6), 779−782.

Krol, J., Loedige, I., Filipowicz, W., 2010. The widespread regulation of microRNA biogenesis, function and decay. Nat. Rev. Genet. 11 (9), 597−610.

Kucherenko, M.M., Barth, J., Fiala, A., Shcherbata, H.R., 2012. Steroid-induced microRNA let-7 acts as a spatio-temporal code for neuronal cell fate in the developing *Drosophila* brain. EMBO J. 31 (24), 4511−4523.

Lagos-Quintana, M., Rauhut, R., Yalcin, A., Meyer, J., Lendeckel, W., Tuschl, T., 2002. Identification of tissue-specific microRNAs from mouse. Curr. Biol. 12 (9), 735—739.

Laneve, P., Gioia, U., Andriotto, A., Moretti, F., Bozzoni, I., Caffarelli, E., 2010. A minicircuitry involving REST and CREB controls miR-9-2 expression during human neuronal differentiation. Nucleic Acids Res. 38 (20), 6895—6905.

Lang, M.F., Shi, Y., 2012. Dynamic roles of microRNAs in neurogenesis. Front. Neurosci. 6, 71.

Le, M.T., Xie, H., Zhou, B., Chia, P.H., Rizk, P., Um, M., et al., 2009. MicroRNA-125b promotes neuronal differentiation in human cells by repressing multiple targets. Mol. Cell. Biol. 29 (19), 5290—5305.

Leucht, C., Stigloher, C., Wizenmann, A., Klafke, R., Folchert, A., Bally-Cuif, L., 2008. MicroRNA-9 directs late organizer activity of the midbrain—hindbrain boundary. Nat. Neurosci. 11 (6), 641—648.

Li, W., Sun, G., Yang, S., Qu, Q., Nakashima, K., Shi, Y., 2008. Nuclear receptor TLX regulates cell cycle progression in neural stem cells of the developing brain. Mol. Endocrinol. 22 (1), 56—64.

Li, X., Cassidy, J.J., Reinke, C.A., Fischboeck, S., Carthew, R.W., 2009. A microRNA imparts robustness against environmental fluctuation during development. Cell 137 (2), 273—282.

Li, Y., Wang, F., Lee, J.A., Gao, F.B., 2006. MicroRNA-9a ensures the precise specification of sensory organ precursors in *Drosophila*. Genes Dev. 20 (20), 2793—2805.

Lin, S.Y., Johnson, S.M., Abraham, M., Vella, M.C., Pasquinelli, A., Gamberi, C., et al., 2003. The *C. elegans* hunchback homolog, hbl-1, controls temporal patterning and is a probable microRNA target. Dev. Cell 4 (5), 639—650.

Liu, K., Liu, Y., Mo, W., Qiu, R., Wang, X., Wu, J.Y., et al., 2011. MiR-124 regulates early neurogenesis in the optic vesicle and forebrain, targeting NeuroD1. Nucleic Acids Res. 39 (7), 2869—2879.

Makeyev, E.V., Zhang, J., Carrasco, M.A., Maniatis, T., 2007. The MicroRNA miR-124 promotes neuronal differentiation by triggering brain-specific alternative pre-mRNA splicing. Mol. Cell 27 (3), 435—448.

Mohammed, J., Bortolamiol-Becet, D., Flynt, A.S., Gronau, I., Siepel, A., Lai, E.C., 2014. Adaptive evolution of testis-specific, recently evolved, clustered miRNAs in *Drosophila*. RNA 20 (8), 1195—1209.

Nishino, J., Kim, I., Chada, K., Morrison, S.J., 2008. Hmga2 promotes neural stem cell self-renewal in young but not old mice by reducing p16Ink4a and p19Arf Expression. Cell 135 (2), 227—239.

Olsen, L., Klausen, M., Helboe, L., Nielsen, F.C., Werge, T., 2009. MicroRNAs show mutually exclusive expression patterns in the brain of adult male rats. PLoS ONE 4 (10), e7225.

Osella, M., Bosia, C., Cora, D., Caselle, M., 2011. The role of incoherent microRNA-mediated feedforward loops in noise buffering. PLoS Comput. Biol. 7 (3), e1001101.

Pasquinelli, A.E., Reinhart, B.J., Slack, F., Martindale, M.Q., Kuroda, M.I., Maller, B., et al., 2000. Conservation of the sequence and temporal expression of let-7 heterochronic regulatory RNA. Nature 408 (6808), 86—89.

Quan, X.J., Hassan, B.A., 2005. From skin to nerve: flies, vertebrates and the first helix. Cell. Mol. Life Sci. 62 (18), 2036—2049.

Rao, P., Benito, E., Fischer, A., 2013. MicroRNAs as biomarkers for CNS disease. Front. Mol. Neurosci. 6, 39.

Rehfeld, F., Rohde, A.M., Nguyen, D.T., Wulczyn, F.G., 2015. Lin28 and let-7: ancient milestones on the road from pluripotency to neurogenesis. Cell Tissue Res. 359 (1), 145—160.

Reinhart, B.J., Slack, F.J., Basson, M., Pasquinelli, A.E., Bettinger, J.C., Rougvie, A.E., et al., 2000. The 21-nucleotide let-7 RNA regulates developmental timing in *Caenorhabditis elegans*. Nature 403 (6772), 901—906.

Richards, G.S., Rentzsch, F., 2014. Transgenic analysis of a SoxB gene reveals neural progenitor cells in the cnidarian *Nematostella vectensis*. Development 141 (24), 4681—4689.

Ristori, E., Lopez-Ramirez, M.A., Narayanan, A., Hill-Teran, G., Moro, A., Calvo, C.F., et al., 2015. A Dicer-miR-107 interaction regulates biogenesis of specific miRNAs crucial for neurogenesis. Dev. Cell 32 (5), 546—560.

Rubio, M., Belles, X., 2013. Subtle roles of microRNAs let-7, miR-100 and miR-125 on wing morphogenesis in hemimetabolan metamorphosis. J. Insect Physiol. 59 (11), 1089—1094.

Sasai, Y., 2001. Roles of Sox factors in neural determination: conserved signaling in evolution? Int. J. Dev. Biol. 45 (1), 321–326.

Saurat, N., Andersson, T., Vasistha, N.A., Molnar, Z., Livesey, F.J., 2013. Dicer is required for neural stem cell multipotency and lineage progression during cerebral cortex development. Neural Dev. 8, 14.

Sempere, L.F., Cole, C.N., McPeek, M.A., Peterson, K.J., 2006. The phylogenetic distribution of metazoan microRNAs: insights into evolutionary complexity and constraint. J. Exp. Zool. B: Mol. Dev. Evol. 306 (6), 575–588.

Sempere, L.F., Martinez, P., Cole, C., Baguna, J., Peterson, K.J., 2007. Phylogenetic distribution of microRNAs supports the basal position of acoel flatworms and the polyphyly of Platyhelminthes. Evol. Dev. 9 (5), 409–415.

Shibata, M., Nakao, H., Kiyonari, H., Abe, T., Aizawa, S., 2011. MicroRNA-9 regulates neurogenesis in mouse telencephalon by targeting multiple transcription factors. J. Neurosci. 31 (9), 3407–3422.

Siciliano, V., Garzilli, I., Fracassi, C., Criscuolo, S., Ventre, S., di Bernardo, D., 2013. MiRNAs confer phenotypic robustness to gene networks by suppressing biological noise. Nat. Commun. 4, 2364.

Smirnova, L., Grafe, A., Seiler, A., Schumacher, S., Nitsch, R., Wulczyn, F.G., 2005. Regulation of miRNA expression during neural cell specification. Eur. J. Neurosci. 21 (6), 1469–1477.

Stappert, L., Borghese, L., Roese-Koerner, B., Weinhold, S., Koch, P., Terstegge, S., et al., 2013. MicroRNA-based promotion of human neuronal differentiation and subtype specification. PLoS ONE 8 (3), e59011.

Tehler, D., Hoyland-Kroghsbo, N.M., Lund, A.H., 2011. The miR-10 microRNA precursor family. RNA Biol. 8 (5), 728–734.

Tsang, J., Zhu, J., van Oudenaarden, A., 2007. MicroRNA-mediated feedback and feedforward loops are recurrent network motifs in mammals. Mol. Cell 26 (5), 753–767.

Van Giau, V., An, S.S., 2016. Emergence of exosomal miRNAs as a diagnostic biomarker for Alzheimer's disease. J. Neurol. Sci. 360, 141–152.

Visvanathan, J., Lee, S., Lee, B., Lee, J.W., Lee, S.K., 2007. The microRNA miR-124 antagonizes the anti-neural REST/SCP1 pathway during embryonic CNS development. Genes Dev. 21 (7), 744–749.

Wei, H., Yuan, Y., Liu, S., Wang, C., Yang, F., Lu, Z., et al., 2015. Detection of circulating miRNA levels in schizophrenia. Am. J. Psychiatry. 172 (11), 1141–1147.

Wheeler, B.M., Heimberg, A.M., Moy, V.N., Sperling, E.A., Holstein, T.W., Heber, S., et al., 2009. The deep evolution of metazoan microRNAs. Evol. Dev. 11 (1), 50–68.

Wulczyn, F.G., Smirnova, L., Rybak, A., Brandt, C., Kwidzinski, E., Ninnemann, O., et al., 2007. Post-transcriptional regulation of the let-7 microRNA during neural cell specification. FASEB J. 21 (2), 415–426.

Zhang, Y., Fan, M., Wang, Q., He, G., Fu, Y., Li, H., et al., 2015. Polymorphisms in microRNA genes and genes involving in NMDAR signaling and schizophrenia: a case-control study in Chinese Han population. Sci. Rep. 5, 12984.

Zhao, C., Sun, G., Li, S., Lang, M.F., Yang, S., Li, W., et al., 2010. MicroRNA let-7b regulates neural stem cell proliferation and differentiation by targeting nuclear receptor TLX signaling. Proc. Natl. Acad. Sci. U. S. A. 107 (5), 1876–1881.

Zhao, C., Sun, G., Li, S., Shi, Y., 2009. A feedback regulatory loop involving microRNA-9 and nuclear receptor TLX in neural stem cell fate determination. Nat. Struct. Mol. Biol. 16 (4), 365–371.

Zhu, Y., Skogerbo, G., Ning, Q., Wang, Z., Li, B., Yang, S., et al., 2012. Evolutionary relationships between miRNA genes and their activity. BMC Genomics 13, 718.

Zou, Y., Chiu, H., Domenger, D., Chuang, C.F., Chang, C., 2012. The lin-4 microRNA targets the LIN-14 transcription factor to inhibit netrin-mediated axon attraction. Sci. Signal. 5 (228), ra43.

MicroRNA AND NEOCORTICAL EVOLUTION

16

Thomas Pratt[1], Tomasz Jan Nowakowski[2] and David Jonathan Price[1]

[1]*University of Edinburgh, Edinburgh, United Kingdom*
[2]*University of California, San Francisco, San Francisco, CA, United States*

INTRODUCTION

Systematic comparative anatomy has given insights into the evolution of the human condition from its primate and more distant ancestors. This has revealed profound changes in the size and structure of the cerebral cortex as well as equally profound changes elsewhere, for example, in limbs for upright walking. The recent sequencing of large numbers of human and nonhuman genomes and transcriptomes, coupled with efforts to correlate DNA sequence (genotype) to function (phenotype), has provided a molecular biology perspective to this classical problem.

MOLECULAR GENETIC CHANGES UNDERPINNING HUMAN EVOLUTION

Genomic DNA can be categorized based on whether it directs the transcription of RNA (promoter and enhancer sequences), serves as a template to be transcribed into RNA, or has no known function ("junk" DNA). Transcribed RNAs can be further subdivided into protein-coding mRNA, comprising translated open-reading frames and untranslated regions (UTRs), and noncoding RNA (ncRNA) which includes pseudogene mRNA, microRNA (miRNA), and long noncoding RNA (lncRNA) (Fig. 16.1A). In humans, $<2\%$ of the 3×10^9 base-pair genome directly codes for protein (Taft et al., 2007). Some estimates suggest that a large proportion of the remaining $>98\%$ is actively transcribed even though it is unlikely that it will ever be translated into protein (Clark et al., 2011). These observations prompt fundamental questions about how much of the genome has biochemical function and is subject to natural selection. Although it is very difficult to accurately count the numbers of mutations driving phenotypic divergence of humans and chimpanzees from their common ancestor, evolutionary genetics calculations incorporating a number of factors including rates that advantageous mutations appear and are fixed in populations and the number of generations since the human lineage diverged from its nonhuman relatives suggest that tens of mutations were sufficient, a significant proportion of which are likely to have contributed to the evolution of the cerebral cortex (Somel et al., 2013). In light of this, it is important to emphasize here that while a handful of mutations in all categories of DNA (protein-coding/regulatory/ncRNA)

Essentials of Noncoding RNA in Neuroscience. DOI: http://dx.doi.org/10.1016/B978-0-12-804402-5.00016-9

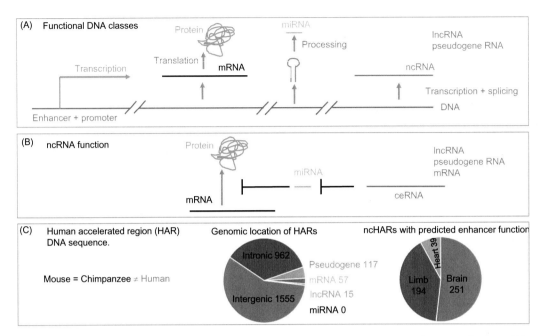

FIGURE 16.1 A summary of various classes of genomic DNA and RNA, and human accelerated regions (HARS)

(A) DNA sequence involved in gene expression can be divided into four functional categories (1) as regulatory elements: enhancers and promoter DNA sequences that drive the transcription of other DNA sequence into RNA; (2) DNA sequence that is transcribed to give mRNA which is translated into protein; (3) DNA sequence that is transcribed and processed to give noncoding miRNA; (4) DNA sequence that is transcribed to give other types of noncoding RNA (ncRNA) which will never be translated into protein. NcRNA is a heterogenous category which includes pseudogenes which have been derived from ancestral protein-coding genes and long noncoding RNAs (lncRNAs) which have not. All transcript categories may also be subject to RNA splicing to remove introns. (B) Functions of ncRNAs in regulating protein translation from mRNA. miRNAs bind to mRNAs and suppress their translation. Competing endogenous RNA (ceRNA) describes an RNA sharing miRNA-binding sites with an mRNA, so competing with it for binding miRNA molecules. CeRNA titrates miRNA away from the mRNA target, so depressing miRNA suppression and positively regulating translation of the mRNA to protein. (C) HARs describe DNA sequences that have been conserved throughout mammalian evolution, e.g., are shared between mouse and chimpanzee, but have changed in the human lineage making them candidates for driving evolution of the human phenotype. Of HARs identified to date, 92% map to intergenic or intronic regions with the remainder mapping to mature pseudogene, mRNA, or lncRNA transcripts. None map to miRNA genes. Bioinformatic analysis predicts that many HARs function as enhancer elements driving transcription in the brain, limb, and heart and experimental validation of a subset of these predicted enhancers found similar proportions driving brain, limb, and heart expression.

Data adapted from Pollard K.S., Salama S.R., Lambert N., et al.: An RNA gene expressed during cortical development evolved rapidly in humans, Nature 443:167–172, 2006; Prabhakar S., Noonan J.P., Paabo S., Rubin, E.M.: Accelerated evolution of conserved noncoding sequences in humans, Science 314:786, 2006; Bird C.P., Stranger B.E., Liu M., et al.: Fast-evolving noncoding sequences in the human genome, Genome Biol. 8:R118, 2007; Bush E.C., Lahn B.T.: A genome-wide screen for noncoding elements important in primate evolution, BMC Evol. Biol. 8:17, 2008; Lindblad-Toh K., Garber M., Zuk O., et al.: A high-resolution map of human evolutionary constraint using 29 mammals, Nature 478:476–482, 2011; Capra J.A., Erwin G.D., McKinsey G., et al.: Many human accelerated regions are developmental enhancers, Philos. Trans. R. Soc. Lond. Ser. B, Biol. Sci. 368: 20130025, 2013.

have been proposed to have driven the evolution of the human phenotype, proof is very difficult to obtain, and it is extremely likely that the bulk of the important changes remain to be discovered.

In the decade that followed the deciphering of the genetic code in the early to mid-1960s, it was proposed that the phenotypic diversity of living organisms might arise as much from differences in systems regulating the levels of production of specific proteins as from differences in the proteins themselves. Britten and Davidson (1969) stated, *"At higher grades of organization, evolution might indeed be considered principally in terms of changes in the regulatory systems."* This view was fueled by the discovery a few years later that the amino acid sequences of homologous proteins from humans and chimpanzees differed by less than 1%, which appeared insufficient to account for the extensive morphological, physiological, and behavioral divergence between these two species (King and Wilson, 1975). King and Wilson argued that a *"relatively small number of genetic changes in systems controlling the expression of genes may account for the major organismal differences between humans and chimpanzees."* The subsequent discovery that high levels of conservation of protein-encoding genes are widespread across the entire animal kingdom has continued to reinforce the view that evolutionary changes in gene regulation were essential for constructing divergent traits using conserved genes as a substrate (Khaitovich et al., 2006; Geschwind and Rakic, 2013; Rakic, 2009; Rakic et al., 2009; Prescott et al., 2015).

Analysis of ancient texts teaches us that for thousands of years much of humanity has considered itself set well apart from all other animals including its closest primate relatives. A more recent version of this hypothesis suggests that the divergence of humans and chimpanzees from their last common ancestor $\sim$6 million years ago was accompanied by accelerated changes to human DNA sequence. Whole-genome sequencing of many species, including humans and nonhuman primates, has allowed regions of accelerated change to be identified, providing a framework for a systematic assessment of the contribution of protein-coding and noncoding sequence changes to human evolution. Sophisticated computational algorithms have been developed to identify genomic loci that exhibit human-specific changes, including large scale human-specific duplications and deletions hypothesized to have altered the regulatory landscape of the human genome (McLean et al., 2011; Cheng et al., 2005; Bailey et al., 2002).

Human accelerated regions (HARs) are genomic sequences conserved between mammalian species, including chimpanzees and mice, but different in the human (Pollard et al., 2006; Lindblad-Toh et al., 2011). The vast majority of HARs, 92%, map to nonprotein-coding regions, strongly suggesting that evolution of regions regulating protein expression made a major contribution to human evolution. Other studies concentrated on identifying accelerated changes in noncoding regions variously termed human accelerated conserved noncoding sequences or accelerated noncoding conserved sequences (ANCs) (Prabhakar et al., 2006; Bird et al., 2007; Bush and Lahn, 2008). For simplicity, we will describe these changes as either HARs (all the human accelerated sequences) or ncHARS (human-accelerated sequences with no known protein-coding function). Merging these lists gives 2615 HARs, 33 of which are common to all studies, that remained conserved throughout mammalian evolution and then evolved rapidly in humans (Capra et al., 2013). Mapping the ncHARS to the human genome (Fig 16.1C) reveals that 962 (36%) are intronic, 117 (4%) overlap a pseudogene, 15 (0.5%) overlap an exon on a lncRNA, and the remaining 59% are intergenic. Of particular relevance to this review is the finding that none of the HAR sequences overlap a miRNA (Capra et al., 2013). They may, however, be involved in regulating miRNA functions; this will be discussed further below. Recent work has suggested a role for these elements in regulating the development of cognitive traits in humans (Doan et al., 2016).

The list of HARs provides a good starting point for understanding changes that drove human evolution. However, ascribing function to DNA sequence elements is not straightforward and may be biased because some functions are easier to investigate experimentally than others. The fact that the vast majority of HARs map to nonprotein-coding regions certainly suggests that human evolution has been driven by changes in gene regulation, as originally proposed over 40 years ago (King and Wilson, 1975). However, not all of this rapid evolution affects the cerebral cortex. HACSN1 is an ncHAR that functions as an enhancer driving gene expression in the developing limb, suggesting a role in the evolution of upright walking rather than in the cerebral cortex (Prabhakar et al., 2008). Even where HAR function maps to cerebral cortex, as in the case of ncRNA HAR1F expressed in Cajal−Retzius neurons during cortical development, the mechanism is not necessarily clear (Pollard et al., 2006). In fact, the largest systematic study of HAR function to date used a combination of bioinformatics and functional assays and found that the bulk of HARs had enhancer activity with similarly high proportions driving gene expression in the developing limb as in the developing brain, a result which satisfyingly mirrored the results of classical comparative anatomy studies (Capra et al., 2013) (Fig 16.1C). Parallel approaches developed to identify and functionally analyze active enhancer marks in the tissues of interest, including developing brain, are beginning to shed further light on the divergent use of enhancer elements that may have been instrumental in evolutionary change (Reilly et al., 2015; Visel et al., 2009; Prescott et al., 2015).

We shall next summarize our understanding of the major species-specific changes in the developmental mechanisms that generate the cerebral neocortex, with particular attention to differences between rodents and primates. These comparisons suggest substrates for evolutionary change. We shall then consider how mutations, including HARs, affecting the sequence and copy-number of nonprotein-coding sequences might have contributed to the evolution of the human cerebral cortex via RNA-dependent mechanisms.

EVOLUTION OF THE MECHANISMS OF CEREBRAL NEOCORTICAL DEVELOPMENT

The six-layered neocortex is unique to mammals. In humans, the neocortex has expanded massively to become the largest component of the brain. The neocortex probably evolved from an earlier reptilian-like dorsal cortex comprising a single layer of excitatory, pyramidal neurons, and intermingled inhibitory neurons subdivided into only a few cortical areas. The cerebral cortex of extant reptiles has three layers, as do those components of the mammalian cerebral cortex that are phylogenetically older than the neocortex, such as the hippocampus. The fossil record suggests that early mammals had small brains and that only a small proportion of their cerebral cortex was six-layered neocortex (Molnár et al., 2014). Primates, which started to evolve along a separate lineage from rodents 65−85 million years ago (O'Leary et al., 2013), showed extensive enlargement of the neocortex and an increase in the number of neocortical areas (Kaas, 2013).

Most of our knowledge of species-specific neocortical developmental mechanisms comes from research on rodents and primates. Despite great differences in their sizes, there are numerous similarities in brain structure and function between members of these orders. They include the conserved laminar structure of the cerebral cortex and its regionalization into major functionally distinct areas with characteristic patterns of connectivity. Many of the fundamental mechanisms of cortical development are also conserved. Newly formed neurons migrate from cortical progenitor zones lining its inner luminal surface (its apical surface in the nomenclature generally applied to epithelia) to adopt cortical positions related to their birthdate. Here they undergo terminal differentiation, elaborating dendrites and extending axons to establish connections and form cortical circuitry (Rakic, 2009).

Important differences between rodents and primates appear from early stages of embryonic development (Rakic, 2009). Before the onset of neurogenesis, the sheet of cortical neuroepithelium already varies greatly in size across species, suggesting tangential expansion of the neuroepithelium as a mechanism for evolutionary expansion of the cortex. In mouse, the production of cortical neurons begins about 10 days after conception and continues for about 8 days (Gillies and Price, 1993; Levers et al., 2001). In humans and other primates such as the macaque monkey, cortical neurogenesis occurs over many weeks, starting at about 35 days postconception and finishing around 105 days postconception (Bystron et al., 2008). One very striking difference between the events that generate the cortex of primates and rodents is the time it takes progenitor cells to go through their cell cycles. Primate cell cycle times, which are very similar in human and nonhuman primates, can be up to five times longer than in rodents at corresponding developmental stages (Kornack and Rakic 1998; Lukaszewicz et al., 2005; Takahashi et al., 1995; Breunig et al., 2011). These differences in cell cycle times are likely to be extremely important in explaining differences between primates and rodents not only because they influence the numbers of neurons generated but also because the length of the cell cycle appears to influence their laminar phenotypes (Pilaz et al., 2009; Dehay and Kennedy, 2007).

Fig. 16.2 compares and contrasts cortical development in rodents and primates (for more detail, see Chapter 5: The Cell Biology of Neural Stem and Progenitor Cells and Neocortex Expansion in Development and Evolution of this book by Huttner and colleagues). The substantial differences in the cellular mechanisms that generate the neocortex of rodents and primates have evolved without significant changes in regulatory proteins that control neocortical cell proliferation, fate, migration, and lamination. Transcription factors such as Pax6 (mouse)/PAX6 (human), Neurog2/NEUROG2, and Tbr2/TBR2 (also known as Eomes/EOMES), which are critical determinants of these processes, are highly conserved. Differences in their patterns of expression are likely to be of more significance. In the rodent cortex, progenitors in the ventricular zone (VZ) express Pax6 and those that give rise to intermediate progenitor cells (IPCs) transiently express Neurog2 (Britz et al., 2006). Pax6 is expressed by very few IPCs in the subventricular zone (SVZ) (<10%−20%) (Fig. 16.2A). Instead, these cells are characterized by their expression of Tbr2 and the postmitotic neurons that they generate express Tbr1 (Englund et al., 2005; Hevner and Haydar, 2012). Thus, sequential Pax6 → Ngn2 → Tbr2 → Tbr1 expression correlates with the transition of VZ progenitors to SVZ progenitors to postmitotic neurons (Telley et al., 2016). In primates, Pax6 is expressed by radial glial cells (RGCs) and a large subset of IPCs in the VZ, inner subventricular zone (ISVZ), and outer subventricular zone (OSVZ) (Hansen et al., 2010; Fietz et al., 2010; Betizeau et al., 2013; Florio and Huttner, 2014; Pollen et al., 2015) with many progenitors coexpressing Pax6 and Tbr2 (Fig. 16.2B). In the macaque primary visual cortex, 60%−80% of VZ progenitors express Pax6 alone during early and midstages of corticogenesis but at later stages 40% of them coexpress Tbr2 (Betizeau et al., 2013; Florio and Huttner, 2014; LaMonica et al. 2013). In the ISVZ of macaque, 60%−80% of progenitors coexpress Pax6 and Tbr2, and only a minority expresses Tbr2 or Pax6 alone. In the OSVZ, Pax6 and Tbr2 are coexpressed by up to 50% of progenitors, and only minorities express Pax6 or Tbr2 alone (Betizeau et al., 2013; Florio and Huttner, 2014). Whether these differences in the molecular profiles of IPCs between rodents and primates relate to important aspects of

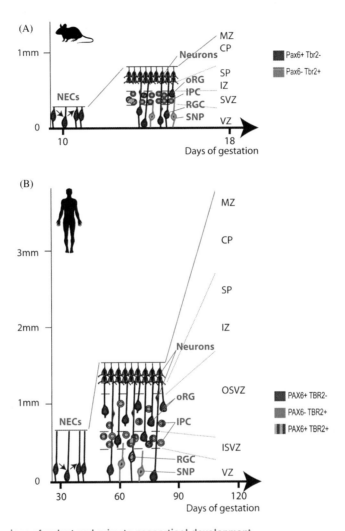

FIGURE 16.2 Comparison of rodent and primate neocortical development

In both rodents and primates at the earliest stages, neuroepithelial progenitor cells (NECs) undergo divisions at the neural tube's inner (or apical) surface to generate two new progenitors, expanding the pool of progenitors. This population of early symmetrically dividing NECs soon transforms and diversifies. Prominent among the new cell types are radial glial cells (RGCs) which, despite their name, are progenitors (Malatesta et al., 2000; Noctor et al., 2001). Progenitor cells that divide at the neural tube's inner surface form a layer known as the ventricular zone (VZ). As forebrain development progresses, an increasing proportion of RGCs divides asymmetrically to produce other cell types (Noctor et al., 2001, 2004). Some daughter cells migrate radially to the pial surface to differentiate into neurons. Many become a new type of progenitor that, instead of dividing at the apical surface of the VZ, divides in a region deep to the VZ called the subventricular zone (SVZ) (Noctor et al., 2004). (A) In rodents, progenitors in the SVZ divide mainly symmetrically to generate two neurons, which migrate into the developing cortex (Noctor et al., 2004; Farkas and Huttner, 2008). (B) Progenitors in the primate SVZ divide

(Continued)

◀ repeatedly and asymmetrically to expand this zone greatly compared to that of rodents (Smart et al., 2002). Primates develop two subventricular proliferative layers, the inner and outer SVZs (ISVZ and OSVZ, respectively) (Smart et al., 2002; Lukaszewicz et al., 2005; Hansen et al., 2010; Florio and Huttner, 2014; Dehay et al., 2015). The ISVZ is densely populated with progenitors that are similar to the majority of those in the rodent SVZ. They are referred to as intermediate progenitor cells (IPCs), and they are distinguished from RGCs by their multipolar morphology, proliferative behavior, and molecular constitution (Noctor et al., 2007). The OSVZ, on the other hand, contains not only IPCs but also many progenitors that have similar properties to RGCs in the VZ, although they lack processes linking them to the apical surface (Hansen et al. 2010; Hevner and Haydar, 2012; Florio and Huttner, 2014; Dehay et al., 2015). These OSVZ progenitors have become known as outer RGCs (oRGs). They undergo proliferative divisions and self-renewing asymmetric divisions to generate one oRG daughter cell and one IPC that can proliferate further (Fietz et al., 2010; Hansen et al., 2010; Florio and Huttner, 2014). They share many features with the closely related RGCs of the VZ, but can be distinguished by their unipolar morphology and their dynamic behavior directly preceding cytokinesis (Hansen et al., 2010). The OSVZ is the major source of neurons for the superficial (or supragranular) cortical layers, which carry out critical functions in intracortical integration and show greater enlargement than other cortical layers in primates (Lukaszewicz et al., 2005; Smart et al., 2002). In addition to RGCs, the VZ of rodents and primates contains another smaller subpopulation of apical progenitors called short neural precursors (SNPs). Unlike RGCs, SNPs do not self-renew and only generate pairs of neurons via symmetric divisions (Gal et al., 2006). Expression and coexpression of the transcription factors Pax6/PAX6 and Tbr2/TBR2 is indicated by green and red shading. Additional abbreviations: *IZ*, intermediate zone; *SP*, subplate.

cortical development remains to be established but, given their roles as high-level developmental regulators, it is likely that they do. Evidence supporting this comes from a recent experiment in which Pax6 expression was sustained in IPCs in the SVZ of rodents, where its expression is usually lost (Wong et al., 2015). This manipulation was sufficient to induce primate-like behavior in rodent IPCs, in that they increased their numbers of divisions resulting in increased upper-layer neuron production. These findings reinforce the overall hypothesis that the evolution of the developmental mechanisms of neocortical development has been achieved through altered regulation of the expression of a relatively small set of highly conserved high-level developmental regulators.

This comparison of primate and rodent cerebral cortex reveals cell and molecular similarities and differences which must have arisen at some point during their divergent evolution from a common ancestor 65 million years ago. Although specific evidence is lacking, it seems likely, given its broad importance, that the posttranscriptional regulation of gene expression by RNA interference (RNAi) involving miRNAs has been a major factor influencing evolutionary changes in the expression of regulators of cortical development.

THE EVOLUTIONARY CONSERVATION OF miRNA PROCESSING SYSTEMS

The RNAi pathway is one of the most highly conserved pathways in living organisms. It is used in many organisms to regulate gene expression by inhibiting gene expression. It directly affects mRNA molecules, by using miRNAs to induce degradation or inhibit the translation of their target mRNAs (Chapman and Carrington, 2007). It is generally agreed that animal and plant miRNA systems evolved along different trajectories having originated from an ancestral proto-RNAi system. Biological functions of endogenous RNAi are diverse and include pathogen defense mechanisms against RNA viruses or transposable elements, a process central for survival. In addition to its role in immunity, RNAi is also responsible for posttranscriptional silencing of gene expression, which is especially manifested during development where endogenous miRNAs have been implicated in the regulation of fate transitions. The tremendous conservation of miRNA processing enzymes Drosha and Dicer highlights the importance of this pathway. Many of these processing enzymes are present in prokaryotes and eukaryotes, although their functions differ (Shabalina and Koonin, 2008). Many simple eukaryotes retain parts of the RNAi-processing machinery, even though they do not use RNAi effectively (Drinnenberg et al., 2009). Argonaute proteins, which are bound and carried to mRNA targets, are also conserved and many organisms express multiple Argonaute family members: there is one in fission yeast, five in *Drosophila* species, eight in humans, and twenty-seven in *C. elegans* (Tolia and Joshua-Tor, 2007).

EVOLUTION OF miRNA SEQUENCE, COPY-NUMBER, AND EXPRESSION

Although the pathways that generate small RNA molecules are highly conserved, there are large variations in the repertoire of miRNAs used by different organisms and, as a general rule, the overall number of identified miRNAs in species drawn from across the animal kingdom increases with morphological complexity (Peterson et al., 2009; Berezikov, 2011; Londin et al., 2015; Kosik 2009). In animals, novel miRNAs arise de novo when DNA mutations create transcribed sequences capable of forming an RNA hairpin that can be recognized and processed into mature miRNAs by Drosha and Dicer. In general, RNAs relatively easily form structures that fold back on themselves in an imperfect way and so the evolution of new miRNAs is more likely than that of protein-coding RNAs (Berezikov, 2011). Based on the observation that phylogenetically new miRNAs are expressed at relatively low levels compared to more ancient miRNAs, it has been suggested that selective pressure is low for many new miRNAs, allowing them to readily acquire new mutations and potentially also new functions (Liang and Li, 2009).

An important contribution to genomic instability and evolutionary changes to the copy-number of genomic elements, including protein-coding and noncoding genes, is made by low-complexity repetitive sequences and transposable elements. At least a half of the human genome is derived from transposable elements (Smit, 1999; Lander et al., 2001; Cordaux and Batzer 2009), which are stretches of DNA that can move from site to site within the genome. They provide a mechanism for the evolution of miRNA genes themselves and for other genes that act in concert with miRNAs to regulate gene expression (discussed below). DNA transposons can excise themselves, move and paste themselves at new genomic sites, whereas retrotransposons duplicate via RNA intermediates that are reverse-transcribed and inserted at new genomic locations. Via this simple mechanism, transposable elements provide a means of duplicating genomic elements that can sometimes carry ncRNAs, including miRNAs (Spengler et al., 2014). Once duplicated, changes can occur independently to alter the function of one or both of the duplicated miRNAs. Transposition can also modulate the expression of miRNAs by bringing them under the control of novel enhancer elements at the site of integration (Cordaux and Batzer 2009; Smalheiser and Torvik, 2005; Piriyapongsa et al., 2007; Yuan et al., 2011).

The most abundant transposable elements in the human genome are Alu and LINE-1 elements, and their potential contribution to the molecular and functional differences between specific cell types across primate species is now testable thanks to the advent of induced pluripotent stem cell technology (Marchetto et al., 2013, Prescott et al., 2015, Franchini and Pollard, 2015). Alu elements are repetitive elements that are typically ∼300 base-pairs long derived from the 7SL RNA gene (7SL RNA is a part of a complex involved in protein trafficking; Alu elements are named after the AluI restriction enzyme site within the consensus Alu sequence). The expansion of Alu elements in the common ancestor of New and Old World monkeys facilitated the expansion of segmental duplications through recombination and can also facilitate frequent local duplications of miRNAs (Zhang et al., 2008, Spengler et al., 2014). The largest human miRNA gene cluster described so far is on chromosome 19 (C19MC). It is a primate-specific miRNA cluster spanning ∼100 kb comprising ∼46 tandemly repeated miRNA genes (Noguer-Dance et al., 2010). The C19MC locus is full of Alu elements. It is possible that in the early stage of primate evolution Alus accumulated in this specific region, thereby promoting rearrangements that led to the formation of the large miRNA family. The end result of this process is that the majority of C19MC miRNAs are sandwiched between Alus that function as promoters of miRNA-producing transcripts. This occurs because the 5′ region of Alus contains an internal RNA polymerase III promoter but Alus do not contain RNA polymerase III termination signals. Thus, Alu transcripts continue into downstream sequence flanking the element until a T-rich terminator sequence is found (Borchert et al., 2009; Richard and Manley, 2009).

RECENTLY EVOLVED miRNAs IN PRIMATES

As discussed above, the repertoire of miRNAs shows an overall expansion with organismal complexity across the animal kingdom. A recent study of this expansion in mammals found a net gain of miRNA families that was about three times higher in placental and marsupial mammals than in monotremes (egg-laying mammals) (Meunier et al., 2013). Evidence from this study suggested that the rates of formation of new miRNA families was even higher in humans and monkeys and that these new families gradually increased their expression levels and altered their mature sequences and target gene repertoires. Many primate miRNAs are not found in the rest of the animal kingdom, and one study identified many miRNAs in humans that are not expressed in chimpazees and vice versa (Berezikov et al., 2006), although some of those differences might relate to developmental stage rather than species as this study compared miRNAs expressed in adult chimpanzee to developing human brain.

A recent study compared interspecies differences in the profiles of mRNA and miRNA expression during the development of selected brain regions including cerebral cortex of macaques, chimpanzees, and humans (Somel et al., 2011). This study related differences in mRNA levels between human, chimpanzee, and macaque brains to developmental trajectories of miRNAs expression. In particular, expression levels of three miRNAs, miR-92a, miR-320b, and miR-454, were found to follow human-specific trajectories in early postnatal development with complementary changes in putative target mRNA expression. It remains unclear whether these differences represent changes in miRNA repertoire with lineage progression in the same cell type or differences in cellular composition with time. Another study in adult human, chimpanzee, and macaque prefrontal cortex identified five miRNAs with human-specific expression signatures: miR-184, miR-487a, miR-383, miR-34c−5p, and miR-299−3p (Hu et al., 2011). One of these, miR-184, which is abundant in the prefrontal cortex and cerebellum of humans but not chimpanzees and macaques, is an important promoter of neural stem cell proliferation (Liu et al., 2010). Of particular interest is miR-941, whose ancestor is found in an intron of the DNAJC5 gene of macaques and is missing from chimpanzees. It has mutated and increased its copy-number in the human lineage between six and one million years ago, and its copy-number remains polymorphic in the human population (2−11 copies). miR-941 is expressed at high levels in pluripotent cells, is repressed upon differentiation, and affects genes involved in neurotransmitter signaling (Hu et al., 2012; Somel et al., 2013).

The primate-specific C19MC cluster, introduced above, has become a subject of intense research. Several studies have shown that abnormal expression of individual miRNAs of the C19MC cluster correlates with tumorigenesis (Flor and Bullerdiek, 2012). Of special interest to brain development, rare embryonal tumors with multilayered rosettes, described as resembling undifferentiated neural tubes, are associated with a fusion between TTYH1 and C19MC leading to extreme ectopic overexpression of the C19MC miRNA cluster under the control of the TTYH1 promoter specifically in RGCs in human cortex. It is thought that these tumors recapitulate events of early neurogenesis. The C19MC cluster is normally expressed only during the earliest stages of embryonic development, and the perpetuation of C19MC expression under the TTYH1 promoter may reactivate a neural developmental program that is normally only active during early fetal development (Noguer-Dance et al., 2010; Archer and Pomeroy, 2014).

THE EVOLUTION OF miRNA TARGETS

Posttranscriptional gene regulation by miRNAs is founded on base complementarity between the miRNA and its target mRNA, and so evolution of the target sequence provides an additional mechanism for modulating miRNA function. Several bioinformatic methods have been developed to predict functional miRNA-target sites, and in some cases predictions have been tested experimentally, but most remain in silico predictions (McLoughlin et al., 2014). 3′UTRs are the main targets of miRNAs, and 3′-UTR length is likely to be an important factor in miRNA-target coevolution (Berezikov, 2011; Bartel, 2009).

In previous sections, we stressed the importance of high-level transcription factors, such as Pax6/PAX6, Neurog2/NEUROG2, Tbr2/TBR2, and Tbr1/TBR1, for the development of the cerebral cortex, and we explained how the expression patterns of these genes differ from rodent to primate. The targeting of transcription factors by miRNAs is recognized to be an important mechanism regulating their expression. There is evidence that transcription factor mRNAs have gained functional binding sites for miRNAs in humans, and this might have contributed significantly to the evolution of human-specific phenotypes. For example, a human-specific single nucleotide change has created a novel miRNA recognition element for miR-183 in the 3′UTR of the human transcription factor FOXO1 (McLoughlin et al., 2014). Only the human FOXO1 3′UTR contains a functional miR-183 site, not found in chimpanzee or mouse 3′UTRs. Human FOXO1 is regulated by miR-183, but mouse FOXO1 is not, and FOXO1-regulated cellular phenotypes, such as cell proliferation and cell migration, are affected by miR-183 targeting only in human cells.

Similarly, a recent study of the miRNAs targeting MECP2 discovered a human-specific miRNA recognition element for miR-483−5b and suggested that expression of miR-483−5 could rescue neurodevelopmental defects in Beckwith−Wiedemann syndrome patients (Han et al., 2013). Further work is required to test whether evolutionary changes in miRNA recognition elements in the mRNAs of the transcription factors that control cortical development account for any of the interspecies differences in the expression of these transcription factors.

THE EVOLUTION OF ncRNA AND ITS ROLE IN miRNA FUNCTION

Most miRNA genes are highly conserved among primates including humans and, in contrast to other ncRNA species including pseudogenes and lncRNAs, the HARS identified by systematic genome comparisons (Fig 16.1C) do not map to miRNAs (Liang and Li, 2009; Capra et al., 2013). This suggests that changes to miRNAs themselves might not be a major factor in human evolution, including the evolution of the human cerebral cortex. A caveat to this conclusion, however, is that existing unbiased HAR searches are underpowered. The discovery of the full sequence of the human genome is far from complete. Recent advances in long read sequencing technologies will enable more precise reconstruction of long stretches of human DNA sequences necessary to complete the discovery of haplotype variation near the hotspots of most recently mutated genomic regions, often characterized by segmental duplications and large deletions. Future studies employing larger numbers of human and nonhuman genomes and cerebral cortex cell type specific miRNA transcriptomes may well identify changes to miRNA sequence associated with human cerebral cortex evolution. Indeed, as described above, a study specifically looking for human miRNA evolution found that miR-941 sequence and copy-number has evolved in the human lineage and may have influenced human cerebral cortex evolution (Hu et al., 2012). The possibility that HARS do not map to miRNAs and that changes to miRNAs themselves might have been less important drivers of human evolution prompts an analogy with the previously discussed paucity of protein-coding changes associated with human evolution. The important changes might not have been in the sequences of miRNAs themselves, but rather in their expression levels by alterations to copy-number or enhancer elements (Hu et al., 2012; Capra et al., 2013; Somel et al., 2013) or, as described in more detail below, indirect changes in the activity of miRNAs through modification of other RNAs that interact with them (Salmena et al., 2011).

Competing endogenous RNA (ceRNA) is a recently described mechanism for regulating miRNA-dependent gene expression independent of changes to miRNA sequence or expression level. A ceRNA can be any RNA species which shares miRNA-binding sites with a given mRNA and acts to regulate the expression of mRNA species indirectly by competing with them for miRNA-binding, exhibiting a "sponge"-like effect. The higher the expression level of the ceRNA, the more of the miRNA it will titrate away from binding to the mRNA, and because miRNA-binding generally inhibits translation, the overall effect of ceRNAs is to desuppress protein expression (Fig. 16.1B) (Poliseno et al., 2010; Tay et al., 2011). Although no role has yet been described for ceRNA function in the development or evolution of the cerebral cortex, it is very tempting to postulate a role for ceRNA sequence evolution as an element modulating gene expression in the cerebral cortex. This idea is strengthened by the results of experiments in which the addition of an exogenous sponge RNA designed to have binding sites for miR92b desuppresses levels of the IPC fate determinant Tbr2, whose mRNA contains miR92b-binding sites, and increases neuronal output (Nowakowski et al., 2013). This suggests that evolutionary changes in ceRNAs, acting in concert with invariant miRNAs, could impact the cerebral cortical phenotype.

Work on ceRNAs has been led by research on the *PTEN* (*Phosphatase and tensin homolog*) gene encoding a phosphatidylinositol-3,4,5-triphosphate 3-phosphatase which is widely expressed in brain and body and inhibits cell signaling pathways including PI3K/AKT, mTOR, and WNT/β-catenin. Consistent with a negative role in proliferative pathways, PTEN functions as a tumor-suppressor with minute changes in PTEN having tumorigenic consequences (Trotman et al., 2003). Particularly relevant to this review, *Pten* suppresses neuronal and glial output in mouse cerebral cortex, and in humans *PTEN* haploinsufficiency is associated with head overgrowth (macrocephaly) and autism spectrum disorders (Crino, 2011; Chen et al., 2015). In light of the critical importance of PTEN dosage in both cancers and in the cerebral cortex, it is tempting to consider mutations affecting PTEN expression as candidates for a role in human cerebral cortex evolution, perhaps by regulating cell number.

In addition to its intrinsic biological interest, *PTEN* serves as an exemplar for the regulation of miRNA function by ceRNAs. PTEN protein level is under tight control, and a major component of this is provided by miRNA regulation. The *PTEN* 3′UTR contains binding sites for multiple miRNAs of the miR-17, miR-21, miR-214, miR19, and miR-26 families. Binding of these miRNAs to the *PTEN* 3′UTR suppresses PTEN protein levels and enhances PI3K/AKT signaling, cell proliferation, and tumor formation in human cell lines (Poliseno et al., 2010; Tay et al., 2011). Research into *PTEN* regulation has shown that ceRNAs, including mRNA and pseudogene species, sharing miRNA-binding sites with *PTEN* are capable of derepressing PTEN expression by competing for miRNAs. *PTENP1* is a processed pseudogene generated by retrotransposition of *PTEN* cDNA sequence into chromosome 9p13.3, where its transcription is regulated by promotor and enhancer elements independently of *PTEN*. Critically, *PTENP1* ncRNA retains the numerous miRNA-binding sites of its *PTEN* ancestral mRNA. A combination of cell-culture and population genetic studies clearly demonstrated that *PTENP1* RNA functions as a ceRNA to positively regulate PTEN protein expression and has tumor-suppressor activity (Poliseno et al., 2010). In addition to sharing miRNA-binding sites with its *PTENP1* pseudogene, *PTEN* mRNA also shares many miRNA-binding sites with other coding mRNAs which act in a network to mutually desuppress each other's inhibition by those miRNAs (Tay et al., 2011).

To conclude, work on *PTEN* has provided strong experimental evidence that ceRNAs are functionally important in regulating the expression of PTEN, the type of gene which could be involved in evolution of cerebral cortex size or cell composition (Salmena et al., 2011). More generally, ncRNA classes which have been demonstrated to have ceRNA function include pseudogenes and lncRNAs. These RNA classes are well represented, unlike miRNAs themselves, in the lists of ncHAR sequences linked to human evolution, as described above (Fig 16.1C). Taken together with the discovery that protein-coding mRNAs themselves act as ceRNAs, allowing evolutionary changes in their enhancer elements to alter both their expression level and also the expression of other genes, we can speculate that HAR mutations might have contributed to the evolution of the human cerebral cortex not by altering miRNA sequence directly but rather by regulating the sequence or expression of mRNAs, lncRNAs, and/or pseudogenes.

FUTURE PROSPECTS

Studies of miRNA gene sequence variation in natural human populations utilizing data available, for example, through the 1000 genomes project (Mills et al., 2011), has the potential to identify statistically significant links between sequence variants and phenotype that have not been possible through other approaches. Links between miRNA variants in natural human populations and disease phenotypes may be especially useful in a clinical context (Sun and Shi, 2015), but will more generally reveal sequences under selection pressures. More extensive interspecies comparisons are needed to understand the possible contributions of miRNA sequence variation to evolutionary changes.

Our current knowledge of miRNAs expressed in the developing primate cerebral cortex is very limited. Several studies have focused on studying miRNA expression in primary primate tissue during neurogenesis stages (Jönsson et al., 2015; Moreau et al., 2013; Arcila et al., 2014), and those studies should promote experimental work testing function. For example, a recent study in macaque cortex found differential expression of miRNAs predicted to target multiple regulators of cell cycle dynamics, which may contribute to the diverse cell cycle kinetic parameters described using time-lapse microscopy (Arcila et al., 2014; Betizeau et al., 2013). Another study identified miRNAs expressed in an area-specific pattern in the early neuroepithelium, which may contribute to early patterning of the neuroepithelium (Jönsson et al., 2015). Although parallel profiling comparison of human, nonhuman primate, and rodent miRNA has not been performed, metaanalyses that would take into account different timing of developmental events, area and cell type specificity, and also strengths and limitations of miRNA profiling approaches could provide useful insights into the conservation of miRNA expression trajectories during development as well as domains in the developing cortex in multiple species (Git et al., 2010).

An important limitation of studies identifying individual differences in miRNA expression profiles or mutations in response elements is that some of these changes could easily be counterbalanced by mutations in other, complementary, pathways. Massively parallel functional assays utilizing the technologies developed for efficient genome editing may provide the scalability necessary to overcome this limitation. Given the recent advances in in vitro differentiation of pluripotent stem cells into cortical progenitor-like cells, such assays could be performed in systems that recreate human specific mutations in human or nonhuman primate cells. Such assays have the potential to provide important insights into the effect sizes of gene expression changes or response-element mutations on physiological phenotypes, such as proliferation rates, which could account for evolutionary differences in brain development.

REFERENCES

Archer, T.C., Pomeroy, S.L., 2014. A developmental program drives aggressive embryonal brain tumors. Nat. Genet. 46, 2–3.

Arcila, M.L., Betizeau, M., Cambronne, X.A., et al., 2014. Novel primate miRNAs coevolved with ancient target genes in germinal zone-specific expression patterns. Neuron 81, 1255–1262.

Bailey, J.A., Gu, Z., Clark, R.A., et al., 2002. Recent segmental duplications in the human genome. Science 297, 1003–1007.

Bartel, D.P., 2009. MicroRNAs: target recognition and regulatory functions. Cell 136, 215–233.

Berezikov, E., Thuemmler, F., van Laake, L.W., et al., 2006. Diversity of microRNAs in human and chimpanzee brain. Nat. Genet. 38, 1375−1377.

Berezikov, E., 2011. Evolution of microRNA diversity and regulation in animals. Nat. Rev. Genet 12, 846−860.

Betizeau, M., Cortay, V., Patti, D., et al., 2013. Precursor diversity and complexity of lineage relationships in the outer subventricular zone of the primate. Neuron 80, 442−457.

Bird, C.P., Stranger, B.E., Liu, M., et al., 2007. Fast-evolving noncoding sequences in the human genome. Genome Biol. 8, R118.

Borchert, G., Lanier, W., Davidson, B.L., 2009. RNA polymerase III transcribes human microRNAs. Nat. Struct. Mol. Biol. 13, 1097−1101.

Breunig, J.J., Haydar, T.F., Rakic, P., 2011. Neural stem cells: historical perspective and future prospects. Neuron 70, 614−625.

Britten, R.J., Davidson, E.H., 1969. Gene regulation for higher cells: a theory. Science 165, 349−357.

Britz, O., Mattar, P., Nguyen, L., et al., 2006. A role for proneural genes in the maturation of cortical progenitor cells. Cereb. Cortex 16 (Suppl 1), i138−i151.

Bush, E.C., Lahn, B.T., 2008. A genome-wide screen for noncoding elements important in primate evolution. BMC Evol. Biol. 8, 17.

Bystron, I., Blakemore, C., Rakic, P., 2008. Development of the human cerebral cortex: Boulder Committee revisited. Nat. Rev. Neurosci. 9, 110−122.

Capra, J.A., Erwin, G.D., McKinsey, G., et al., 2013. Many human accelerated regions are developmental enhancers. Philos. Trans. R. Soc. Lond. Ser. B, Biol. Sci. 368, 20130025.

Chapman, E.J., Carrington, J.C., 2007. Specialization and evolution of endogenous small RNA pathways. Nat. Rev. Genet. 8, 884−896.

Chen, Y., Huang, W.C., Sejourne, J., et al., 2015. Pten mutations alter brain growth trajectory and allocation of cell types through elevated beta-catenin signaling. J. Neurosci. 35, 10252−10267.

Cheng, Z., Ventura, M., She, X., et al., 2005. A genome-wide comparison of recent chimpanzee and human segmental duplications. Nature 437, 88−93.

Clark, M.B., Amaral, P.P., Schlesinger, F.J., et al., 2011. The reality of pervasive transcription. PLoS Biol. 9, e1000625.

Cordaux, R., Batzer, M.A., 2009. The impact of retrotransposons on human genome evolution. Nat. Rev. Genet. 10, 691−703.

Crino, P.B., 2011. mTOR: a pathogenic signaling pathway in developmental brain malformations. Trends Mol. Med. 17, 734−742.

Dehay, C., Kennedy, H., Kosik, K.S., 2015. The outer subventricular zone and primate-specific cortical complexification. Neuron 85, 683−694.

Dehay, C., Kennedy, H., 2007. Cell-cycle control and cortical development. Nat. Rev. Neurosci. 8, 438−450.

Doan, R.N., Bae, B.I., Cubelos, B., 2016. Mutations in human accelerated regions disrupt cognition and social behaviour. Cell 167, 341−354.

Drinnenberg, I.A., Weinberg, D.E., Xie, K.T., et al., 2009. RNAi in budding yeast. Science 326, 544−550.

Englund, C., Fink, A., Lau, C., et al., 2005. Pax6, Tbr2, and Tbr1 are expressed sequentially by radial glia, intermediate progenitor cells, and postmitotic neurons in developing neocortex. J. Neurosci. 25, 247−251.

Farkas, L.M., Huttner, W.B., 2008. The cell biology of neural stem and progenitor cells and its significance for their proliferation versus differentiation during mammalian brain development. Curr. Opin. Cell Biol. 20, 707−715.

Fietz, S.A., Kelava, I., Vogt, J., et al., 2010. OSVZ progenitors of human and ferret neocortex are epithelial-like and expand by integrin signaling. Nat. Neurosci. 13, 690−699.

Flor, I., Bullerdiek, J., 2012. The dark side of a success story: microRNAs of the C19MC cluster in human tumours. J. Pathol. 227, 270–274.

Florio, M., Huttner, W.B., 2014. Neural progenitors, neurogenesis and the evolution of the neocortex. Development 141, 2182–2194.

Franchini, L.F., Pollard, K.S., 2015. Genomic approaches to studying human-specific developmental traits. Development 142, 3100–3112.

Gal, J.S., Morozov, Y.M., Ayoub, A.E., et al., 2006. Molecular and morphological heterogeneity of neural precursors in the mouse neocortical proliferative zones. J. Neurosci. 26, 1045–1056.

Geschwind, D.H., Rakic, P., 2013. Cortical evolution: judge the brain by its cover. Neuron 80, 633–647.

Gillies, K., Price, D.J., 1993. The fates of cells in the developing cerebral cortex of normal and methylazoxymethanol acetate-lesioned mice. Eur. J. Neurosci. 5, 73–84.

Git, A., Dvinge, H., Salmon-Divon, M., et al., 2010. Systematic comparison of microarray profiling, real-time PCR, and next-generation sequencing technologies for measuring differential microRNA expression. RNA 16, 991–1006.

Han, K., Gennarino, V.A., Lee, Y., et al., 2013. Human-specific regulation of MeCP2 levels in fetal brains by microRNA miR-483-5p. Genes Dev. 27, 485–490.

Hansen, D.V., Lui, J.H., Parker, P.R., Kriegstein, A.R., 2010. Neurogenic radial glia in the outer subventricular zone of human neocortex. Nature 464, 554–561.

Hevner, R.F., Haydar, T.F., 2012. The (not necessarily) convoluted role of basal radial glia in cortical neurogenesis. Cereb Cortex 22, 465–468.

Hu, H.Y., Guo, S., Xi, J., et al., 2011. MicroRNA expression and regulation in human, chimpanzee, and macaque brains. PLoS Genet. 7, e1002327.

Hu, H.Y., He, L., Fominykh, K., et al., 2012. Evolution of the human-specific microRNA miR-941. Nat. Commun. 3, 1145.

Jönsson, M.E., Nelander Wahlestedt, J., et al., 2015. Comprehensive analysis of microRNA expression in regionalized human neural progenitor cells reveals microRNA-10 as a caudalizing factor. Development 142, 3166–3177.

Kaas, J.H., 2013. The evolution of brains from early mammals to humans. WIREs Cog. Sci. 4, 33–45.

Khaitovich, P., Enard, W., Lachmann, M., Pääbo, S., 2006. Evolution of primate gene expression. Nat. Rev. Genet. 7, 693–702.

King, M.C., Wilson, A.C., 1975. Evolution at two levels in humans and chimpanzees. Science 188, 107–116.

Kornack, D.R., Rakic, P., 1998. Changes in cell-cycle kinetics during the development and evolution of primate neocortex. Proc. Natl. Acad. Sci. USA 95, 1242–1246.

Kosik, K.S., 2009. MicroRNAs tell an evo-devo story. Nat. Rev. Neurosci. 10, 754–759.

LaMonica, B.E., Lui, J.H., Hansen, D.V., Kriegstein, A.R., 2013. Mitotic spindle orientation predicts outer radial glial cell generation in human neocortex. Nat. Commun. 4, 1665.

Lander, E.S., Linton, L.M., Birren, B., et al., 2001. Initial sequencing and analysis of the human genome. Nature 409, 860–921.

Levers, T.E., Edgar, J.M., Price, D.J., 2001. The fates of cells generated at the end of neurogenesis in developing mouse cortex. J. Neurobiol. 48, 265–277.

Liang, H., Li, W.H., 2009. Lowly expressed human microRNA genes evolve rapidly. Mol. Biol. Evol. 26, 1195–1198.

Lindblad-Toh, K., Garber, M., Zuk, O., et al., 2011. A high-resolution map of human evolutionary constraint using 29 mammals. Nature 478, 476–482.

Liu, C., Teng, Z.Q., Santistevan, N.J., et al., 2010. Epigenetic regulation of miR-184 by MBD1 governs neural stem cell proliferation and differentiation. Cell Stem Cell 6, 433–444.

Londin, E., Loher, P., Telonis, A.G., et al., 2015. Analysis of 13 cell types reveals evidence for the expression of numerous novel primate- and tissue-specific microRNAs. Proc. Natl. Acad. Sci. USA 112, E1106–E1115.

Lukaszewicz, A., Savatier, P., Cortay, V., et al., 2005. G1 phase regulation, area-specific cell cycle control, and cytoarchitectonics in the primate cortex. Neuron 47, 353–364.

Malatesta, P., Hartfuss, E., Gotz, M., 2000. Isolation of radial glial cells by fluorescent-activated cell sorting reveals a neuronal lineage. Development 127, 5253–5263.

Marchetto, M.C., Narvaiza, I., Denli, A.M., et al., 2013. Differential L1 regulation in pluripotent stem cells of humans and apes. Nature 503, 525–529.

McLean, C.Y., Reno, P.L., Pollen, A.A., et al., 2011. Human-specific loss of regulatory DNA and the evolution of human-specific traits. Nature 471, 216–219.

McLoughlin, H.S., Wan, J., Spengler, R.M., et al., 2014. Human-specific microRNA regulation of FOXO1: implications for microRNA recognition element evolution. Hum. Mol. Genet. 23, 2593–2603.

Meunier, J., Lemoine, F., Soumillon, M., et al., 2013. Birth and expression evolution of mammalian microRNA genes. Genome Res 23, 34–45.

Mills, R.E., Walter, K., Stewart, C., et al., 2011. Mapping copy number variation by population-scale genome sequencing. Nature 470, 59–65.

Molnár, Z., Kaas, J.H., de Carlos, J.A., et al., 2014. Evolution and development of the mammalian cerebral cortex. Brain Behav. Evol. 83, 126–139.

Moreau, M.P., Bruse, S.E., Jornsten, R., et al., 2013. Chronological changes in microRNA expression in the developing human brain. PLoS ONE 8, e60480.

Noctor, S.C., Flint, A.C., Weissman, T.A., et al., 2001. Neurons derived from radial glial cells establish radial units in neocortex. Nature 409, 714–720.

Noctor, S.C., Martinez-Cerdeno, V., Ivic, L., Kriegstein, A.R., 2004. Cortical neurons arise in symmetric and asymmetric division zones and migrate through specific phases. Nat. Neurosci. 7, 136–144.

Noctor, S.C., Martínez-Cerdeño, V., Kriegstein, A.R., 2007. Contribution of intermediate progenitor cells to cortical histogenesis. Arch. Neurol. 64, 639–642.

Noguer-Dance, M., Abu-Amero, S., Al-Khtib, M., et al., 2010. The primate-specific microRNA gene cluster (C19MC) is imprinted in the placenta. Hum. Mol. Genet. 19, 3566–3582.

Nowakowski, T.J., Fotaki, V., Pollock, A., et al., 2013. MicroRNA-92b regulates the development of intermediate cortical progenitors in embryonic mouse brain. Proc. Natl. Acad. Sci. USA 110, 7056–7061.

O'Leary, M.A., Bloch, J.I., Flynn, J.J., et al., 2013. The placental mammal ancestor and the post-K-Pg radiation of placentals. Science 339, 662–667.

Peterson, K.J., Dietrich, M.R., McPeek, M.A., 2009. MicroRNAs and metazoan macroevolution: insights into canalization, complexity, and the Cambrian explosion. Bioessays 31, 736–747.

Pilaz, L.J., Patti, D., Marcy, G., et al., 2009. Forced G1-phase reduction alters mode of division, neuron number, and laminar phenotype in the cerebral cortex. Proc. Natl. Acad. Sci. USA 106, 21924–21929.

Piriyapongsa, J., Mariño-Ramírez, L., Jordan, I.K., 2007. Origin and evolution of human microRNAs from transposable elements. Genetics 176, 1323–1337.

Poliseno, L., Salmena, L., Zhang, J., et al., 2010. A coding-independent function of gene and pseudogene mRNAs regulates tumour biology. Nature 465, 1033–1038.

Pollard, K.S., Salama, S.R., Lambert, N., et al., 2006. An RNA gene expressed during cortical development evolved rapidly in humans. Nature 443, 167–172.

Pollen, A.A., Nowakowski, T.J., Chen, J., et al., 2015. Molecular identity of human outer radial glia during cortical development. Cell 163, 55–67.

Prabhakar, S., Noonan, J.P., Paabo, S., Rubin, E.M., 2006. Accelerated evolution of conserved noncoding sequences in humans. Science 314, 786.

Prabhakar, S., Visel, A., Akiyama, J.A., et al., 2008. Human-specific gain of function in a developmental enhancer. Science 321, 1346–1350.

Prescott, S.L., Srinivasan, R., Marchetto, M.C., et al., 2015. Enhancer divergence and cis-regulatory evolution in the human and chimp neural crest. Cell 163, 68–83.

Rakic, P., Ayoub, A.E., Breunig, J.J., Dominguez, M.H., 2009. Decision by division: making cortical maps. Trends Neurosci. 32, 291–301.

Rakic, P., 2009. Evolution of the neocortex: a perspective from developmental biology. Nat. Rev. Neurosci. 10, 724–735.

Reilly, S.K., Yin, J., Ayoub, A.E., et al., 2015. Evolutionary genomics. Evolutionary changes in promoter and enhancer activity during human corticogenesis. Science 347, 1155–1159.

Richard, P., Manley, J.L., 2009. Transcription termination by nuclear RNA polymerases. Genes Dev. 23, 1247–1269.

Salmena, L., Poliseno, L., Tay, Y., et al., 2011. A ceRNA hypothesis: the Rosetta Stone of a hidden RNA language? Cell 146, 353–358.

Shabalina, S.A., Koonin, E.V., 2008. Origins and evolution of eukaryotic RNA interference. Trends Ecol. Evol. 23, 578–587.

Smalheiser, N.R., Torvik, V.I., 2005. Mammalian microRNAs derived from genomic repeats. Trends Genet. 21, 322–326.

Smart, I.H., Dehay, C., Giroud, P., et al., 2002. Unique morphological features of the proliferative zones and postmitotic compartments of the neural epithelium giving rise to striate and extrastriate cortex in the monkey. Cereb. Cortex 12, 37–53.

Smit, A.F., 1999. Interspersed repeats and other mementos of transposable elements in mammalian genomes. Curr. Opin. Genet. Dev. 9, 657–663.

Somel, M., Liu, X., Khaitovich, P., 2013. Human brain evolution: transcripts, metabolites and their regulators. Nat. Rev. Neurosci. 14, 112–127.

Somel, M., Liu, X., Tang, L., et al., 2011. MicroRNA-driven developmental remodeling in the brain distinguishes humans from other primates. PLoS Biol. 9, e1001214.

Spengler, R.M., Oakley, C.K., Davidson, B.L., 2014. Functional microRNAs and target sites are created by lineage-specific transposition. Hum. Mol. Genet. 23, 1783–1793.

Sun, E., Shi, Y., 2015. MicroRNAs: small molecules with big roles in neurodevelopment and diseases. Exp. Neurol. 268, 46–53.

Taft, R.J., Pheasant, M., Mattick, J.S., 2007. The relationship between non-protein-coding DNA and eukaryotic complexity. Bioessays 29, 288–299.

Takahashi, T., Nowakowski, R.S., Caviness Jr, V.S., 1995. The cell cycle of the pseudostratified ventricular epithelium of the embryonic murine cerebral wall. J. Neurosci. 15, 6046–6057.

Tay, Y., Kats, L., Salmena, L., et al., 2011. Coding-independent regulation of the tumor suppressor PTEN by competing endogenous mRNAs. Cell 147, 344–357.

Telley, L., Govindan, S., Prados, J., et al., 2016. Sequential transcriptional waves direct the differentiation of newborn neurons in the mouse neocortex. Science 351 (6280), 1443–1446.

Tolia, N.H., Joshua-Tor, L., 2007. Slicer and the argonautes. Nat. Chem. Biol. 3, 36–43.

Trotman, L.C., Niki, M., Dotan, Z.A., et al., 2003. Pten dose dictates cancer progression in the prostate. PLoS Biol. 1, E59.

Visel, A., Blow, M.J., Li, Z., et al., 2009. ChIP-seq accurately predicts tissue-specific activity of enhancers. Nature 457, 854–858.

Wong, F.K., Fei, J.F., Mora-Bermúdez, F., et al., 2015. Sustained Pax6 expression generates primate-like basal radial glia in developing mouse neocortex. PLoS Biol. 13, e1002217.

Yuan, Z., Sun, X., Liu, H., Xie, J., 2011. MicroRNA genes derived from repetitive elements and expanded by segmental duplication events in mammalian genomes. PLoS ONE 6, e17666.

Zhang, R., Wang, Y.Q., Su, B., 2008. Molecular evolution of a primate-specific microRNA family. Mol. Biol. Evol. 25, 1493–1502.

Index

Note: Page numbers followed by "*f*" and "*t*" refer to figures and tables, respectively.

Printed in the United States
By Bookmasters